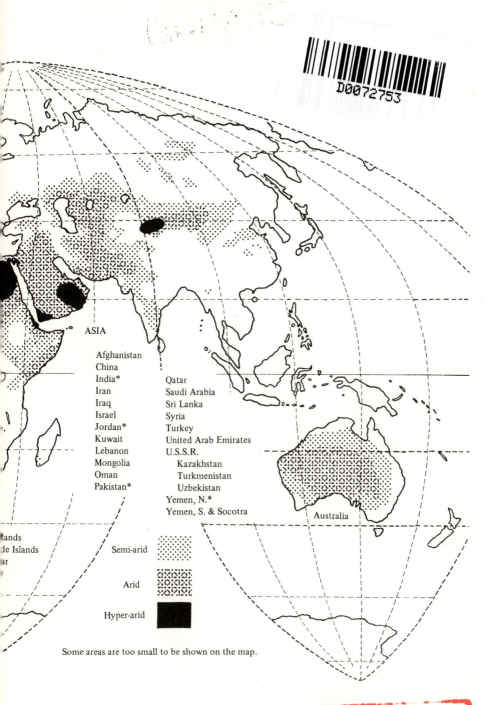

ASIA

Afghanistan
China
India*   Qatar
Iran   Saudi Arabia
Iraq   Sri Lanka
Israel   Syria
Jordan*   Turkey
Kuwait   United Arab Emirates
Lebanon   U.S.S.R.
Mongolia   Kazakhstan
Oman   Turkmenistan
Pakistan*   Uzbekistan
                   Yemen, N.*
                   Yemen, S. & Socotra

lands
de Islands
ar

Semi-arid

Arid

Hyper-arid

Some areas are too small to be shown on the map.

Australia

Kew International Conference on Economic Plants for Arid Lands

Frontispiece The exposed, edible, tuberous roots of *Ceiba parvifolia* (pochote) growing in the Tehuacan Valley, Puebla, Mexico. Photograph by R.S. Felger, November 1983.

PLANTS FOR ARID LANDS

Proceedings of the Kew International Conference
on Economic Plants for Arid Lands held in the
Jodrell Laboratory, Royal Botanic Gardens, Kew,
England, 23–27 July 1984

Editors

G. E. WICKENS   *Royal Botanic Gardens, Kew,*
Richmond, Surrey

J. R. GOODIN   *Texas Tech University,*
Lubbock, Texas, USA

D. V. FIELD   *Royal Botanic Gardens, Kew,*
Richmond, Surrey

London
GEORGE ALLEN & UNWIN
Boston   Sydney

George Allen & Unwin (Publishers) Ltd,
40 Museum Street, London WC1A 1LU, UK

George Allen & Unwin (Publishers) Ltd,
Park Lane, Hemel Hempstead, Herts HP2 4TE, UK

Allen & Unwin Inc.,
Fifty Cross Street, Winchester, Mass. 01890, USA

George Allen & Unwin Australia Pty Ltd,
8 Napier Street, North Sydney, NSW 2060, Australia

First published in 1985

British Library Cataloguing in Publication Data

   Kew International Conference on Economic Plants
for Arid Lands *(1984: Kew)*
Plants for arid lands: proceedings of the Kew
International Conference on Economic Plants for
Arid Lands held in the Jodrell Laboratory, Royal
Botanic Gardens, Kew, England, 23–27 July 1984.
1. Arid regions agriculture
I. Title   II. Wickens, G.E.   III. Goodwin, J.R.
IV. Field, D.V.
630′.915′4   S613
ISBN 0–04–581019–2

Library of Congress Cataloging in Publication Data applied for

Printed in Great Britain
by Mackays of Chatham Ltd

KEW INTERNATIONAL CONFERENCE ON ECONOMIC PLANTS FOR ARID LANDS

The Conference Organizers are grateful to the following for their financial support without which the Conference would not have taken place:

American Airlines
Anglo-Jewish Association
Bentham & Moxon Trust
British Council
Commission of the European Communities
Commonwealth Foundation
Dulverton Trust
Imperial Chemical Industries PLC (Plant Protection Division)
International Foundation for Science
Midland Bank Ltd
Shell International Petroleum Company Ltd
United Nations Environment Programme

Foreword

Economic plants have been defined by SEPASAT as those plants that are utilised either directly or indirectly for the benefit of Man. Indirect usage includes the needs of Man's livestock and the maintenance of the environment; the benefits may be domestic, commercial or aesthetic. Economic plants constitute a large and so far uncalculated percentage of the quarter of a million higher plants in the World today. However, it has been calculated that 10% (25 000) of these species are now on the verge of extinction and extinction means that a genetic resource that could be of benefit to Man will be lost for ever. Furthermore, for every species lost an estimated 10–30 other dependent organisms are also doomed.

Fewer than 1 per cent of the World's plants have been sufficiently well studied for a true evaluation of the potential floral wealth awaiting discovery, not only in the rain forests, which man is now actively destroying at a rate of 20 ha a minute, but also in the very much neglected dry areas of the World.

More than one third of the earth's land area ($c.$ 49 million km^2) is arid, supporting more than 785 million people, or 17.7 percent of the world's population, a population that is expanding at a faster rate than the environment can effectively support. The pressure on the environment and food production is such that 50–70 thousand km^2 of useful productive land is being lost each year through desertification. An estimated 3.8 million km^2 of the arid and semi-arid lands are also saline; in the Indian subcontinent alone some 400 km^2 are lost annually to agriculture because of increasing salinity, while in the USA with its far more sophisticated agriculture the area lost is 800–1200 km^2.

At this present time, because of the depressed world economy and unprecedented drought, it is the people of the undeveloped countries and arid and semi-arid environments that are particularly effected. There is real need for these people, with the active support of the more fortunate developed countries, to develop their plant (and other) resources to the full, and not over-exploit them.

The Royal Botanic Gardens, Kew have a long historical association with economic botany stretching back over two centuries to the times when Kew started distributing plants around the world. In recent years Kew has become more concerned with the plight of the developing countries and three years ago, with the financial support of OXFAM, set up a project, known as the Survey of Economic Plants for Arid and Semi-arid Tropics (SEPASAT) to record and ultimately advise on plants for these areas. It is intended that this should be the foundation of what is hoped eventually will become a world-wide data bank on economic plants, bringing together the scattered published information and the unpublished into a usable source of information.

While accepting the need for Kew to play a part in helping the undeveloped countries of the arid and semi-arid tropics, utilising much of its hitherto untapped resources, there was felt to be a need to examine the role of like-minded organisations and countries throughout the world. This resulted in the Kew International

Conference on Economic Plants for Arid Lands (KICEPAL) and it is hoped that it will be the first of a series of conferences to be held at fairly regular intervals and hosted by various nations. It is essential that we learn to share our knowledge, political barriers notwithstanding, utilise our plant resources properly and to the full, and by so doing, preserve the environment for future generations. The Royal Botanic Gardens, Kew will play their part by honouring the recommendations of the Conference to act as a neutral International Data Base for the collection and exchange of information among all interested parties.

Professor E A Bell
Director
Royal Botanic Gardens, Kew

Preface

The last decade or so has seen an increasing awareness of the value of many under-exploited plants in the arid and semi-arid tropics, and the need to manage these potential resources for the benefit of local peoples and as an aid to combating arid land desertification.

As a result of this awareness, accentuated by the effects of the Great Drought of the late 60's, many countries have directed much research into studying their economic plant resources. This has been reinforced by increased action and funding by some International Agencies.

Because of this increased activity and a general need to improve public awareness of what is being done and by whom, we believe that it is time to take stock; to review the problems being faced, both nationally and internationally, and to see what action is necessary to derive the best results from all this research. The Kew International Conference on Economic Plants for Arid Lands, co-sponsored by the International Center for Arid and Semi-Arid Land Studies (ICASALS), Texas Tech University, was convened to discuss and assess this theme.

This volume contains 28 out of the 30 invited papers for the Conference with two mini-papers substituted for manuscripts not received in time for publication. A further 20 mini-papers were also presented by various contributors on subjects of their own choosing; a selection of these will be published in the Journal of Arid Environments by kind permission of the editor, Professor J.L. Cloudsley-Thompson.

Where necessary the senior editor has taken the liberty of amending the botanical names to those currently accepted at Kew; where appropriate synonyms used in the original manuscript are shown in parenthesis. This has been done to ensure uniformity of taxonomic treatment throughout the volume.

G.E. Wickens	J.R. Goodin	D.V. Field
SEPASAT	ICASALS and	Herbarium
Royal Botanic Gardens	Department of Biological	Royal Botanic Gardens
Kew, Richmond	Sciences, Texas Tech Univ.,	Kew, Richmond
Surrey	Lubbock, Texas	Surrey

Kew International Conference on Economic Plants for Arid Lands

Preamble

Recognising the diversity of the botanical wealth of the Arid and Semi-Arid areas of the World as highlighted at the Kew International Conference on Economic Plants for Arid Lands (July 1984),

Fearful of the threats to these resources by lack of appreciation of their actual and potential value to aid the inhabitants of these severely stressed regions – these threats were identified as:
  (a) Actual species loss and loss of genetic diversity within a species,
  (b) Total loss of habitat or severe degradation to the point of non-recoverability,
  (c) Change of land use so fulfilling (a) or (b) or so restricting long term options for ecosystem rehabilitation
(One or more of these threats may be caused when severe droughts force desert peoples to turn even more than usual from 'agriculture' to gathering native species in the wild in order to sustain their own life and that of their beasts),

Mindful of the need for the amelioration of the severe restrictions imposed by the arid environment, and for the development of and better support for arid land communities,

the Conference highlighted the importance of addressing the needs of the people and ensuring that a greater awareness of economic plant resources and their development is undertaken before they are irretrievably lost.

Recommendations

This conference *urges* decision makers everywhere to support actively the implementation of the following recommendations of the Conference.

(a) *Public Awareness and existing Local Knowledge.* The value of arid lands plants should be identified and publicised. Local knowledge of these plants is frequently profound but often ignored; it is recommended that all such information should be gathered nationally and internationally where appropriate. This information is part of the cultural heritage of any nation as well as of direct value to arid land utilisation and management. Major educational programmes, from University level to local improvement schemes for rural agriculturalists in their own languages, were seen as essential. Many arid land problems could be solved by using arid land plants!

(b) *Scientific data and Co-ordination.* Besides the knowledge held at this level there is a vast amount of scientific data already in existence which needs to be evaluated, organised and gathered together for universal exchange. The Conference requests the Royal Botanic Gardens, Kew, to act as a neutral International Data Base for the collection and exchange of all such information (whether scientific or ethno-botanical) among all interested parties.

(c) *Research and Training.* While it is clear that there is a wealth of data available which can be coordinated, there were many spheres where further research was essential, from new technologies through domestic and industrial implementation to the socio-economic consequences. This research should be seen as part of a holistic or integrated approach. Other major areas highlighted as clearly needing further and active research included:– nitrogen fixation, sand-dune stabilisation, saline surviving and multi-purpose plants, also research into rain-fed and irrigated agricultural methods for arid lands besides a major increase in arid lands seed collection and distribution. The need in many countries to develop appropriate research and training institutes was also identified.

(d) *Funding.* Whilst much information can be coordinated for relatively little cost the potential to aid the arid and semi-arid lands using their own plants is immense, and all International and National funding agencies are strongly urged to support food, fodder, fuelwood, fibre, medicinal and industrial resource plant research and technology for development.

Contents

1   The needs of the people

Mary Cherry

OXFAM, 274 Banbury Road, Oxford OX2 7DZ

It must be unusual, possibly unique, for a scientific conference to start with a paper which is essentially not scientific but social. I applaud the organisers most heartily for their choice of keynote subject. Scientific research can never be an end in itself. It must always have as its objective the satisfying of a human need, directly or indirectly, perceived or as yet unperceived by the beneficiaries.

It is a sad fact of our modern life that specialisation, often minutely focussed and intensely concentrated, while bringing scientific and technological advances unprecedented in the history of mankind, can fail disastrously in human terms. Narrowness of vision and understanding can mean that a breakthrough in the laboratory leads to a breakdown, or at least a let-down, for the people who should have benefited and who often suffer from unforeseen repercussions. And then, of course, there is the waste of time and resources that could have been employed for good.

At this conference our attention is focussed on arid lands, which means that often we will be thinking about the needs of many of the world's poorest people. There are, of course, arid lands in some rich countries. Some of you will come from these; from Australia, the USA and so on. But it is estimated that about 43 per cent of the world's land surface is arid or semi-arid and a high proportion of that is in developing countries. Furthermore, within those countries, it is the poorer people who scratch a living, or fail to scratch a living, in drought-prone conditions where irrigation facilities are meagre, unreliable or non-existent.

It is such people that OXFAM tries to help in Asia, Africa, the Middle East and Latin America. It is such people that you can help through your work. OXFAM is supporting Kew's Survey of Economic Plants for Arid and Semi-Arid Tropics (SEPASAT) project because we believe that this is of real relevance to these environmentally and economically disadvantaged people of the Third World. That is why I am here today and asking you to focus your work on the needs of the people.

So where are these people and what are their needs? The end papers list countries which totally or in part fall within the definition of hyper-arid to semi-arid tropics or subtropics. The asterisks indicate countries in which there are OXFAM-supported projects or where we have given some help in recent years. It does not follow that all projects would be in the semi-arid or arid parts of those countries, but many are. A feature of these environments is that, very often, they are deteriorating and that deterioration is related to the scant and poor vegetation which, in turn, owes its condition to the pressure of human need in an environment of climate and soil that already cannot sustain it. We know that within recorded time many of our arid areas had good vegetative cover, either natural or cultivated. In some areas the deterioration from tree-covered slopes or useful food-producing lands to a barren landscape has taken place within our lifetime and still continues. Somehow this downward spiral has to be reversed. It is already a disaster for millions and it will soon be a disaster for millions more.

Scientists working with economic plants for arid lands have a vital part to play in turning this spiral around. To quote from the first issue of the SEPASAT Newsletter:

'Economic plants may be defined as those that are utilised either directly or indirectly for the benefit of Man. Indirect usage includes the needs of Man's livestock and the maintenance and improvement of the environment; the benefits may be domestic, commercial or aesthetic.'

Given availability, poor people will make use of a wide range of plant species and will put them to much more complete and varied use than will richer people. These plant materials may be highly suitable for the needs, environment and particularly the social-economic requirements of the people. The intervention of other cultures bringing alternative and more modern materials may or may not be desirable, but it is certainly undesirable that the plant materials and the knowledge of how to use them should be lost. Unfortunately, much traditional knowledge is already lost or on the verge of being lost.

Sadly, it is all too common today for people to suffer either natural or manmade disasters which throw them back again upon simple, natural resources. OXFAM is very often quickly on the scene at those disasters. After playing our part in relieving initial distress, we see our main role as helping people restart their economy. For example, field staff frequently assist in getting seed to an area quickly to enable people to plant a crop to replace one destroyed by drought or flood. OXFAM may also fund simple co-operative food or seed stores to enable people advantageously to utilize any salvaged crops.

OXFAM follows in the tradition of an organisation that started during the Second World War as the Oxford Committee for Famine Relief to help hungry refugees, regardless of where they were geographically or politically, and it continues to respond to conditions of famine and other disasters (Fig. 1.1). However, the main thrust of the organisation's work is development, to help people, whose normal circumstances are an on-going disaster and to improve the quality of their life a little. In discussing the main categories of need in relation to the use of plants in arid and semi-arid conditions, I will mention some OXFAM-funded projects.

To start with the fundamental resource, the soil: in many parts of the world the people for whom we are concerned either have an inadequate area of land on which

Figure 1.1 Father and son in Niger collecting edible leaves after local crops have failed. Famine foods are vital for poor people in arid lands.

to support their family or they are landless and yet dependent upon the land. The tragedy is that often they are in the very areas where the resource is getting less due to soil erosion, land degradation (loss of structure and fertility or development of salinity) or encroaching desert. Other papers at this conference will be dealing with tree species and other vegetation for soil protection and restoration. Usually the necessary scale of projects of this kind makes them appropriate for governments to tackle with assistance from large international organisations such as World Bank and UN agencies. However, non-government organisations, such as OXFAM, have various smaller-scale roles to play. These are likely to be with those people who tend to be overlooked because they are remotely-situated or who are difficult to assist because of their traditions and lifestyle. Non-governmental organisations also assist in the motivation of people to protect their own environment. This may mean helping them to develop alternative means of livelihood rather than those which are endangering the soil, such as cutting firewood for sale. They may also give them the support which they may need to protect themselves from side-effects and repercussions of large-scale projects, e.g. afforestation, dam building, etc., which may be planned and executed in ways which lack understanding of, or sympathy for, the indigenous people of the area.

Notably in Somalia, but also elsewhere, we have helped in the development of techniques for sand dune stabilisation. These involve, for example, putting in lines of brushwood perpendicular to the prevailing winds and planting *Commiphora* on the dunes. Communication is as important as developing the technique and so we fund seminars, at an appropriate level, for this purpose.

Similarly, we have been working with local organisations in one of India's major arid zones, the Rajasthan desert. It is worth pointing out that Indian arid areas are probably unique in the degree to which they are populated. The Rajasthan desert, which extends into parts of Gujarat and Haryana States, covers about 0.29 million km^2. Over 14 million people and more than 80 million head of domestic livestock live in the area, according to figures from the Central Arid Zone Research Institute, Jodhpur. This institute has done a great deal of valuable work in stabilising dunes, demonstrating the use of micro-windbreaks of locally-available brushwood, and researching and recommending appropriate tree and agricultural species and techniques for the environment, recognising that the sheer pressure of human and animal need is a prime cause of soil deterioration and desertification.

One local organisation near Jodhpur, which OXFAM supports, has been doing modest but useful work in agricultural promotion in arid areas. An interesting part of their work has been the popularising of the growing of ber, or jujube trees (*Ziziphus mauritiana*). Wild species, which are extremely drought-hardy and are a predominant component of the natural vegetation of the desert, have long been used for fodder, fuel and fencing. Now varieties which have been much improved for yield and fruit quality are grafted on to the wild species and provide a reasonable cash crop from land which is unsuitable for cultivation.

Many of the local organisations around the world which OXFAM assist are planting, or encouraging the planting of *Leucaena leucocephala,* not just for its soil-holding and fertility-building characteristics, but because of its many other uses which are so valuable for poor people – food, fodder and fuel. In Maharashtra, India, I have seen *Leucaena* literally create topsoil – an oasis of vegetation in a barren landscape – and there one of our projects is very successfully inter-cropping with grain.

One does not think of Indonesia as an arid country, but on Sawu Island of Nusa Tenggara Timur rainfall averages from as low as 250 mm to 1400 mm per year and the number of days with rain in a year ranges from 14 to 86. There, an OXFAM-assisted project is regreening the island with *Leucaena,* coconut, cashew and banana. We give support to the Green Deserts organisation in Sudan, for example, where some of their work involves trying to provide better vegetation for the use of nomadic people. Live-fencing, too, is something which comes well within the scope of OXFAM-type projects because of its good land utilisation and multi-purpose use. Tanzania is just one of many countries where we are encouraging its use.

Of all the categories of need which plant materials can supply, I suppose the most urgent is that of fuel. A shortage of fuel has so many desperate affects for poor people and particularly for women. If there is no fuel there is no cooked meal; that may tip the balance of an already marginally adequate diet or put the people at risk from dangerous components of their staple diet which are normally destroyed or removed by cooking. Similarly, if there is no possibility of boiling water, that method of sterilisation is removed. If there is no longer a supply of wood near the homestead, the women, already heavily burdened with children, household chores and probably agricultural work too, will have to walk further and further each day to glean what they can.

An OXFAM initiative in Upper Volta five or six years ago has become something of a classic. The Field Director there at that time had already had several years of experience in the Sahel working on a well-digging programme with the American Peace Corps. He already knew a lot about drought and the social problems of the people in that part of the world where the women were having to trek further and further to collect firewood. Convinced that something must be done to encourage more tree-growing and water conservation, he took some of his leave to visit the Negev area of Israel. There he saw trees flourishing in desert conditions, each with its own micro-catchment area. He returned to Upper Volta and with great determination introduced that very appropriate simple form of water-harvesting.

Within a relatively short time, with the assistance of a woman forester and the advice of soil scientists, they had trials underway with 20 tree and bush species and 300 micro-catchments spread through eight villages. Meanwhile, they decided to carry out a survey by questioning the women who did the wood cutting, about the types of species least-favoured by (and therefore least vulnerable to) goats, sheep and cattle. That sounds like a logical thing to have done. But how many tree-planting projects, even village fuel-wood projects, consult the women during the planning stage? Very few, I believe. The answers given were very interesting. Several of the species mentioned were among those already selected for trials as probably being appropriate, but many on the women's list were not. They were more often shrubs than trees because the women found shrubs easier to reach and cut and because of their rapid rate of regeneration. Many were said to be fast-drying and others were preferred because they burned well with few sparks and little smoke.

The women, with generations of experience handed down to them, certainly knew their firewood. As a result, the following nine species were selected for use in the programme, of which five marked with an asterisk were chosen because animals do not like them even during the dry season: *Cassia sieberiana**†, *Piliostigma reticulatum**†, *Acacia macrostachya*, *Pterocarpus lucens* (a tree), *Combretum glutinosum*†, *Combretum micranthum**†, *Combretum nigricans**†, *Guiera senegalensis**†, *Mitragyna inermis* (a tree). Those marked with a dagger are especially suitable for bare degraded land near villages, together with *Boscia angustifolia*.

Figure 1.2 An Ethiopian farmer working his dry soil with a plough made of wood. Poor people depend on materials of plant origin for virtually all their needs.

This is an example of how vital social aspects, that might so easily have been overlooked, were focussed upon and became a major reason for the success of this project. It is still continuing and has the full co-operation of local government and other organisations. It is vitally important that in all project planning the needs of the indigenous people, particularly the so-called 'voiceless' people be understood and catered for. In the case of tree projects, if there is to be any interference with the natural forest or there is to be new planting, especially with exotics, it has to be realised that forest people are often utterly dependent upon many forest resources, such as plants, animals, wild honey, etc., besides the actual wood itself.

Much will be said at this conference, I am sure, about drought- and salt-tolerant species and varieties of food crops. I will merely touch on famine food and fodder (Fig. 1.2, p. 6). In areas where famine arising from drought is a frequent experience it is advantageous if people can be helped by technical and economic means to grow and store enough food to sustain them during the worst periods. However, anyone who has worked in such areas will know that this is usually easier said than done. In such areas a crop which may yield less at good times but continue to provide something at bad times is likely to be highly desirable. Often wild fruits or grass seeds are the poor people's salvation and it is important that they are not lost. For example, in dry parts of Orissa, India, the fruits of mahua (*Madhuca longifolia* var. *latifolia*) save many people from starvation. Sometimes the famine foods which people depend upon are dangerous. A good – or bad – example is that of the Indian vetch or grass pea, *Lathyrus sativus,* which causes lathyrism, a serious neurological disease causing paralysis of the lower limbs in both man and animals, if consumed for any length of time. Regrettably, in parts of India, landless labourers have little option but to eat the grass pea in times of famine arising from drought. Steeping the seeds in water and then sun-drying or parboiling reduces the toxin content substantially, but also removes nutrients. Now low-toxin types have been selected and are being multiplied. I am sure there is great scope for scientists, particularly plant breeders, to work to improve some of these traditional famine foods which have the merit of surviving harsh dry conditions.

OXFAM is also concerned with browse species and other famine fodder for livestock, particularly in areas populated by nomadic peoples. The human social repercussions when a traditional nomadic system breaks down can be quite horrifying. For example, we are sponsoring trials of perennial browse species in the Turkana area of Kenya and are concerned with the delicate balance between available feed, available water and livestock numbers in a number of other areas in which OXFAM staff or project holders are working.

Whereas food, fuel, fodder and water must be the greatest needs of people, there are a multitude of other uses for plant materials, e.g. the use of pulverised *Moringa oleifera* seeds as water purifiers in one of our projects in Nusa Tenggara Timur, Indonesia. For poor people plants provide their human and veterinary medicines; their plant nutrients, and pesticides; materials for housing, furniture, dress, tools, materials for crafts such as basketry and woodwork and so on.

In India last year a forester-turned-village-social-worker drew my attention to some categories of need which are liable to be overlooked as, quite rightly, we turn our attention to alleviating poverty and hunger. But for everyone, including the poor, there should be more to life than bread alone. My Indian friend pointed out that if a tree or any plant has beauty, perhaps in the form of colourful blossoms, in addition to fulfilling utilitarian purposes, it will add to the quality of life of the people living round about. Another cultural need is for a shaded meeting place, provided by a clump of two or three large trees. In some cultures a grove of village trees designated specifically to provide funeral wood is also important. A merit of these trees, which are grown for such cultural reasons, is that no one would cut or harm them without the authority to do so.

At the far end of the spectrum from these plants, which relate to the culture and tradition of people, are plants which have been developed for industrial uses. Very often these have been known and used by indigenous desert-dwellers for genera-tions and have only recently been researched and commercially exploited. I am thinking, for example, of jojoba (*Simmondsia chinensis*) which yields oil equivalent in quality to that of the sperm whale, and guayule (*Parthenium argentatum*), which is a source of natural rubber. Both are native to northern Mexico and the south-west of the USA and were little-known elsewhere until reported on by a panel of the U.S. National Academy of Sciences. Such plants scarcely fall within the terms of this paper since they are more likely to meet the needs of governments and of commercial concerns than the needs of ordinary people. However, two cautionary notes might be appropriate. One is that publicity given to the potential of such new crops tends to lead to premature commercial exploitation before agronomic and other aspects have been fully researched. The second is that the development of new plantations, particularly in poor countries, can be to the detriment of the poorer, indigenous people of the area who may lose traditional rights to the land or suffer other social repercussions.

In conclusion, the poorer people of the arid and semi-arid areas of the world have a multitude of needs which are, or could be, satisfied by plants ranging from the largest trees to the tiniest herbs. Not least of the needs of such people today is the assistance of concerned scientists because arid lands and the people who live on and by them have long been the 'Cinderellas' of research. Plant breeding, seed production, agronomy, management systems appropriate to the environment and the economic circumstances and traditions of the people, all need a tremendous input by research workers followed up by wise extension officers.

However, any amount of input in money and research workers' time will be at best wasted and at worst disastrous if scientists do not start by listening to the people they are trying to help, the farmers and villagers who have lived and worked in the area for generations, and by studying and understanding their traditional methods. Terrible mistakes have been made in the past by the unwise imposition of so-called improved methods in unsuitable circumstances. For real progress to be made in improving the vegetation of arid lands and thus improving the productivity and usefulness of dry land plants to meet the needs of the people, a true spirit of partnership is required between scientists and those who work on and live by the land.

2 The arid environment

A. T. Grove

Downing College, Cambridge CB2 1DQ, UK
or, African Studies Centre, Free School Lane, Cambridge, UK

Introduction

One person in every seven lives in an arid environment. The dry lands produce one fifth of the world's food supplies; they yield one half of the world's precious and semi-precious metals; they contain the bulk of the world's reserves of oil and natural gas (Heathcote 1983). Between 1960 and 1984 the populations of arid and semi-arid lands have more than doubled.

Climatic classification

Such impressive figures depend on the way in which the arid environment is defined. Normally it is on a basis of climate, especially the relation of precipitation input to evaporation loss. A definition very commonly used is that of Meigs (1953). Based on Thornthwaite's (1948) aridity coefficient, it is derived from the relation between mean annual precipitation, potential evapo-transpiration, water surplus and water replenishment calculated for individual months throughout the year. Indices distinguish the semi-arid from the arid lands; extremely arid lands are defined as those where 12 successive months have been recorded without rain falling. According to this convention and slight modifications that have been made to it, about 5 per cent of the land surface of the globe is distinguished as extremely arid, about 15 per cent as arid, and about 15 per cent semi-arid; in total about one third of the terrestrial world.

Hare (1977) used an alternative index in his review of climate and desertification for the UN Conference on Desertification (UNCOD) held in Nairobi in 1977. This was Budyko's radiation index of dryness (D) which compares the radiation balance (R) with the energy required to evaporate the mean annual precipitation (P), the latent heat of vaporisation of water being L:

$$D = R/LP$$

This index has the disadvantage that it is troublesome to calculate the radiation balance since it is necessary to have values for albedo. However, as Hare demonstrates, the aridity indices of Thornthwaite, Meigs and Budyko are roughly interchangeable, at least in low latitudes.

Too much weight should not be placed on indices of this kind which do not take into account the distribution of rainfall within the rainy season; one should not expect precise correlation between annual index, biotic distributions and irrigation requirements. Such indices provide a useful guide to the distribution and intensity of aridity over the globe. In the semi-arid environments, rain-fed agriculture is possible in most years; in the arid regions seasonal pastoralism may be possible but irrigation is needed for cultivation.

For special purposes, such as studies in connection with the transferability of plants from one region to another, additional climatic parameters have to be included in the classification process. Agglomerative methods of classification are available such as those used by Russell and Moore (1970, 1976). They compared climates in Africa and Australia by taking 16 monthly attributes for 300 stations, giving 192 attributes for each station and analysing the data for the whole year, for the summer and also for the winter seasons, to produce dendograms grouping the stations into 20, 30 and 40 sets that allowed climatic regions to be delineated. Once the data have been assembled and stored on a computer a very adaptable tool is available for drawing up classifications for particular purposes. Individual attributes can be included, excluded or weighted and maps produced for particular years or sets of years.

Climatic variability

It is possible, but not always easy, to allow for variability of precipitation in such classifications. In arid environments, where rain is commonly associated with local convectional storms, even annual totals may differ greatly over short distances. Interannual variability is high, with standard deviations as compared with mean values higher than in more humid regions. Furthermore, the importance of the large coefficients of variability of the precipitation is enhanced by evaporation rates which are also high and tend to increase as precipitation diminishes. The nature and causes of the variability differ from one part of the world to another and have varied through time.

Meteorologically, climatic aridity is associated with atmospheric subsidence. This is most widespread in high pressure cells which are functions of the global circulation and of the distribution of land and sea. The disposition of the sub-tropical high pressure cells explains the location of the sub-continental deserts of northern and southern Africa and Australia. The eastward extension of the Saharan—Arabian deserts into central Asia is associated with subsidence in the lee of mountain ranges and the dryness of air that has travelled thousands of kilometres from the main sources of water vapour, the low latitude oceans. Lack of weather systems giving rain helps to explain the aridity of regions like Texas. On the west sides of continents, deserts extending along the tropical coasts have special characteristics, relatively low temperatures, morning fog and heavy dewfall associated with the presence offshore of cool, upwelling water brought to the surface by winds blowing towards the equator on the eastern sides of the sub-tropical high pressure cells.

Coastal desert environments present extreme examples of rainfall variability as has been demonstrated quite recently in Peru. In most years very little rain falls, but from time to time when the warm equatorial counter-current, El Niño, spreads east and south displacing the cold Humboldt current, air moving inshore is no longer chilled and stabilised in its lower layers and heavy rains give rise to severe flooding. Such fluctuations in ocean currents are associated with very large-scale perturbations of the general circulation. Widespread drought and even excessive rain are also influenced by processes operating on a global scale. The Southern Oscillation, which can be compared to an atmospheric tide over the Pacific with a quasi-periodicity of a few years, is probably associated with some of the more important departures from mean conditions.

The sub-tropical high pressure cells moving north and south with the sun inhibit rain in summer on the poleward sides of the tropical deserts and in winter on the equatorward sides. The range of seasonal movement of the cells, their dimensions and intensity have an important bearing on rainfall and plant growth in the semi-arid regions. In the semi-arid tropics, a pause in the early rains can result in crops dying because at the beginning of the growing season the soil is dried out to a great depth after the long dry season. The consequent importance of mid-season corrections to cropping programmes in India, for instance, has often been stressed. On the poleward sides of the deserts, in lands with mediterranean climates, the risks are less because the main growing season commences during the winter rainy season when crops can rely on water stored in the soil, and the amount present is known to farmers and advisory services. Prior knowledge of soil moisture for the growing season is most readily available in those mid-latitude regions where crop growth depends heavily on prior winter snowfall.

The distribution through time of wet and dry years in arid environments has a random element but dry years and wet years cluster together. Not only is there persistence in the time series but there are also steps involving long-term changes in mean values of precipitation. Kraus (1955) pointed to a step down in the precipitation at many semi-arid stations which took place about the end of the last century. It caused a diminution in the mean discharge of the Nile and a fall in level of the great lakes of Africa. The lakes recovered to some degree in the 1950s and early 1960s and some of them remained high while the sahelian and sudan zones on the south side of the Sahara were experiencing their worst longterm drought ever recorded (Grove 1983).

The sahelian drought, which began in 1968 and appeared to culminate about 1973, is still with us. In scarcely any of the years since 1970 has the rainfall reached the mean values of 1954—70 (Lamb 1982). It seems likely that 1983—4 was one of the driest years of this dry period with the discharge of the Senegal and the rainfall in the vicinity of Lake Chad being the lowest ever recorded this century. Drought has returned to Ethiopia and the levels of artificial lakes in Nubia, Kainji, Bandama and Volta have fallen to unprecendentedly low levels. At present we have no good explanation for this disastrous situation, nor any idea how long it is likely to continue.

Much effort is being made to discover means of predicting such events. One approach which has long been popular is to search for periodicities in the climatic or hydrologic record that might allow extrapolation into the future. Faure and Gac (1981) have pointed to the regularity of an oscillation with twin peaks in the seven-year running mean values of the annual discharge of the Senegal and also of the Niger and Chari rivers in West Africa in the course of this century and have ventured to predict that sahelian drought will be alleviated after 1984 though it might be expected to return in the early years of the next century.

In southern Africa, rainfall records compiled and analysed by Tyson *et al.* (1975) showed that the four rainfall regions display different rainfall periodicities. In the summer rainfall region extending from northern Natal across the Transvaal into Botswana, years with deficient rain clustered at intervals of 18—20 years. The last of the drought periods had been in the early 1960s; partly because of the indication that the next drought might occur in the not too distant future, a conference was held in Gaborone in 1978 to consider appropriate responses to the hazard (Hinchey 1978). The next year was dry and so was 1981—2; a severe drought has now entered its fourth year and Tyson and his colleagues might well feel that they were too diffident about the predictive value of their analysis. But such orderliness in annual rainfall time series is exceptional and other approaches to prediction, notably atmospheric modelling, give greater promise of success.

Modelling the three dimensional circulation of the atmosphere, or a part of it, involves accumulating climatic data partly derived from satellite monitoring, processing it according to recognised physical laws, and obtaining an output that may be of some predictive value. Natural conditions such as sea surface temperatures and the distribution of snow cover and sea-ice, variables that may have an important bearing on the occurrence of departures from mean conditions of arid climates can be incorporated as inputs. In addition it is possible to investigate the effects of man's activities on the turbidity or dustiness of the atmosphere, the reflectivity of the ground surface, and the air's carbon dioxide content (Gates 1981).

Arid climates as affected by man

Bryson and Baerreis (1967) drew attention to the possibility that, in north-west India as an instance, increased wind erosion and dust transport might result in reduced transmission of solar radiation with a resulting increase in subsidence, suppression of convectional activity, and reduction of rainfall. Later studies have yet to confirm his conclusions but they deserve continued attention.

Otterman (1974), noticing the greater reflectivity of the surface on the heavily grazed Egyptian side of the border in northern Sinai as compared with the well vegetated Israeli side, reasoned that destruction of vegetation would reduce surface temperatures, lower sensible heat flux to the atmosphere, and suppress convective shower formation. Modelling studies over the last decade have tended to confirm Otterman's findings that higher albedo suppresses convective cloud and rainfall in arid regions, though with the possible exception of the Great Plains of North America (Sud & Fennessy 1982). In other words, heavy grazing, cultivation and collecting wood fuel may cause climatic as well biotic deterioration; an expanding desert thus feeds on itself.

The possible consequences of increased loading of the atmosphere with CO_2 released by the consumption of fossil fuel are still very uncertain. It is possible that the rise in temperature which was the main feature of global climate change in the first half of this century and which has been resumed in this its last quarter is a consequence of the greater absorption in the atmosphere of long-wave radiation emitted from the surface. Such a rise in temperature, by increasing evaporation losses on land, would increase aridity. However, the higher temperatures and possible greater energy of the winds would increase evaporation from the ocean surface, thereby promoting higher precipitation on land. Whether such a change in precipitation is taking place would be very difficult to detect on account of the high interannual variability of precipitation. Global circulation models have been constructed in an attempt to predict the distribution of changing values of the difference between precipitation and evaporation resulting from the enhanced 'greenhouse effect', but the discrepancies between the maps that have been produced do not at present inspire much confidence in the results.

Long term climatic history

There is a good deal of uncertainty as to the age of present aridity. The high degree of adaptation of plant and animal life to the aridity of the Namib suggests that a desert has been in existence in south-west Africa since the Miocene, some 15 million years ago, when cold bottom water started to accumulate in the ocean basins and began to upwell off the west coasts of the tropical continents. Evidence of early aridity is provided by evaporites and lithified dunes in sedimentary rocks distributed over all the continents and represented in rock strata from all the geological periods (Frakes 1979). Dust of aeolian origin in deep-sea cores tells of desert expansion in late Cenozoic times (Street 1981). Pollen in Quaternary sediments gives indications of increasing desiccation in the tropics early in the Ice Age (Bonnefille 1983). But the most striking evidence of marked alternations of changing environments in lands now arid comes from satellite imagery of continental plains and basins which exhibit features developed over the last 20 000 years of geological times, the last one per cent of the Quaternary period.

On imagery of the steppe, savanna and currently cultivated fields south of the Sahara we can easily discern the remains of ancient continental dunefields extending over millions of square kilometres; alongside are irrigated croplands on alluvial sediments bounded by the strandlines of former lakes. The soils, the vegetation cover and the entire landscape in its essentials are the outcome of climates alternately wetter and drier than those of the present day. Such features are clearest in zones a thousand kilometres wide at the margins of the tropical deserts where the spectacular changes in environmental conditions continued until less than 5000 years ago; the last great change of climate, involving desiccation which has persisted into the present, was in progress when the Pharaohs were already ruling a united Egypt.

In North America the main changes in climate took place much earlier, as the ice retreated c. 10 000 years ago, about the time when people first occupied the Great Plains. Satellite images of the southwest USA show strandlines left behind by Bonneville, Lahontan and a host of other pluvial lakes that dried up about that time, at the end of the Pleistocene; further north can be distinguished the long wave pattern of dunes and other aeolian features, shaped by the westerlies before the climate became warmer and wetter and the forest biomes shifted polewards (Wells 1983).

Elsewhere arid environments, it seems, were much more extensively developed in the closing stages of the last glaciation about 20 000 to 12 000 years ago than they are today, both in the tropics and in higher latitudes. The boreal forests and the equatorial forests were confined to much smaller areas than at present. Loess originating from the outwash of ice sheets and mountain glaciers was deposited by westerly winds over millions of square kilometres of the continental interiors in mid-latitudes, while desert dunes extended hundreds of kilometres nearer the equator than they do today.

The extensive spreads of aeolian sediments that accumulated in the glacial periods are the parent materials of soils that today produce much of the world's cereals and pulses. They obscure stony or old lateritised material beneath and are rather uniform and easily cultivable by machinery but they are also susceptible to renewed mobilisation by the wind and to erosion by gullies extending headwards from water courses and roadside ditches.

Whereas broad interfluves are the areas where aeolian sediments were commonly deposited and now provide the soils for rainfed agriculture, basin depressions in arid environments are the places where lakes and stream-borne sediments accumulated in the wetter periods of the Late Quaternary. The irrigated Gezira in the Sudan Republic is an immense sediment cone of the Blue Nile veneered with clay in early Holocene times (Williams & Adamson 1982). The flood-plain sediments of the Nile in Egypt accumulated as sea-level rose at the end of the Pleistocene. The South Chad scheme in north-east Nigeria irrigates black clays that accumulated in the early Holocene, about 12 000 to 4500 years ago, on the floor of a lake that occupied an area of c. 320 000 km^2. This compares with the present lake, from which the irrigation water is pumped, which has had an area of *c.* 20 000 km² for much of this century, but has shrunk in the last decade to a fraction of its normal size (Grove 1984).

The range of the climatic changes in lands now arid or semi-arid within the last twenty thousand years have been greater, especially in the tropics, than is sometimes realized. In semi-arid tropical Africa, at least, it seems likely that precipitation increased in many areas by as much as 600—800 mm between about 12 000 and 9500 years ago. Since about 5000 years ago mean annual values have diminished by about 300 mm (Grove 1984a). Over much of the world, plants adapted to aridity found conditions suited to them much extended at the height of the last glaciation and then much curtailed as conditions became wetter, favouring less tolerant competitors, about 10 000 years ago. The early Holocene wet period in the tropics was interrupted for several centuries about 10 000 and again about 7400 years ago, before conditions approaching present aridity became established (Street-Perrott & Roberts 1983). During this time, men were becoming pastoralists and cultivators and before long they were needing charcoal for smelting; anthropogenic desertification was set in train.

Cultivation involved the selection of plants for economic purposes, the transfer of plant material from one region to another, and deliberate as well as inadvertent change in the arid environment, notably the application of water to cultivated crops to make up for natural deficiency.

The irrigated arid environment

The world's irrigated area has increased from less than 10 million hectares at the beginning of the nineteenth century to more than 200 million hectares today. It constitutes about 13 per cent of the world's arable lands and uses 1400 km³ of water, about 15 times the Nile's annual discharge into Lake Nubia but little more than three per cent of the total volume brought down to the sea by the world's rivers. Over 80 per cent of the irrigated land is in arid environments, mostly in China, India and Pakistan, the USA and USSR (Worthington 1977).

There are two great dangers to be faced with regard to irrigation in the arid environment. Firstly, the water supply must be reliable. Sub-surface water is being exploited all over the world; almost invariably extraction exceeds supply to the aquifers and the level of water tables sinks with the consequent need for deeper wells and more powerful pumps. In many arid regions it is now known that the water being extracted originally accumulated in wetter periods in the Late Quaternary and is not being replenished. With industry and other urban users being prepared to pay high prices for water at present being used to grow crops, considerable areas in places like Texas and Arizona are likely to go out of cultivation over the next few years, or less water-demanding crops will have to be grown.

Much the same story can be told about the old semi-arid lands of southern Europe. The growth of tourism and the expansion of industry in coastal settings has attracted people from the interior uplands to the coastal cities. Traditional subsistence agriculture involving pastoral transhumance and the cultivation of cereals, olives, figs, almonds and vines, is giving way to commercial production of salad vegetables, melons, peaches and citrus fruit that require far more water. The competing demands for water from towns, hotels and industry in the flat, sandy coastal strips are extracting excessive amounts of water from aquifers of no more than moderate capacity with resultant incursions of salt water from the sea (Grove 1984b).

Irrigation is more generally dependent on the storage of rainy season stormwater in reservoirs; this is the long-established means of coping with the demands of agriculture in arid lands. There are opportunities for expansion but the best and cheapest storage sites are already being utilised. Furthermore, many of them are being destroyed and will no more be usable on account of sediment filling the reservoirs. Semi-arid lands have higher sediment yields per unit area than any other environments and reservoirs in semi-arid areas need very large catchments if they are to be supplied with any considerable volume of water. As a result, reservoirs receive enormous loads of sediments annually and their storage capacity diminishes accordingly, while the surface area exposed to evaporation remains little altered and the water they can supply rapidly shrinks.

The greatest threat to irrigation agriculture in the arid environment is salinisation. Most of the salt is supplied to the soil by irrigation water and from the underlying sediments. Precipitation of chlorides, sulphates and carbonates of sodium takes place in the soil layer when water tables are allowed to rise within a metre or two of the surface and upward capillary movement takes place into the root zone producing toxic salt levels. The threat varies greatly from one area to another. Waters draining from ancient lateritised upland plains in Africa have low solute loads and water from the upper Niger led over its Inland Delta may remove as much soluble material as it supplies. In the Gezira, the soils are sufficiently clayey to prevent much upward movement of water. But in many, probably most irrigated areas, salt is accumulating in soil profiles. The collapse of the Sumerian civilisation about 2000 BC has been attributed to salting of the soils of Mesopotamia. Over the world as a whole as much land is falling out of cultivation as a result of salinisation and waterlogging as is being reclaimed by irrigation. The solution clearly involves controlling the amounts of water applied to the land, preventing leaks from canals, and ensuring adequate drainage to keep soil water levels low. In the intermediate stages, the growth of salt tolerant plants may assist matters (Kovda 1977).

Heathcote (1983) draws attention to the fact that irrigation in arid environments of the USA requires enormous volumes of high-quality water that could often be used more profitably for agricultural or non-agricultural purposes in the humid areas from which it originates and concludes that the evidence from the USA points to a restricted future for irrigation in the arid lands. Outside the advanced countries, objections to irrigation are often linked to high social as well as economic costs.

The ecological changes associated with irrigation schemes range from the increased demand for fuel by settlers on the schemes (Hughes 1982), resulting in the denuding of nearby areas, to pests and diseases associated with permanent water alongside people (Grove 1984c). The vectors of malaria and schistosomiasis, mosquitoes and certain species of snails, are particularly difficult to eradicate. Locusts and quelea birds are provided with congenial living and breeding areas. Above all, capital intensive irrigation schemes remove from the people control over their own lands and livelihoods (Adams 1984).

In both developed and underdeveloped countries probably the highest priority in the future is for increasing the productivity of existing irrigation facilities through renovation, reducing water losses, providing drainage and improving management of water and labour. Not only are high yielding varieties of crops needed but also varieties resistant to drought and with minimum water requirements, and crops that are wanted by the people growing them.

Needs for the future

Of the enormous area distinguished as arid, only about one tenth is under cultivation, either irrigated or rainfed. By far the greater part of the arid environment is unused except by grazing and browsing animals and, as far as can be seen at present, is unlikely to be used for any other purpose. Rainfed agriculture is subject to drought and partly as a consequence people are tending to concentrate on those areas where rainfall is more reliable. In oil-rich countries with arid environments people are concentrating in cities near the coast and are abandoning the desert interiors. Much the same is true of lands around the Mediterranean and in the future it may be true of lands elsewhere. It seems possible that great areas of the arid lands that have been degraded may be less heavily used in the future and thus have the opportunity to recover.

For the future it might be suggested that economic plants will mainly be needed for urban areas where people in arid environments are congregating, for more intensive rainfed and irrigated cultivation and for the much more extensive areas, in part grazed or browsed but largely left as wilderness.

References

Adams, W.M. 1984. River control in West Africa. In *The Niger and its neighbours: environmental history and hydrobiology, human use and health hazards of the major West African rivers,* A.T. Grove (ed.), in press. Rotterdam: Balkema.

Berger, A. (ed.) 1981. *Climatic variations and variability: facts and theories.* Dordrecht: Reidel.

Bonnefille, R. 1983. Evidence for a cooler and drier climate in the Ethiopian uplands towards 2.5 Myr ago. *Nature* 303: 487–491.

Bryson, R.A. and D. Baerreis 1967. Possibilities of major climatic modifications and their implications: north-west India, a case for study. *Bull. Am. Met. Soc.* 48: 136–142.

Frakes, L.A. 1979. *Climates throughout geological time.* Amsterdam: Elsevier.

Faure, H. and J-Y. Gac 1981. Will the Sahelian drought end in 1985? *Nature* 291: 475–478.

Gates, W.L. 1981. The climatic system and its portrayal by climate models: a review of basic principles. I. Physical basis of climate, II. Modelling of climate and climatic change. In *Climatic variations and variability: facts and theories,* A. Berger (ed.): 3–19. Dordrecht: Reidel.

Grove, A.T. 1983. Evolution of the physical geography of the East African Rift Valley Region. In *Evolution, time and space,* R.W. Sims, J.H. Price and P.E.S. Whalley (eds): 117–156. New York: Academic Press.

Grove, A.T. 1984a. Changing climate, changing biomass and changing atmospheric CO_2. *Progress in Biometeorology* 3: 5–10.

Grove, A.T. 1984 b. Climate before the historical period. In press.

Grove, A.T. (ed.) 1984c. *The Niger and its neighbours: environmental history and hydrobiology, human use and health hazards of the major West African rivers.* Rotterdam: Balkema.

Hare, F.K. Climate and desertification. In *Desertification: its causes and consequences,* Secretariat, U.N. Conference on Desertification, Nairobi: 63–120. Oxford: Pergamon.

Heathcote, R.L. 1983. *The arid lands: their use and abuse.* London: Longman.

Hinchey, M.T. (ed.) 1979. *Proceedings of the symposium on drought in Botswana.* Gaborone: Botswana Society in association with Worcester, Mass. Clark University Press.

Hughes, F.M.R. 1982. Provision of fuelwood for irrigation schemes: the case of the Bura Irrigation Settlement Project, Kenya. In *Problems of the management of irrigated land in areas of traditional and modern cultivation,* H.G. Mensching (ed.): 50–60. Hamburg: University Geography Department.

Kovda, V.A. 1977. Arid land irrigation and soil fertility: problems of salinity, alkalinity, compaction. In *Arid land irrigation in developing countries,* E.B. Worthington (ed.): 211–236. Oxford: Pergamon.

Kraus, E.B. 1955. Secular changes of tropical rainfall regimes. *Q. J. R. Met. Soc.* 81: 138–210.

Lamb, P.J. 1982. Persistence of subsaharan drought. *Nature* 299: 46–48.

Meigs, P. 1953. World distribution of arid and semi-arid homoclimates. In *Reviews of research on arid zone hydrology. Arid Zone Programme* 1: 203–210. Paris: UNESCO.

Otterman, J. 1974. Baring high-albedo soils by overgrazing: a hypothesised desertification mechanism. *Science* 186: 531–533.

Russell, J.S. and A.W. Moore 1970. Detection of homoclimates by numerical analysis with reference to the Brigalow region (Eastern Australia). *Agric. Met.* 7: 455–479.

Russell, J.S. and A.W. Moore 1976. Classification of climate by pattern analysis with Australasian and southern African data as an example. *Agric. Met.* 16: 45–70.

Street, F.A. 1981. Tropical palaeoenvironments. *Progress in Physical Geography* 5: 157–185.

Street-Perrott, F.A. and N. Roberts 1983. Fluctuations in closed-basin lakes an an indicator of past atmospheric circulation patterns. In *Variations in the Global Water Budget,* A. Street-Perrott, M. Beran and R. Ratcliffe (eds): 331–345. Dordrecht: Reidel.

Sud, Y.C. and M. Fennessy 1982. A study of the influence of surface albedo and July circulation in semi-arid regions using the GLAS GCM. *J. Clim.* 2: 105–125.

Thornthwaite, C.W. 1948. An approach toward a rational classification of climate. *Geog. Rev.* 38: 55–94.

Tyson, P.D., T.G.J. Dyer and M.N. Mametse 1975. Secular changes in South African rainfall: 1880–1972. *Q. J. R. Met. Soc.* 101: 817–833.

Wells, G.L. 1983. Late glacial circulation over central North America revealed by aeolian features. In *Variations in the global water budget,* A. Street-Perrott, M. Beran and R. Ratcliffe (eds): 317–330. Dordrecht: Reidel.

Williams, M.A.J. and D.A. Adamson (eds) 1982. *A land between two Niles.* Rotterdam: Balkema.

Worthington, E.B. (ed.) 1977. *Arid land irrigation in developing countries.* Oxford: Pergamon.

3 Wild desert relatives of crops: their direct uses as food

Gary P. Nabhan[1] and Richard S. Felger[2]

[1]Office of Arid Lands Studies,
University of Arizona, Tucson, Arizona 85719, USA

[2]Office of Arid Lands Studies and
Instituto de Ecologia, Mexico City, Mexico

Introduction

Arid lands worldwide offer a rich variety of wild and indigenous cultivated food plants adapted to low and variable precipitation regimes. For example, the number of edible, ethnographically documented food plant species of the Sonoran Desert in southwestern North America is approximately 450, roughly 20 per cent of that desert's vascular flora – revised from preliminary estimate in Felger and Nabhan (1978). Estimates of similar magnitude can be made for several other deserts and semi-arid regions around the world, including those in India and Africa, e.g. Bhandari (1978) and Becker (1983).

Relative to the floras of other biomes, arid lands of the New World are particularly rich in certain kinds of food plants. These include CAM metabolism succulents such as agaves and columnar cacti with fructose-rich fruit, herbaceous and woody perennials with large starch-rich tuberous roots, and ephemeralized annuals with a high harvest index of protein- and oil-rich seeds. North American deserts have a high percentage of herbs with mucilaginous seeds with great hydroscopic capacities (Young & Evans 1973). The Sonoran and other tropical-derived New World hot deserts are vegetationally and floristically rich in woody legumes with nutritionally significant seeds and pods.

It is ironic that much of the modern agricultural developments in arid zones depend on temperate or tropical crop species that are not well-adapted to high heat, low soil moisture and low humidity. These plants require large amounts of irrigation water as well as micro-environmental modifications to be economally productive. Although these crops are good yielders per unit area given supplements of groundwater or river-diverted irrigation, these strategies are costly, both economically and energetically. Irrevocable groundwater depletion in arid zones is becoming a common tragedy worldwide. For example, groundwater extraction in the USA tripled between 1950 and 1975, and the mining of groundwater greatly exceeds recharge (Office of Technology Assessment 1983). In regions of Arizona, as well as Sonora and Baja California, Mexico, the costs of pumping poor quality groundwater from depths of 100 m or more are not met by the economic yields of conventional crops, so hundreds of thousands of hectares of farmland are being abandoned each year (Foster *et al.* 1980).

River diversion likewise has created profound and irreversible resource degredation, as well as conflicts of use, both nationally and internationally. Because of this degredation, destruction of the river systems and competition for what water remains, the irregular stream flows of the arid and semi-arid USA no longer adequately supply farmers in two-thirds of the irrigated districts in the western USA (Office of Technical Assessment 1983).

Yet even with the expensive inputs to the transplanted tropical/temperate crops, we are not likely to see much in the way of significant increases in the near future. Most of these crops have reached their physiological ceilings for yields under optimal conditions (Evans 1980). Because more than 3 million hectares of prime U.S. farmland has been converted to nonfarm uses, including more than ½ million hectares in the cornbelt, we can expect more marginal lands to have to pick up this lost productivity (Sampson 1981). Yet humid-adapted crops need 20–40 per cent more water to produce the same yields in arid zones as they do in temperate zones (Pimental *et al.* 1982). When not buffered from the extremes of the desert, these humid-adapted conventional crops perform poorly when compared to desert-adapted minor crops and wild species. This has been demonstrated by comparisons of seed set in heat intolerant pinto beans (*Phaseolus vulgaris*) with domesticated tepary (*P. acutifolius*) and wild desert beans (*P. filiformis*) in the extremely arid Pinacate Desert region of Sonora under runoff agriculture (Nabhan 1983).

Accordingly, eminent plant breeders such as Borlaug (1983) have suggested that we have the most to gain in world food production by the utilization of germ plasm better adapted to environmental stresses in marginally productive zones. Borlaug (1983) sees this gain primarily accomplished by incorporating into conventional crops hardy genes from wild species adapted to stress conditions, i.e. the indirect use of wild genetic resources.

However, other scientists have favoured the development of altogether new crops for arid zones (Felger & Nabhan 1978, National Academy of Sciences 1975). When most people talk about 'new' crops they are often including ancient or underexploited but already domesticated crops such as the grain amaranths and domesticated tepary beans. We feel that the concept of new crops in the true sense should be restricted to the process of taking undomesticated wild species with economic potential and selecting or breeding the germplasm to bring into cultivation superior strains.

Whereas some new desert crops, such as jojoba, fit this description, many do not. Most such new crop inventories include a certain number of:

a.   Minor crops that have long been domesticated but which are currently under-utilized.

b.   Wild species closely related (congeneric) to one or more domesticated or economic horticultural crops. For such plants there often already exists considerable genetic, eco-physiological, chemical data, as well as cultivation, propagation, processing, and preparation information.

We will examine this latter group of plants to evaluate its relative potential for the development of new food crops for arid lands. In the following discussion we refer to these plants as 'wild food species in the economic genera', after Ames's (1939) discussion of economic annuals and human cultures. We emphasize that these are but one subgroup of potential new crop species deserving further evaluation. In general, we advocate screening arid-adapted plants of several life forms for integration into desert agro-ecosystems (Fig. 3.1, p. 22) that would be structurally complex and generically diverse (Felger 1979, Nabhan 1985). The plants we discuss share the following characteristics:

a. They apparently have more tolerance or resistance to abiotic or biotic stresses than do most congeneric commercial cultivars.

b. They offer either a greater range of economic products or one product of superior chemical content when compared to conventionally cultivated counterparts.

c. Their breeding systems, genes, physiological pathways, and chemical content are often better known than those of other potential new crops, and there is already cultural acceptance.

d. There is a rich history of uses and knowledge of these plants by indigenous peoples recorded in ethnographic literature and/or learnable from present day local people.

We have identified a number of economic genera that have wild desert-dwelling relatives in the Western Hemisphere. For these purposes we have surveyed the vascular plant genera, mostly New World, with domesticates and horticultural clonal varieties of considerable antiquity as noted in Dressler (1953), Zevon and de Wet (1982), and others. We have then consulted regional North American floras for wild species in these genera occurring in deserts. From this inventory we have selected examples of those species which we feel have superior adaptations to arid lands, economic products, and adaptability to cultivation.

To highlight their potential value we provide brief sketches of selected crop candidates below. Surprisingly, many of these wild species remain locally utilized and commercially marketed, and some are or have been exported beyond their natural geographic ranges.

The number of wild relatives of domesticated or long-cultivated plants is probably relatively large. For example there are several dozen species of these wild relatives in the Sonoran Desert alone, many of which have agronomic promise. In this paper we discuss a span of life-forms in such diverse genera as *Agave* (Agavaceae), *Capsicum* (Solanaceae), *Ceiba* (Bombacaceae), *Cnidoscolus* (Euphorbiaceae), *Cucurbita* (Cucurbitaceae), *Lycium* (Solanaceae), *Opuntia* (Cactaceae), and *Phaseolus* (Leguminosae). These range from those which are being brought into cultivation and known to be cross-compatible with well-known domesticated relatives (e.g. *Agave angustifolia* and *Capsicum annuum* var. *aviculare*), to those which have not yet been cultivated and have underexploited congenerics which are little known outside their local areas of domestication or cultivation (e.g. *Cnidoscolus* and *Lycium*).

Sonoran Desert Agrisystem

Intercropping plants of several lifeforms

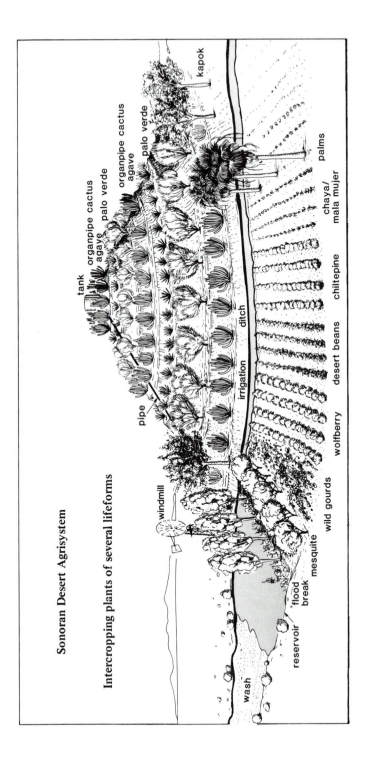

Figure 3.1 Diagrammatic representation of a desert water-harvesting agro-ecosystem with diverse perennial lifeforms, as conceptualized by Mirocha and Nabhan. High yielding plants not discussed in the text are the palms and two legumes, mesquite and palo verde. Sonoran Desert palms with economic potential include two species of *Washingtonia* and several species of *Brahea* (= *Erythea*) which have edible fruit. Mesquite, *Prosopis* spp. section *Algarobia*, is well-known for fuel wood and edible pods and seeds. Several species of palo verde, *Parkinsonia* (= *Cercidium*) spp., produce nutritionally significant seeds under very hot, arid conditions.

copyright Nabhan/Mirocha 1984

Agave angustifolia (mescal bacanora, dwarf sisal)

This monocarpic perennial succulent is the most wide-ranging and variable of the approximately 140 species of *Agave*. It ranges through much of the arid and semi-arid zones of Mexico, from 150 km south of the U.S. border in Sonora southward to Guatemala, and eastward to Tamaulipas (Gentry 1982). Its rosettes of sword-like leaves are usually 0.8–1.2 m long, but occasionally achieve a length of 2 m. Basal stolons produce clonal offsets, and this vegetative suckering is the plant's primary mode of reproduction. After 8–15 years each rosette sends up a 3–5 m tall panicle of flowers, which are usually bat-pollinated. These reproduce tens of thousands of seeds, and then the rosette dies. However, under cultivation the life cycle of Sonoran plants can be shortened to about six years.

The primary use of the wild plant within its natural range, especially in north-western Mexico, is for the production of a distilled alcohol known as mescal bacanora, or simply, mescal. Prehistorically the pit-roasted 'heart' or 'cabeza' (stem and leaf bases, was eaten as a sweet, calorie-rich food, or fermented into a mild beer, rather than being fermented and then distilled as done today. Bahre and Bradbury (1980) describe the cottage industry based on this plant in Sonora, where the local product is preferred over commercial tequila made from its close relative, *Agave tequiliana.* The Sonora state government recently has encouraged experimental plantings near Sahuaripa and Onavas, Sonora, with the hope of producing enough alcohol from field plantings to augment the supply of this regional beverage. It is estimated ½ million plants of this species are harvested from wild stands in Sonora each year (Nabhan 1984b), such that natural populations are seriously being depleted.

The other major economic use of *A. angustifolia* is as a source of hard long-fiber. It has been predicted that a hybrid, known as #11648 – a product of the backcross (*A. amaniensis* ✕ *A. angustifolia*) ✕ *A. amaniensis* – will replace common sisal, *A. sisalana,* in African fiber production. The advantage of this hybrid is due to its superior leaf configuration, fiber quality and absence of lateral spines on the leaves (Wienk 1976). However, it is vulnerable to *Phytophthora* root rot.

Within the wide-ranging *A. angustifolia* gene pool there is considerable variation in economic characters and environmental adaptations. *A. angustifolia* also has been hybridized with *A. amaniensis, A. cantala,* and *A. sisalana* (Osborne & Singh 1980), and we believe that it is a progenitor of *A. tequiliana.* However, plant breeders have made little use of desert ecotypes to improve heat tolerance or water use efficiency of these crops. Wild *A. angustifolia* produces well in frost-free areas of Sonora, where mean annual precipitation is as low as 235 mm, and survives in these sites in years with as little as 75 mm rainfall (Hastings & Humphrey 1969). Such localities typically suffer July mean ambient temperatures of over 32°C.

Although we know of no data on the water use efficiency of this species, agaves are reported to have transpiration ratios which are among the lowest of any plants in the world (Nobel 1976). This measure of water use efficiency per unit weight of plant biomass harvested indicates that agaves are generally twice as efficient as maize, with transpiration ratios of 71:136 (Ehrler 1975).

We believe that a one-product effort of agave fiber or sugar production will remain relatively uneconomic for some time to come. However, alcohol and food production, with secondary extraction of fiber and steroids could prove highly lucrative. Steroids account for less than 5 per cent of the leaf weight, but are extractable for many pharmaceutical uses (Gentry 1982). They may also be the active ingredient in an agave leaf extract that promotes bacterial growth to reduce sewage odours in barnyard ponds and sewage holding tanks (R. McDaniel, pers. comm.). Such a product has already been patented and marketed in California.

Capsicum annuum var. *aviculare* (chiltepine, chile del monte, chile pequin)
As the most widely distributed wild taxon in the genus *Capsicum,* the chiltepin naturally ranges as far north as mountains along the Mexican border in southern Arizona and extends southward into Colombia. This frost-sensitive perennial shrub is fast-growing once established, and obtains heights of 2–3 m in the understory of riparian desertscrub and thornscrub vegetation. Usually regarded as a tropically-adapted species, the chiltepin ranges well into the Sonoran and Chihuahuan Deserts where pond evaporation may be in excess of 1500 mm. We consider such outlying populations to be desert ecotypes deserving of greater scientific investigation.

Throughout its range the chiltepin produces fiery red, bird-dispersed fruit. The high capasicin content of the fruit (2600 parts per million) undoubtedly serves as an irritant to mammals that would be ineffectual seed dispersers (Maga 1975). Hundreds of fruit are produced per m³ of canopy each year, with yields peaking in late August through early October in the North American deserts.

Rich in vitamins C, A, riboflavin, fiber, and sodium (Maga 1975), these fruits are highly esteemed by desert people for their taste, pungency, and usefulness as an anti-oxidant food preservative. They are used in an immature, green form as a condiment, sold fresh or pickled with vinegar, garlic and wild orégano (*Lippia palmeri*). They are also sold in their mature red form. This dried spice currently sells for $28–32 per lb in the USA and almost a third of that in Mexico. Chiltepines are the major ingredient in several extremely popular brands of hot sauce in Sonora and Sinaloa (e.g. 'Yaqui Salsa Brava'), and are now the featured ingredient in one California brand.

Wild harvests of these small chiles have been as high as 15 tonnes per year in Sonora. However, overgrazing, overharvesting and occasional droughts or freezes have combined to reduce population densities (Nabhan 1984b). Within the last three years, Alfredo Noriego of Hermosillo, Sonora, has developed vegetative propagation of three superior-yielding plants, and transplanted several thousand cuttings into a five hectare field; 10 000 plants/ha can produce one tonne of fruit. He estimates that wholesale value of the annual production of one hectare may exceed $3500 USA dollars at current prices. However, more than 15 ha will likely saturate the current local market (Alfredo Noriego and colleagues, pers. comm.) and a wider market will be needed.

Sr. Noriego irrigates plants six times per year in an extremely arid area near Kino

Bay, Sonora. However, this plantation, with its narrow genetic base, has suffered from high mortality due to root disease. Chiltepines are cross-compatible with most domestic jalapeño and bell peppers, and have been used as a source of resistance to three strains of virus (Villalon 1983).

Ceiba acuminata and *C. parvifolia* (pochote, kapok)

C. acuminata occurs in semi-arid subtropical scrub (thornscrub) of western Mexico and ranges northward to the southern margins of the Sonoran Desert in Sonora, Mexico. It is a fast-growing, openly-branched, drought-deciduous tree, and has large, blunt spines on the trunk, and large, bat-pollinated flowers. The 20 cm-long kapok-filled fruit has been occasionally commercially harvested. The seeds are edible and have a nutlike flavour (Gentry 1942). The closely related *C. parvifolia* from semi-arid regions of Puebla and Oaxaca in Mexico produces large quantities of huge tuberous roots, locally called camotes (see fontispiece), and *C. acuminata* seems to be essentially similar in this regard.

The Yaqui and other Indian groups relied heavily on the tuberous roots of *C. acuminata*. The Yaqui preferred to harvest them at the end of the summer rainy season while the camotes were turgid and succulent (Roman Borbón, pers. comm.). The roots commonly were roasted in hot coals. There are several closely related species in Mexico. In the Tehuacán Valley, in the state of Puebla, the large camotes of *C. parvifolia* have served local people as a wild-harvested staple for thousands of years (Callen 1973), and are still occasionally harvested (see frontispiece).

Ceiba acuminata and *C. parvifolia* should be investigated as a perennial root crop for hot, semi-arid regions. Propagation is by seed and perhaps also by cuttings. The plants might be coppiced to maintain them as shrubs, and the young, tender tuberous roots harvested annually. The younger roots are suitable as food for humans, while the older and larger, more fibrous ones might find use as animal food. While we do not yet have yield data, *C. parvifolia* produces prodigious quantities of large tubers, and the present ethnographic data indicates substantial yields for *C. acuminata*. Preliminary analysis of large, tuberous roots of *C. parvifolia* from the Tehuacán Valley indicated that the nutritional value is not high. However, these were old, and somewhat woody camotes harvested during the dry season.

The floss from the seeds of *C. pentandra* is the primary commercial kapok fiber, but it also occasionally has been harvested from the fruit of various other species, including *C. acuminata* (Uphof 1968). The young leaves of *C. pentandra* are used as a pot herb, the young fruit is eaten raw, and the petals, germinated seeds, and oil from the seed are eaten (Watt 1889, Burkill 1935, Dalziel 1937). This economically important tree has long been cultivated and domesticated forms have been developed in tropical Asia and elsewhere (Burkill 1935, Ochse & van den Brink 1980).

Cnidoscolus palmeri (mala mujer, ortiguilla)

This little-known genus offers an incredible range of agronomic possibilities. There are about 60 species restricted to warm regions of the New World, some of which yield rubber, spinach-like leaves, protein- and oil-rich seeds, and potato-like roots. These attributes are known among nine of the species and some of the other species are expected to share similar characteristics. Hybridization potentials remain unknown.

The most striking features of the genus is the presence of obnoxious stinging hairs covering the plants which can inflict a painful rash. However, there are domesticated forms without stinging hairs or essentially lacking them, such as *C. chayamansa* or chaya. Unknown in the wild, it has been cultivated as a major leaf vegetable among the lowland Maya and others since ancient times in southern Mexico and Central America (Berlin *et al.* 1974, National Academy of Sciences 1975). Another hairless form, a selection of *C. phyllacanthus,* is known from Brazil (Braga 1960).

Two species of *Cnidoscolus* are known to have edible roots. *C. palmeri,* endemic to the Gulf of California region, has delicious potato-like tuberous roots. This woody shrub, 1−1.5 m high, grows in rock crevices on steep granite slopes along the Gulf Coast of Sonora, Baja California Sur, and several islands in the Gulf.

Each shrub has several dozen fleshy tuberous roots 5−20 cm or more in length. These roots are potato-like, but often compressed as they grow wedged between rocks. Several tubers develop at irregular intervals along each slender lateral root which may extend 1−2 m from the centre of the plant (Felger & Moser 1984). A single shrub may have 15 kg or more fresh weight of tubers.

The tubers are edible raw or cooked. Fresh, raw tubers are crisp, succulent and white like a raw potato and taste somewhat like jícama or yam bean, *Pachyrhizus erosus.* Cooked they are like potatoes. Starch content (dry weight or 'moisture free' basis) measured 82.2% and protein 7.5% (Felger *et al.* 1981). This compares favourably with the potato, although the whole protein content is higher for *C. palmeri* tubers.

These tubers are suitable for harvesting at any time of the year and apparently store well (Felger & Moser 1984). The tubers were roasted in hot coals or on a grill. The blackened bark, or skin, was removed and the inside eaten, often with honey. The young tubers were said to be best.

The other species with an edible root is a small, herbaceous perennial in Florida, aptly named *C. stimulosus,* or tread-softly. It produces a single, parsnip-like edible root (Morton 1977). Many other species of *Cnidoscolus* produce fleshy or tuberous roots, but it is not known if they are edible.

C. maculatus, ortiguilla, produces substantial quantities of relatively large edible seeds rich in protein and oil content (Felger *et al.* 1981). It is endemic to the southern part of the Baja California peninsula. In addition other species of *Cnidoscolus* produce rubber as well as edible seeds (Williams 1962).

Cucurbita digitata and C. sororia (coyote gourd)

The *C. digitata* species complex of perennial gourds represents the most arid-adapted *Cucurbita* taxa in the Western Hemisphere (Bemis & Whitaker 1969). *C. digitata* ranges from extremely arid areas of Baja California Norte across the Sonoran Desert into the northwestern area of the Chihuahuan Desert in Texas and Chihuahua. Cross-compatible taxa include *C. palmata* in the Colorado River Delta area, where it naturally hybridizes with *C. digitata, C. cylindrata,* and *C. cordata* (Bemis & Whitaker 1965). These latter taxa are largely restricted to the Baja California peninsula of northwestern Mexico. Members of this xerophytic species complex thrive in natural conditions of less than 150 mm of annual rainfall, and might produce economic harvests on 250 mm.

Members of the taxonomically confused *C. sororia* complex are not usually considered xerophytic. However, a population of this species complex (resembling *C. palmeri* and *C. sororia*) at the northern limits of this group's natural range, extends into the southern part of the Sonoran Desert where annual rainfall is *c.*350 mm. There this wild cucurbit freely hybridizes with cultivated *Cucurbita mixta* (Merrick & Nabhan 1984). Striped cushaw land races of *C. mixta* are the most xerophytic cultivated squashes in the world (Whitaker 1968). We believe that all of the above species are better adapted to hot low deserts than *C. foetidissima,* the xerophytic gourd which has received considerable agronomic attention (Bemis *et al.* 1978; Bemis *et al.* 1979).

These xerophytic gourds all have certain characteristics in common: the production of 200–300 seeds per fruit, which contain roughly 30% oil and 30% protein (Bemis *et al.* 1967); and the presence of oxygenated tetracyclic triterpenes known as cucurbitacins in their foliage, cotyledons, fruit, and roots (Metcalf *et al.* 1982). Seed of *C. cordata* contain 16.1% oil, *C. digitata* 20.2%, *C. palmeri* 34.5%, *C. palmata* 27.9–31.6%, and *C. sororia* 34.1% (Bemis *et al.* 1976). However, *C. digitata* accessions have a higher percentage of conjugated unsaturation in fatty acids relative to other cucurbits. Since there is correlation between polyunsaturated oils and coronary disease, they may be more suited to consumer acceptance.

The *C. sororia* complex could be a key candidate for Latin American seed-derived products such as 'mole pipian', which are currently derived from *C. mixta* land races. For these products the fleshy part of the squash is commonly discarded or fed to livestock. By producing a smaller gourd and more fruit per plant, seed production in wild *C. sororia* complex is potentially greater than that of *C. mixta* land races which maximize fruit size. Natural hybrids and backcrosses of *C. sororia* with *C. mixta* are sometimes sweet with intermediate seed sizes (Nabhan 1984a) and would make ideal material for selecting high seed production per plant.

If oil were the only product of these wild cucurbits and their derivitives, there would be difficulty in competing with many other seed crops in the marketplace. However, proteinaceous seed meal and cucubitacins for biological control of *Diabrotica* beetles in conventional crops may soon stimulate commercial production. *C. palmata* and *C. cylindrata* contain cucurbitacins E and I and a glycoside form, with the quantity in the roots being quite substantial. *C. sororia* and *C. palmeri* produce cucurbitacins B and D. When extracted, these chemicals can be used either as sprays for trap crops to attract beetles away from vulnerable crops, or when mixed with small quantities of insecticides, and sprayed on the whole crop, as a way to specifically kill *Luperini* beetle pests while hardly affecting other insects (Metcalf *et al.* 1982). The economic market for effective control of corn (maize) rootworm alone may stimulate commercial production of wild cucurbits.

Lycium fremontii (desert wolfberry, tomatillo)
This genus contains about 100 species, mostly in tropical and warm regions of the world; many are desert species and/or adapted to alkaline or saline soils. The domesticated Chinese boxthorn or Chinese wolfberry, *L. chinense,* is widely culti-vated in tropical regions of southeast Asia for its edible leaves and vitamin-rich fruit (Burkill 1935, Ochse & van den Brink 1980, Zee & Hui 1981). Several other species of *Lycium* have long been cultivated as ornamentals. Most, if not all lyciums are readily propagated by seed and cuttings.

There are about a dozen species in southwestern United States, most of which have edible fruit. One of these, *L. fremontii,* seems to have agronomic promise. It is a thorny shrub native to southern Arizona and the Gulf of California region in northwestern Mexico. It thrives on desert soils, upper beaches and semi-saline and alkaline flats both near the coast and on inland deserts. The sweet and mildly tart fruit, produced in great quantity in spring, is a significant source of vitamin C and iron (Greenhouse 1979).

The fruit has long been esteemed by the desert people in southwestern North America (Castetter & Bell 1942, Bean & Saubel 1972, Felger & Moser 1976). The Seri, Pima, and other Indians still pick the berries in spring. A common method of preparation is to boil the fruit and consume it plain or sweetened with sugar or honey. The fruit was also dried and stored, later to be mixed with water and then consumed. Today the Pima sometimes prepare it as jam or preserve (Greenhouse 1979). The Seri, Pima, and other Indians still pick the berries in spring. A common method of preparation is to boil the fruit and consume it plain or sweetened with sugar or honey. The fruit was also dried and stored, later to be mixed with water and then consumed. Today the Pima sometimes prepare it as jam or preserve (Greenhouse 1979). The Seri know of some bushes which yield fruit so sweet that it does not require sweetening with sugar or honey.

Each shrub commonly produces prodigious quantities of juicy, bright orange-red berries. Plants harvested by the Seri Indians in Sonora, Mexico, have berries averaging 12.5 mm long, with a range of 7−18 mm (Felger & Moser 1984). The fruit ripens in March and April.

The desert lyciums seem to be easy to cultivate, and this species should lend itself to development as an arid-adapted fruit crop. It will certainly tolerate poor quality water, and perhaps brackish water. The Arizona populations certainly are tolerant of moderate amounts of freezing weather in winter, and the shrub will survive even prolonged drought. In the absence of later winter rainfall, probably one or two irrigations would insure a good fruit crop. It could be developed as a berry-producing bush in the desert, and harvested by hand or with machinery like that developed for raspberries and grapes.

Opuntia fulgida (jumping cholla)

This common cactus proliferates across the desert floor throughout most of the mainland region of the Sonoron Desert. It has one or more woody trunks and commonly reaches heights of 1.5–2 m. Pendulous chains of fruit hang from the upper branches, and may remain attached for several years. There is great variation in spininess condition within the species, and even within populations at a single site. For example var. *fulgida* is characterized by dense spines, so dense that the surface of the plant cannot be seen through the spines, whereas var. *mammillata* has many fewer spines and the green stem is readily visible. These varieties and many minor variations are often sympatric because the plants commonly reproduce clonally by fallen joints or even fruits, which root and grow.

The fruit is fleshy, green even when ripe, and lacks spines but has many glochids. There is considerable variation in fruit size, which seems to be genetically controlled. At least along or near the coast of Sonora in the Seri Indian region, plants with extra large fruit occur among both varieties in widely scattered places. Throughout most of its range the fruit of this species is about 2.5–4 cm in length, and although it was on occasion eaten by the native people, it is of little merit. However, the Seri Indians harvest fruit from specific stands and plants which consistently produce fruit several times larger than usual. These larger-fruited plants are wholly sympatric with the smaller-fruited plants. The exceptionally large cholla fruit, which is tart but sweet, compares well with that of the economically important and long-domesticated meso-American prickly pears such as the *O. ficus-indica*/*O. megacantha* complex*. Disadvantages of the cholla fruit are that the ovary wall is relatively thick and the seeds rather large, both parts of which are generally discarded. However, there is so much variation within wild populations that even preliminary selections should be able to identify superior stock. There are advantages for developing large chollas rather than large prickly-pears for very hot and arid regions. The large chollas with their cylindrical stems have a much lower surface-volume ratio than do the large prickly-pears with their pancake-shaped stems- 'pads' or cladodes (Nobel 1981). Large prickly pears do not extend into extreme desert due to constraints placed on them by their physiological capabilities (Felger 1980).

Phaseolus ritensis and *P. filiformis* (desert beans)

Phaseolus ritensis is the most arid-adapted perennial bean of the eight species of *Phaseolus* in the metcalfei section, a group of wild beans centered in the Sierra Madre Occidental of western Mexico (Piper 1926, Nabhan *et al.* 1980). It is a summer emergent that sends out vines 4–5 m long from a woody, tuberous root with foliage dying back in fall. In the northern part of its range this species ranges from 1200–2500 m elevation where annual rainfall is as little as 300 mm. It has

---

*Most authorities regard *O. megacantha* as a synonym of *O. ficus-indica,* the former being the spiny form, the latter spineless (G.E.W.).

been successfully crossed with the distantly related common bean, *P. vulgaris* (Braak & Kooistra 1975) and with the relatively closely related lima bean, *P. lunatus* (Baudoin 1981, Buhrow pers. comm.).

The seed of *P. ritensis* has been used by local people for food, and the roots as glue and as a fermenting agent, and the roots are still sold for medicinal use (Nabhan *et al.* 1980). Seed protein content ranges from 20.5–30.9% and it has a favourable cystine content. The seed is about the size of the common garden pea. The species is resistant to bean golden mosaic virus (Baudoin 1981), which is a serious plague for all beans in both Africa and South America. In the mountains of northwestern Mexico the Mountain Pima Indian farmers have brought it into cultivation.

P. filiformis will grow in conditions too arid for any other *Phaseolus,* ranging into areas with less than 80 mm annual rainfall. In extremely arid regions it is a short-lived ephemeral, and can germinate either with hot or cool weather rainfall, but can persist up to seven months with adequate soil moisture. It is rather small seeded, but extremely drought and heat tolerant (Nabhan 1983). The mature seeds and immature pods were eaten by the Sand Papago and Mexicans who lived in the extremely arid Pinacate Region of northwestern Sonora, Mexico. Protein content of the seed is as high as 26.1% (Ariffin 1984). Caches of seeds of this desert bean have recently been recovered from pre-pottery archaeological sites in southern Arizona which are probably at least 2000 years old. This species also has resistance to bean golden mosaic virus. As a light-frost-hardy legume that can both tolerate and escape drought (the latter by virtue of its telescoped life cycle) this species has remarkable potential in marginal lands. However, attempts to cross it with *P. vulgaris* have run into reproductive barriers (Maréchal & Baudoin 1978).

If superior selections are developed from the primary gene pools of these legumes, it may be possible to augment bean yields in hot and extremely arid areas now lacking significant production. In this way, these wild species may follow in the steps of the domesticated tepary (*P. acutifolius*), which is undergoing a revival in arid lands. Bean breeders have a wealth of variation to draw upon for selection and improvement.

Conclusions

Arid lands harbor a vast reservoir of potential new crops. Plant breeders are becoming increasingly aware of wild, desert-dwelling, congeneric relatives of cultivated species as sources of genes useful in crop improvement for water use efficiency, drought, heat and salt tolerance, pest and disease resistance and chemical content. However, these genetic resources have been used directly as food for millenia and continue to offer benefits to mankind. In addition, hybridization and inovative selection techniques offer powerful means for further development of these desert-adapted economic plants. Some are cross-compatable with domestic species, while for others the breadth of useful gene pools remain unknown.

For some of these genera, we already have a basic understanding of their genetic, physiological and agronomic characteristics. The desert plants discussed here have eco-physiological advantages over their agronomic congenerics when grown under conditions of limited water availability. All survive better under drought conditions, and a few may also have superior water use efficiency ratios than conventional crops. The ethnobotanical record has demonstrated that their food products are not inferior in taste, nutritive quality or utility. Some of these North American arid land species are still wild-harvested or under incipient cultivation on a cottage-industry scale.

Scientists have much to learn from native peoples conversant with these plants, and if properly re-oriented, scientists have much to offer humanity in the development of arid land resources. We can build on the millenia-old knowledge to help transform arid land agriculture to be more productive while also retaining water and energy efficiency.

References

Ames, O. 1939. *Economic annuals in human cultures.* Cambridge, Mass.: Harvard University.

Ariffin, R. 1984. *Proximate analysis of Sonoran Desert food plants.* Tucson: University of Arizona thesis.

Bahre, C.J. and D.E. Bradbury 1980. Manufacture of mescal in Sonora, Mexico. *Econ. Bot.* 34(4): 391–400.

Baudoin, J.P. 1981. Observations sur quelques hybrides interspécifiques avec *Phaseolus lunatus* L. *Bull. Rech. Agron. Gembloux* 16(4): 273–286.

Bean, L.J. and K.S. Saubel 1972. *Temalpakh.* Banning, Ca.: Malki Museum Press.

Becker, B. 1983. The contribution of wild plants to human nutrition in the Ferlo (Northern Senegal). *Agroforestry Systems* 1: 257–267.

Bemis, W.P. and T.W. Whitaker 1965. Natural hybridization between *Cucurbita digitata* and *C. palmata. Madroño* 18: 39–47.

Bemis, W.P. and T.W. Whitaker 1969. The xerophytic *Cucurbita* of northwestern Mexico and southwestern United States. *Madroño* 20(2): 33–41.

Bemis, W.P., J.W. Berry and C.W. Weber 1979. The buffalo gourd – a potential arid land crop. In *New agricultural crops*, G.A. Ritchie (ed.): Boulder, Co.: 65–87. Westview Press.

Bemis, W.P., J.W. Berry, C.W. Weber and T.W. Whitaker 1978. The buffalo gourd: a new potential crop. *HortScience* 13: 235–240.

Bemis, W.P., J.W. Berry, M.J. Kennedy, D. Woods, M. Moran, and A.J. Deutchman, Jr. 1967. Oil composition of *Cucurbita. J. Amer. Oil Chem. Soc.* 44: 429–430.

Berlin, B., D.E. Breedlove, and P.H. Raven 1974. *Principles of Tzeltal plant classification.* New York: Academic Press.

Bhandari, M.M. 1978. *Flora of the Indian Desert.* Jodhpur: Scientific Publishers.

Borlaug, N.E. 1983. Contributions of conventional plant breeding to food production. *Science* 219(4585): 689–694.

Braak, J.P. and E. Kooistra 1975. A successful cross between *Phaseolus vulgaris* L. and *Phaseolus ritensis* Jones with the aid of embryo culture. *Euphytica* 24: 669–679.

Braga, R. 1960. *Plantas do Nordeste, especialmente do Ceará,* 2nd ed. Ceará, Brazil: Impresa Oficial, Fortaleza.

Burkill, I.H. 1935. *Dictionary of the economic products of the Malay Peninsula.* London: Crown Agents for the Colonies.

Callen, E.O. 1973. Dietary patterns in Mexico between 6500 BC and 1580 AD. In *Man and his foods*, C.E. Smith (ed.): 29–49. Alabama: University of Alabama Press.

Castetter, E.F. and W.H. Bell 1942. *Pima and Papago Indian agriculture.* Albuquerque: University of New Mexico Press.

Dalziel, J.M. 1937. *The useful plants of West Tropical Africa.* London: Crown Agents for the Colonies.

Dressler, R.L. 1953. The pre-columbian cultivated plants of Mexico. *Bot. Mus. Leafl., Harv. Univ.* 16(6): 115–172.

Ehrler, W.L. 1975. Environmental and plant factors influencing transpiration of desert plants. In *Environmental physiology of desert organisms*, N.F. Hadley (ed.): 52–67. Stroudsburg, Pa.: Dowden, Hutchinson & Ross.

Evans, L.T. 1980. The natural history of crop yields. *Amer. Scient.* 68: 388–397.

Felger, R.S. 1979. Ancient crops for the twenty-first century. In *New agricultural crops*, G.A. Ritchie (ed.): 5–20. Boulder, Co.: Westview Press.

Felger, R.S. 1980. Vegetation and flora of the Gran Desierto, Sonora, Mexico. *Desert Plants* 2(2): 87–114.

Felger, R.S., L.S. Leigh, S.L. Buchmann, D.O. Cornejo, M.A. Dimmitt, D. Johnson Gordon, C. Nagel, L. Ratener and C.A. Stigers 1981. Inventorying the world's arid lands for new crops: a model from the Sonoran Desert. In *Proceedings of arid lands resource inventories workshop*, H.G. Lund (ed.): 106–116. USDA For. Ser. Gen. Tec. Rpt. WO–28.

Felger, R.S. and M.B. Moser 1976. Seri Indian food plants: Desert subsistence without agriculture. *J. Ecol. Food Nutr.* 5(1): 13–27.

Felger, R.S. and M.B. Moser 1984. *People of the desert and sea: Ethnobotany of the Seri Indians.* Tucson: University of Arizona Press.

Felger, R.S. and G.P. Nabhan 1978. Agroecosystem diversity: a model from the Sonoran Desert. In *Social and technological management in dry lands*, N. Gonzalez (ed.): 129–149. Boulder, Co.: Westview Press.

Foster, K.E., R.L. Rawles, and M.M. Karpiscak 1980. Biomass potential in Arizona. *Desert Plants* 2(3): 197–201.

Gentry, H.S. 1942. *Rio Mayo plants.* Washington, D.C.: Carnegie Inst. Publ. 527.

Gentry, H.S. 1982. *Agaves of continental North America.* Tucson: University of Arizona Press.

Greenhouse, R. 1979. *The iron and calcium content of some traditional Pima foods and the effects of preparation methods.* Tempe: Arizona State University thesis.

Hastings, J.R., and R.R. Humphrey (eds) 1969. Climatological data and statistics for Sonora and northern Sinaloa. *Tech. Rep. Meteorol. Climatol. Arid Regions*, No. 19. Tucson: Univ. of Arizona Inst. Atmos. Phys.

Maga, J.A. 1975. Capsicum. *Critical reviews in food science.* Cleveland, Ohio: Chemical Rubber Company.

Maréchal, R. and J.P. Baudoin 1978. Observationes sur quelques hybrides dans le genre *Phaseolus.* IV. L'hybride *Phaseolus vulgaris* × *Phaseolus filiformis. Bull. Rech. Agron. Gembloux* 13(3): 233–240.

Merrick, L.C. and G.P. Nabhan 1984. Natural hybridization of wild *Cucurbita sororia* group and domesticated *C. mixta* in southern Sonora, Mexico. *Cucurbit Genetics Cooperative* 7: 73–75.

Metcalf, R.L., A.M. Rhodes, R.A. Metcalf, J. Ferguson, E.R. Metcalf and P-Y Lu 1982. Cucurbitacin contents and diabroticite (Coleoptera: Chrysomelidae) feeding upon *Cucurbita* spp. *Environm. Entomol.* 11(4): 931–938.

Morton, J.F. 1977. *Wild plants for survival in South Florida*, 4th edn. Miami: Fairchild Tropical Garden.

Nabhan, G.P. 1983. *Papago fields: arid lands ethnobotany and agricultural ecology.* Tucson: University of Arizona dissertation.

Nabhan, G.P. 1984a. Evidence of gene flow between cultivated *Cucurbita mixta* and a field edge population of wild *Cucurbita* at Onavas, Sonora. *Cucurbit Genetics Cooperative* 7: 76–77.

Nabhan, G.P. 1984b. Genetic resources of the U.S./Mexico borderlands: conservation, management and use. In *Bioresources and environmental hazards of the United States – Mexico Borderlands,* P. Ganster, H. Walter, and H. Applegate (eds). Berkeley: University of California Press (in press).

Nabhan, G.P. 1985. Replenishing desert agriculture with native plants and their symbionts. In *Meeting the expectations of the land,* W. Jackson and B. Colman (eds). San Francisco: North Point Press (in press).

Nabhan, G.P., J.W. Berry and C.W. Weber 1980. Wild beans of the greater southwest: *Phaseolus metcalfei* and *Phaseolus ritensis. Econ. Bot.* 34(1): 68–85.

National Academy of Sciences 1975. *Underexploited tropical plants with promising economic value.* Washington, D.C.: National Academy of Sciences.

Nobel, P.S. 1976. Water relations and photosynthesis of a desert CAM plant, *Agave deserti. Plant Physiology* 58: 576–682.

Nobel, P.S. 1981. Influences of photosynthetically active radiation on cladode orientation, stem tilting, and height of cacti. *Ecology* 62: 982–990.

Ochse, J.J. and R.S. van den Brink 1980. *Vegetables of the Dutch East Indies.* Amsterdam: Asher.

Office of Technology Assessment 1983. *Water-related technologies for sustainable agriculture in U.S. arid/semiarid lands.* Washington, D.C.: U.S. Government Printing Office.

Osborne, J.F. and D.P. Singh 1980. Sisal and the other long fibre agaves. In *Hybridization of crop plants,* W.R. Fehr and H.H. Hadley, (eds): 564–575. Madison, Wis.: American Society of Agronomy/Crop Science of America.

Pimental, D., S. East, W.L. Chao, E. Stuart, O. Dintzis, G. Einbender, W. Schlappi, D. Androw and K. Broderick 1982. Water resources in food and energy production. *Science* 32(11): 861–866.

Piper, C.V. 1926. Studies in American Phaseolinae. *Contr. U.S. Nat. Herb.* 22: 663–701.

Sampson, R.N. 1981. *Farmland or wasteland: a time to choose.* Emmaus, Penn.: Rodale Press.

Uphof, J.C. 1968. *Dictionary of economic plants.* Lehre: J. Cramer.

Villalon, B. 1983. 'Tam midl jalapeno-1' pepper. *HortScience* 18: 492–493.

Watt, G.A. 1889–1893. *A dictionary of the economic products of India.* London: W.H. Allen.

Wienk, J.F. 1976. Sisal and relatives. In *Evolution of crop plants,* N.W. Simmonds (ed.): 1–4. London: Longman.

Whitaker, T.W. 1968. Ecological aspects of cultivated *Cucurbita. HortScience* 3: 9–11.

Williams, L. 1962. Lactiferous plants of economic importance II. Mexican chilte (*Cnidoscolus*): a source of gutta-like material. *Econ. Bot.* 16(2): 53–70.

Young, J.A. and R.A. Evans 1973. Mucilaginous seed coast. *Weed Sci.* 21(1): 52–54.

Zee, S.Y. and L.H. Hui 1981. *Hong Kong food plants.* Hong Kong: Urban Council.

Zeven, A.C. and J.M.J. de Wet 1982. *Dictionary of cultivated plants and their regions of diversity.* Wageningen: Centre for Agricultural Publishing and Documentation.

4 Crops for arid lands

J. M. Wilson[1] and J. R. Witcombe[2]

[1]School of Plant Biology, University College of North Wales, Bangor, Gwynedd LL57 2UW, UK

[2]ICRISAT, Patancheru P.O., Andhra Pradesh, India 502 324

Introduction

The semi-arid regions of the world rely on a relatively small number of staple food crops, most of which are also grown in the temperate humid climates. The Consultative Group for International Agricultural Research (CGIAR) has established two international research centres to work on the major food crops of these regions. The International Centre for Agricultural Research in the Dry Areas (ICARDA) is based in Aleppo, Syria, and was established in 1977. It researches on forage crops, cereals and food legumes. The cereal crops are durum wheat, bread wheat, barley and triticale and the food legumes are lentils, faba bean and kabuli chickpeas. The work on kabuli chickpeas is a collaborative effort with ICRISAT (see below). The region to which the research of ICARDA is directed is the Mediterranean region and south-west Asia. It extends from Morocco in the west to Pakistan in the east and from Turkey in the north to Sudan in the south. The area includes a large high altitude zone where both cold and drought stress are encountered.

The International Crops Research Institute for the Semi-Arid Tropics (ICRISAT) is based in Hyderabad, India. Its research is centred on sorghum, pearl millet, pigeon pea, desi and kabuli ckickpea and groundnut. The research is directed towards the semi-arid tropics which covers an area of 20 million km² which includes much of India and two large belts of Africa and the Sahel. Other semi-arid tropical regions are found in south-east Asia, northern Australia, Mexico and central South America. The semi-arid tropics include some of the most highly populated areas of the world and the region produced over half of the world's sorghum, over 80 per cent of the pearl millet, over 90 per cent of the chickpea and pigeon pea and over 60 per cent of the groundnut.

Between them ICARDA and ICRISAT research into crops that sustain the lives of many millions of people in a large area of the world. It is strikingly clear that in the semi-arid regions it is the conventional crops that sustain life, and research on these is the most likely to have encouraging results. Recent progress on some of these crops is briefly reviewed.

Temperate cereals

Triticum aestivum (bread wheat)

Bread wheat ranks first among food crops in the ICARDA region and provides the principal food for the majority of the population. Over 90 per cent of bread wheat is grown on 250 to 650 mm of rainfall and half the area receives less than 400 mm annual precipitation. Because many of the modern high yielding varieties are more suitable for irrigated and high fertility conditions ICARDA is working with the International Centre for Corn and Wheat Improvement (CIMMYT) in Mexico to develop varieties and techniques suitable for the low-rainfall zones. It is hoped that the ICARDA/CIMMYT bread wheat project will lead to the development of germplasm that is tolerant to drought, cold, diseases and insects. Improved varieties of bread wheat such as Mexipak are already gaining ground on the more drought tolerant durum wheats. However, although good high yielding wheat varieties have been developed and released in the ICARDA region, improved agronomic practices which are necessary to obtain good yields have not been widely adopted.

Triticum turgidum (durum wheat)

Durum wheat occupies the largest cereal area in many countries in North Africa and the Middle East as it is better adapted to arid environments than bread wheat. In spite of this, little work has so far been done on durum wheat improvement. However, at ICARDA the yield of durum wheat is steadily approaching the yield of the best bread wheat lines, even though bread wheat has received a more intensive research input over a longer period of time. One of the main objectives of the ICARDA programme is to develop durum wheat varieties that have the genetic potential to perform satisfactorily on limited soil moisture, with the capacity also to respond well when water and nutrition become less limiting. Selection for nutrient and moisture use efficiency are also important.

Hordeum vulgare (barley)

Barley covers about 10 million ha in the Middle East, West Asia and North Africa. It is a more dependable crop in very dry regions than wheat. Barley has an important role in the agriculture of some regions as a forage crop during the winter months. After winter grazing it can be left for grain. This alternative use of barley explains its cultivation in very arid regions where grain yield is highly unpredictable – in drier years in some areas the crop is grazed continuously with no grain harvested.

Although large areas are grown, and in many regions it is the major crop, the average yields are extremely low, not least because it is grown in the driest zones (Table 4.1), usually in soils of medium or low fertility (Weltzein & Srivastava 1981).

Table 4.1 Moisture levels under which barley is grown as indicated by national programmes. Areas as a per cent of total for each region (after Weltzien & Srivastava 1981).

Region	Irrigated	Rainfall mm					
		500	400–500	300–400	300	200	Residual
Near and Middle East	6	4	14	35	35	6	—
South and Far East	10	10	—	—	5	—	75
North Africa	8	—	28	45	19	—	—

Yield trials conducted by ICARDA show the large potential for the improvement of the crop, largely by improved agronomic practices (Table 4.2). In the driest regions a more than two-fold increase can be obtained. Moreover, in some seasons a genetic improvement of the order of 25 per cent over local control varieties has been demonstrated. In general a marked genetic improvement has been produced in the less dry areas, indeed in areas with less than 250 mm rainfall per annum local varieties are superior.

Table 4.2 Barley yields in farmers' fields in Aleppo Province, Syria, compared with ICARDA trials (1980/1) (from ICARDA 1981).

Rainfall zone (mm/yr)	Yields farmers' fields (kg/ha)	Yields ICARDA trials (kg/ha)	Potential increase (%)
⟩350	3150	4100	30
250–300	1211	3610	198
⟩250*	1106	2400	117
200–250	674	1710	254

* in half the years

Triticosecale spp. (triticale)

Triticale is a cross between wheat and rye and has been developed in an attempt to combine the tolerance of rye to frost, drought and diseases and the nutritional quality of its grain with the yield potential, milling and baking quality of wheat. In

ICARDA trials improved high yielding triticale lines gave greater yields than both bread and durum wheat (ICARDA 1982). However, the high yields obtained may be due mainly to high disease 'resistance' since it is grown in only a small area and there has been no opportunity for the build-up of pathogens. Research efforts are being directed towards improving the acceptability of the grain, in particular its colour, weight and baking quality. Much interest has been shown in the use of triticale as an animal feed, especially for the poultry industry.

Cereals of the semi-arid tropics

The term 'millets' refers to any of the small seeded cereal and forage grasses used for food, feed or forage. In earlier times sorghum and even maize were included in the 'millets and minor cereals' category before receiving separate status. Today the term 'millets and minor cereals' is considered to embrace ten genera and at least fourteen species, some of which may receive separate status as they gain in importance in the future as did the sorghums (Rachie 1975).

Millet is the fifth most important cereal in the world with Asia being the primary producer as India and China together produce about 70 per cent of the world's millet. Collectively the millets can tolerate a very wide range of conditions and there are types for practically every situation and need. Species and varieties are available for high temperatures and low humidities or vice versa; heavy, light, stony or infertile soils, varying moisture conditions and short or long seasons. They also have very desirable characteristics such as high nutritional value, relatively few insect and disease pests and drought resistance. In spite of this millet grain yields are frequently low due to the marginal conditions under which they are grown. It is a cereal of the poorest countries and the poorest peoples as well as being one of the main sources of calories in Africa. This paper will review the species of millet which have the most promise for arid regions.

Pennisetum glaucum (pearl millet)

Pearl millet is the most important of all the millets. It can grow in soils and rainfall areas that will not support the growth of other cereals and give an economic but low yield. Pearl millets are cultivated in sandy soils of the Sahelian zone with annual rainfall of less than 250 mm.

Pearl millet is an outbreeder in which cytoplasmic male sterility has been discovered and exploited for the production of hybrid varieties. Large areas of hybrid varieties are cultivated in India. Furthermore, recent plant breeding at ICRISAT has produced synthetic varieties that can outyield the best hybrids.

Setaria italica (foxtail or Italian millet)

Mostly grown in China where exceptionally high yields of 11 000 kg/ha are

claimed. It is drought resistant and grows at high elevations (to 200 m) on a variety of loamy, alluvial or clayey soils. The world collection of 660 entries are maintained in India and this extreme diversity should be valuable in future breeding programmes.

Panicum miliaceum (common or proso millet)

Proso millet is a relatively short duration emergency or quick season irrigated crop with low moisture requirements. This millet is mainly grown in Russia, China and northern India.

Panicum antidotale (little millet)

Little millet is similar to proso millet and also matures quickly and withstands drought well but is little known outside India. There is less genetic diversity in this species than in foxtail millet.

Paspalum scrobiculatum (kodo millet)

Kodo millet is reported to be extremely hardy, drought resistant and to grow on stony or gravelly soils which do not support other crops. It is grown mainly in India but its main disadvantage is that is takes a long time (5–6 months) to mature compared to 2–4 months for other millets.

Eleusine coracana (finger millet)

A high yielding millet which withstands a wide range of temperature and moisture conditions. However, in India it is mainly grown on irrigated lands where moisture is insufficient for rice. The world collection contains over 1000 entries and great variations in grain qualities, disease resistance, maturities and lodging resistance exist.

Sorghum bicolor (sorghum)

Sorghum, along with millet, is a major cereal of rain fed agriculture in the semi-arid tropics. Much of the breeding work has been carried out in the USA and hybrid varieties were produced in the 1950s using cytoplasmic male sterility. At ICRISAT population breeding has also produced elite lines with high yield potential, although none of the elite varieties outperform hybrid controls at all the locations tested.

Hybrids have greater yield potential than varieties and it has been found that yields of hybrids are more likely to be sustained under harsh semi-arid conditions. ICRISAT is therefore increasing its efforts to identify good seed parents for hybrids, especially for use in Africa. In addition, problems associated with breeding for drought resistance are receiving increased attention, especially for the low rainfall areas (650 mm or less) where stress may occur intermittently and the crop grows mainly on residual moisture. Grain quality, especially mould resistance, is also being improved by breeding at ICRISAT. There is no apparent reason why much more efficient sorghum plants should not be developed since the crop has a C_4 photosynthetic pathway and a good translocation system. Population pressure in the semi-arid tropics necessitates the rapid development of sorghum and hopefully the concentrated research effort at ICRISAT will greatly improve the sorghum crop in due course.

Major food legumes

Lens culinaris (lentil)

Asia accounted for 68 per cent of the world production of lentil in 1977. The highest producing countries were India, Turkey and Syria. In general, the areas where lentils are most important are those with the lowest yields. In India, lentils are generally grown as a component of a double cropping system on unirrigated fields. In areas of lower rainfall lentil is the sole crop of the year and is planted after the monsoon fallow. In areas with Mediterranean-type environments lentils are mostly grown in regions of low annual rainfall (often less than 300 mm) (Saxena 1981).

The lentil crop is extremely sensitive to above optimal irrigation and water-logged soils. It has a poor efficiency of water utilization (Summerfield 1981). Nevertheless, it is capable of producing a crop of high quality grain under low rain-fall conditions, and the lentil straw has a high feed value.

Lentil is an under-researched crop and a breeding programme at ICARDA has clearly demonstrated that even selected landraces can substantially outyield those that are widely grown locally (Table 4.3).

Table 4.3 Seed yield of selected land races of lentil compared to local ones at various sites in ICARDA 1980/1 regional trials (from Erskine 1984).

Selection number	Seed yield (kg/ha)		
	Syria	Lebanon	Jordan
78S 26003	1643	1141	1724
76TA 66088	1670	933	2004
Jordan local	1572	841	1697
Lebanon local	1011	918	1150
Syria local	1354	971	1393
Standard error	±263	±461	±332

Lentil is an extremely labour-intensive crop since it is hand harvested. By selecting for plants which are tall and erect, high yielding varieties have been produced which are suitable for mechanical harvesting.

Cicer arietinum (chickpea)

Chickpea is the third most important pulse in the world. About 11 million ha of chickpea are grown, of which about 85 per cent are the desi type (small seeded, angular) and the rest are kabuli type (large seeded, grain shaped). The desi type is grown in India and Ethiopia and the kabuli type in the Mediterranean region and Latin America.

Singh and Malhotra (1984) have described the breeding of a winter sown variety of chickpea for the Mediterranean region which has double the yield of local spring sown varieties. The variety was selected from a germplasm collection without hybridisation. Winter-sown varieties need to be both cold tolerant and resistant to ascochyta blight. Winter-sown varieties can be grown in drier regions than was previously possible since they mature two weeks earlier than spring-sown material and avoid more extreme levels of heat during their reproductive phase.

Vicia faba (faba bean)

Although faba bean is widely grown either in temperate humid environments or with irrigation, it is grown, nevertheless, in semi-arid environments as a rainfed crop. The principal regions where it is grown as such are the Mediterranean Basin, Ethiopia, Iran, Afghanistan, India, Nepal, China and Latin America. China is by far the world's largest producer of faba bean although the area under its cultivation is in marked decline. In Latin America it is mostly cultivated in regions which are too cold for the *Phaseolus* bean, and most of the Latin American production is in Brazil.

Although recent breeding efforts at ICARDA have produced disease resistant varieties, notably for ascochyta blight, major genetic improvements have not yet been made (El-Sayed 1984). The faba bean varieties do not tend to have wide adaptation, and population breeding has not yet produced the desired breakthough in yield.

Cajanus cajan (pigeon pea)

Pigeon pea is used in the semi-arid tropics as a subsistence crop by poor farmers. About 95 per cent of the pigeon pea crop is grown in India. Pigeon pea under most conditions is predominately inbreeding but shows marked heterosis. At ICRISAT hybrid seed can be produced using a genetic male sterile. Hybrids yield 30 per cent more than improved cultivars.

Early maturing varieties of pigeon pea can be followed by winter wheat. Lines have been developed that mature in less than 100 days. Traditional late varieties mature in over 250 days.

Pigeon pea is frequently used in an intercropping system and overall yields of the mixtures are considerably higher than the pure stands. Intercrops used are *Setaria*, mung bean (*Vigna mungo*), millet sorghum and groundnut. Examples of yields under intercropping are shown in Table 4.4.

Table 4.4 Grain yields (kg/ha) of pigeon pea grown alone or intercropped with either setaria or millet on a black soil (ICRISAT 1975–76).

	pigeon pea	pigeon pea + setaria	setaria
pigeon pea	2530	2530	–
setaria	–	3330	3290
	pigeon pea	pigeon pea + millet	millet
pigeon pea	2530	1970	–
millet	–	3390	3090

A further promising development is the use of pigeon pea as a perennial crop. The crop is ratooned at the first harvest, i.e. the stems are cut to leave a stubble which regrows. Most of the ratooned plants survive the dry period before the monsoon rains because they are deep-rooted. A second harvest is taken from the same plants about four months later. Substantial quantities of firewood may be removed from the crop after the third harvest. To use this system a crop that is resistant or tolerant of wilt and sterility mosaic is essential (ICRISAT 1982).

Arachis hypogea (groundnut)

Groundnut production is the most important legume of the semi-arid tropics. About 70 per cent of the world production is in the semi-arid tropics but yields are generally low (around 800 kg/ha) and are unpredictable. Increases in yield of about 30 per cent have been obtained from improved cultivars (ICRISAT 1982). Early maturing lines display drought avoidance and high yield. There is considerable potential to improve the crop using relatives of *Arachis hypogea*, particularly for disease resistance characters. Groundnut is frequently used in an intercrop system, often with pearl millet.

Vigna unguiculata (cowpea)

Cowpeas are an important source of protein in the subsistence and peasant farming communities of semi-arid Africa and Asia. They are also an important source of fodder. Seed production for human use is probably greatest in India and West Africa but they are not important in world trade. A review of the literature on cowpeas has been made by Summerfield *et al.* (1974).

Minor legumes

In addition to the main food legumes considered above there are many other plants of the family Leguminosae which are at present greatly underexploited. These plants can be valuable as root crops, pulses, forage, timber and for their fruits. In many parts of the world these crops are virtually unknown and yet their drought tolerance and other attributes suggest that they could become major crops in the future. In the present paper some of the pulse crops which have interesting potential in arid lands are considered. Restricted space means that species such as lablab bean (*Lablab purpureus*) and tarwi (*Lupinus mutabilis*) are not considered here and the reader is referred to BOSTID report No. 25 on 'Tropical Legumes: Resources for the Future' for further information on these and other promising species (National Academy of Sciences 1979).

Phaseolus acutifolius (tepary bean)

Teparies are one of the most drought tolerant beans and gives good yields in arid regions that are too dry for other beans. Yields of up to 4000 kg/ha have been obtained under minimal irrigation in California, clearly outproducing most other field legumes grown under similar conditions. The bean has a high protein content (23–25%) and yet the plant is virtually unknown outside North America.

The main limitations to the adoption of this plant in the semi-arid regions is its comparatively low resistance to diseases and pests and sensitivity to salty soils and waterlogging. The beans can be eaten in the same way as other dried beans but their small size, tendency to cause flatulence and longer cooking time make them less popular than other bean species. For further information see Nabhan (1983).

Vigna subterranea (bambara groundnut)

This is an African pulse crop which can thrive in poor arid soils where groundnuts, corn and sorghum often fail. Like groundnut it forms pods on or just beneath the ground. Bambara seeds have less oil and protein than groundnuts but nevertheless have a high nutritional value. The seeds cannot be eaten raw as they are too hard, although the immature seeds are soft enough to eat raw. Little research has been done on this crop despite its considerable promise. It appears to be less susceptible to pests and diseases than groundnuts and with good management the yields compare favourably with other crops even on very poor soils.

Bambara groundnut is an important nutritional source for the poor peoples of Africa and research is needed on improving the agricultural management of the crop and adapting commercial methods of groundnut farming in developing this species as a large scale field crop.

Tylosema esculentum (marama bean)

Marama bean is a native of the Kalahari regions of southern Africa and has not yet been cultivated. However, as it grows in areas where few other crops can survive and produces seeds of similar quality to groundnut it appears to have considerable potential to feed the indigenous peoples of these regions. The plant is a prostrate vine with a large underground tuber which produces new stems after dieback during the cooler months. The tuber can be eaten when cooked and is an important emergency source of water for humans and animals. So far it has only been culti- vated as an ornamental in South Africa and much research is needed on nearly all agricultural practices such as germination, spacing, weeding and harvesting.

Vigna aconitifolia (moth bean)

Moth bean is reputed to be the most drought tolerant pulse crop grown in India and is cultivated in India's driest state, Rajasthan. It thrives under high temperature conditions, on poor sandy soils and the moisture remaining in the soil near the end of the rainy season is often enough for the crop to complete seed formation. The seeds are small but high in protein. Young pods can be eaten as a table vegetable and the foliage is a good livestock feed and can be made into hay. Moth bean has good resistance to pests and diseases and the major problem associated with the further adoption of this plant is the lack of recent published guidelines for farmers. Seedbed preparation is very important as the small seeds find it difficult to penet- rate even a thin soil crust. Future research should focus on the breeding of upright types which are easier to harvest and types which have greater resistance to yellow mozaic leaf virus and nematodes.

Cordeauxia edulis (ye-eb)*

Ye-eb is a wild legume of the arid (often as low as 150–200 mm rainfall) desert regions of Somalia and Ethiopia and has received very little domestication. The shrub produces a chestnut-like flavour nut which is similar to macadamia and pistachio nuts. During the Sahelian drought of 1973–6 the ye-eb was over- exploited and protection of the native stands is now required. In its native habitat the plant grows on extremely poor red sandy soils and takes 3 to 4 years before bearing the fruit pods. Very few other plants can survive in the ye-eb's native habitat and its potential in other hot dry regions of low uncertain rainfall deserves investigation.

---

*Yicib is the correct spelling in modern Somali, although the 'c' is almost silent (G.E.W.).

Canavalia ensiformis (jackbean) and *C. gladiata* (swordbean)

Both species are extremely hardy and can grow well on a variety of poor, leached, acid soil with as little as 700 mm annual rainfall and temperatures which range from 14 to 27°C. Jackbean is a New World species common to drought areas of Arizona and Mexico, whilst swordbean is Old World in origin and is cultivated mainly in India and humid parts of Africa. Both beans are highly productive and on fertile soils their yields are similar to other legume pulse crops. They often thrive where cowpea and other beans fail.

Novel crops

Unfortunately restricted space permits the inclusion of only 3 crops under this heading, grain amaranth, buffalo gourd and jojoba. For further information on little known tropical plants that with further research could become important cash and food crops in the future the reader is referred to BOSTID report No. 16 on 'Underexploited Tropical Plants with Promising Economic Value' which described the potential of 36 different species, many of which are drought tolerant (National Academy of Sciences 1975).

Amaranthus ssp. (grain amaranth)

Amaranth is perhaps best known as a leafy vegetable which is widely grown in S.E. Asia. However, its attractiveness as a crop for the future stems from the grain types which are found mainly in the semi-arid, seasonally wet areas of the tropics, e.g. Nepal and India. There are three cultivated species of grain amaranth and *Amaranthus hybridus* subsp. *hypochondriacus* is the most important as it forms the bulk of the Asiatic crop. *A. hybridus* subsp. *cruentus* and *A. caudatus* are the other two grain amaranths. The grain is of high protein quality with an amino acid composition very similar to the optimum balance required in the human diet. Amaranth is a C_4 plant that can produce rapid growth in semi-arid conditions. The high degree of genetic variability will enable breeders to produce varieties capable of good yields in arid regions on land that was previously unusable. Plant breeding programmes are aimed at: (a) increasing seed size (amaranth seeds are very small – often 1000 per gram, tetraploid lines exhibit a 2.5 fold increase in seed weight); (b) producing shorter, stockier plants with increased resistance to lodging; (c) improving the harvest index, and (d) producing uniform ripening and increased resistance to shattering.

Growing problems are mainly associated with the small size of the seed which has to be placed close to the surface to ensure germination. Rains and predation can rapidly reduce the number of seed so that overseeding followed by thinning is frequently practised. Alternatively, transplanting is practised in northern India and Nepal and can be intercropped with finger millet (*Eleusine coracana*). However, this method is not practical for commercial grain production and it has been shown that transplanting can decrease yields. Grain amaranths can be mechanically harvested until the canopy closes. Most of the research on amaranth is being conducted in the USA at the Vegetable Crops Department of Cornell University and at the Organic Gardening and Farming Research Center in Pennsylvania. The reader can find more details about amaranths in a review by Feine *et al.* (1979).

Cucurbita foetidissima (buffalo gourd)

The buffalo gourd has evolved in the semi-arid regions of western North America and this wild perennial has considerable potential for feeding the indigenous peoples of these regions. The plant is perennial in habit and has an asexual mode of reproduction by vigorous vine growth which is capable of rooting at each node. Three different parts of the plant have nutritional value. Most important is the yield of oil (30–40%) and protein (30–35%) from the seed of the fruits. In addition the very large roots (up to 40 kg in three or four seasons' growth) contain approximately 20% starch and finally the vines have potential as forage for domestic animals (Bemis *et al.* 1979).

Much research remains to be done on the domestication of this plant. For instance, the wide variation in the number of seeds per fruit and the number of fruits per plant provides the opportunity to improve yield by selection. There is a vast store of genetic variation in this species from which other desirable characteristics could be selected. The buffalo gourd grows well on light soils, has low water requirements and appears to be relatively free from diseases and insects. The buffalo gourd is, however, frost sensitive, the vines being killed at temperatures below 4°C but the roots can withstand air temperatures of up to -25°C if the soil is covered with an insulating layer of snow. Trials are being made to establish the buffalo gourd by direct seeding and harvesting the fruits mechanically at the end of the first season by a type of vine thresher. At the end of the second season, after harvesting the fruits, alternate swaths of the field can be mechanically dug for the roots. Thus the buffalo gourds would be thinned by harvesting at the end of the second year's growth. In summary, this plant shows considerable promise as a novel crop for the semi-arid tropics.

Simmondsia chinensis (jojoba)

Jojoba has received a great deal of publicity in recent years as its fruit contains 40–60% liquid wax of high value as it is similar to sperm whale oil and has an expanding list of uses from engine lubricants to cosmetics. Jojoba wax is valuable for its stability, purity, simplicity, lubricity and can be modified by partial dehydrogenation to produce a variety of soft white waxes and creams for use in industry. Unfortunately the remaining residue after wax extraction cannot be used as animal feed due to the presence of an unusual toxin in the meal.

The attraction of this crop for people in the warmer arid regions of the world is its potential to grow and produce good yields with comparatively little water (Hogan 1979). It grows in native stands where rainfall is less than 120 mm per year and it can also grow on saline soils with saline water. Once established the plant can have a net positive photosynthesis with water potentials as low as -7000 k Pa. However, it grows best and produces highest yields at between 380—500 mm of moisture per year.

Most interest in this plant has been shown by the USA, Israel, Mexico and Australia with both large and small farming organisations involved in the development of this crop. Whether these plantations are profitable depends on predictions of the price of jojoba oil in future years (jojoba does not yield seed until 5 years after planting) and on whether claims for the seed yield of the varieties planted are fulfilled. The seed yield of plants grown from seed may vary from 0.6—5.4 kgm/ bush. Furthermore, many of the areas where jojoba was first planted may be unsuitable as it is now realised that jojoba is more sensitive to cold than was commonly thought (temperature should not fall below -4°C), and, although the established plant is very drought tolerant, an adequate supply of water (600—750 mm/year) is often necessary during the first two years of establishment. In addition jojoba is not free of disease problems and is attacked by many different pathogens, e.g. *Phytophthora* and *Pythium* species and also insects.

In conclusion it is unlikely that the 'miracle' plant jojoba will enable many people to get rich quickly. Research is needed on the selection and breeding of jojoba for productivity, cold resistance, early maturation, disease resistance, hermaphroditism and an upright shape that facilitates mechanical harvesting. Harvesting by hand is very expensive and it is considered that jojoba can only capture the high-volume low-cost lubricant market if it is plantation grown and machine harvested. Until many of the problems of growing and harvesting jojoba in plantations are overcome and sound, established markets are created for the wax, jojoba can only be considered as a speculative investment.

Mechanisms of drought resistance

The purpose of this section is to review briefly what is known about the morphological, biochemical and physiological basis of drought resistance in plants and to attempt to determine whether breeding for these attributes leads to improved yield or yield stability under stress. Table 4.5, p. 48 shows that the mechanisms of drought resistance can be divided into three primary types. First, drought escape is the ability of the plant to complete its life cycle before a serious plant water deficit develops. Second, drought tolerance at high tissue water potential is primarily a

Table 4.5   Mechanisms of drought resistance (after Jones *et al.* 1981).

---

1 Drought escape
    (a) Rapid phenological development
    (b) Developmental plasticity

2 Drought tolerance with high tissue water potential
    (a) Maintenance of water uptake
        (i) increased rooting
        (ii) increased hydraulic conductance
    (b) Reduction of water loss
        (i) reduction in epidermal conductance
        (ii) reduction in absorbed radiation
        (iii) reduction in evaporative surface

3 Drought tolerance with low tissue water potential
    (a) Maintenance of turgor
        (i) solute accumulation
        (ii) increase in elasticity
    (b) Dessication tolerance
        (i) protoplasmic resistance

---

drought avoidance mechanism by avoiding tissue dehydration. Third, drought tolerance at low tissue water potential by mechanisms which enable the maintenance of turgor and/or increase the tolerance of the protoplasm to dessication.

DROUGHT ESCAPE

Drought escapers, or ephemerals, appear rapidly after heavy rain in desert communities. They grow quickly and produce seed before the soil water supply is depleted and it is considered that they have no special physiological, morphological or biochemical mechanisms to cope with water deficites. It is, however, known that many summer ephemerals have a Kranz-type anatomy typical of C_4 pathway of photosynthesis which increases water use efficiency (Mulroy & Rundel 1977).

    Ephemerals appear to survive arid environments due to their rapid growth and developmental plasticity. Under dry conditions little vegetative growth occurs and there are few flowers and seeds, yet in wet years their indeterminate growth habit enables large amounts of seed to be produced. The selection for rapid development has been one of the most rewarding approaches that plant breeders have used. For instance, in cereals, such as wheat and barley, varieties which flower early are less affected by drought but a disadvantage is that these varieties do less well in years of adequate rainfall. Plant breeders have also produced cereal varieties that are less sensitive to water deficits at the most sensitive stages of development such as around flowering time.

DROUGHT TOLERANCE AT HIGH TISSUE WATER POTENTIAL

Plants in this category can maintain high tissue water potential by mechanisms which will maintain water uptake or reduce water losses under dry conditions. Water uptake can be maintained either by an increase in root density or by growing deeper. The resistances in the roots to water flow may also vary between species but their importance in relation to drought resistance is not known. However, chilling temperatures rapidly descrease the permeability of the roots of tropical and sub-tropical species to water and this combined with the 'locking open' of the stomata of many crop species at 5°C can lead to rapid leaf dehydration and injury (Wilson 1983).

Mechanisms for reducing water loss include regulation of the stomatal aperture and the development of an impermeable cuticle. Unfortunately, as CO_2 enters via the stomata a reduction in water loss by decreased stomatal aperture also results in reduced photosynthesis and eventually plant yield. Water loss can however be reduced by leaf movements and changes in the reflectance of leaves which decreases the amount of radiation absorbed. Wilting, the production of hairs, surface wax or salt all increase leaf reflectance. Finally, shedding of the older lower leaves can reduce the rate of water loss by many crop species.

DROUGHT TOLERANCE AT LOW TISSUE WATER POTENTIAL

Turgor can be maintained as water potential decreases due to the accumulation of solutes such as sugars which lowers the osmotic potential or it can be maintained by high tissue elasticity. As few measurements of tissue elasticity have been made and it seems unlikely that fully expanded tissues would increase in elasticity in response to low water potentials, the maintenance of turgor appears to be primarily by solute accumulation. Munns *et al.* (1979) found that the major solutes accumulating in wheat leaves during drought stress were sugars and amino acids. In other plants organic acids, potassium and chloride are accumulated to significant levels. Much attention has also been given to the role of proline and betaine accumulation in plants during water stress. (Wynn Jones & Storey 1981). The accumulation of proline and glycinebetaine may be a direct response of the plant to the stress or possibly a by-product of the stress by its effect on cell metabolism such as protein synthesis. Therefore, it does not follow that the rate or degree of accumulation of proline or glycinebetaine is an index of drought tolerance which can be used by plant breeders.

Protoplasmic resistance to water deficits is best demonstrated in the 'resurrection' plants and in rapidly drying mosses such as *Tortula ruralis* which can resume normal cell metabolism within a few minutes of re-hydration. Very little research has been done on changes in the chemical composition of these plants during dehydration and re-hydration. It is possible that proline, glycinebetaine or other cytoplasmic non-toxic solutes may accumulate which protects the membranes and macromolecules from severe dehydration. For a comprehensive review of mechanisms of drought tolerance the reader should consult Levitt (1979).

BREEDING FOR DROUGHT RESISTANCE

Drought resistance is a problem area in crop breeding because it is difficult to establish uniform and repeatable drought resistance screening nurseries for field selection. In addition, we lack proof that breeding for biochemical or physiological changes such as proline accumulation during adaptation to stress leads to improved yield or yield stability under stress. It is not therefore surprising that few plant breeders will breed for physiological mechanisms. Indeed, the only way to test this approach is to breed for the mechanism and test the results.

In this short review I have attempted to show that our knowledge of the morphological and physiological basis of drought resistance has increased markedly over the last 30 years. However, more research is needed on assessing the importance of the various mechanisms of adaptation to stress in relation to improved yield. At the present time our understanding of the mechanisms of adaptation is not sufficient to be able to predict the interaction and integration of the various mechanisms that may be operating. We need to be able to assess the relative importance of the various mechanisms in a similar genetic background. Only then will it be possible to state whether it is better to search for morphological or biochemical adaptations to water stress.

Conclusions

From this brief review of recent advances in some of the major food crops of the semi-arid regions a picture emerges of crops that have been under-researched in the past, and that have an encouraging potential of producing higher yields with improved varieties and management. Indeed, with the adoption of existing technology it is feasible in many crops to double existing yields. It is recognised by the international centres that often the barriers to increasing yields are socio-economic. For example, in northern Syria barley yields are limited by a land tenure system which results in a very low input agriculture that depletes soil fertility (ICARDA 1982). If such problems exist to introducing existing technology to conventional crops, then it is likely that the advantages of novel crops need to be truly spectacular for them to receive rapid adoption.

Despite the enormous potential of the conventional food crops there are strong ecological arguments in favour of perennial cropping systems that protect the soil from erosion. In the low rainfall zones forage crops, agro-forestry and orchard crops such as grape, olive and pistachio, may be a better alternative to annual crops that expose the soil to erosion and that are more unpredictable in yield.

References

Bemis, W.P., J.W. Berry and C.W. Weber 1979. The Buffalo gourd: a potential arid land crop. In *New agricultural crops*, G.A. Ritchie (ed.): 65–87. Boulder, Colorado: Westview Press.

El Sayed, F.A. 1984. Evaluation and utilization of faba bean germplasm in an international breeding program. In *Advances in agricultural biotechnology. Genetic resources and their exploitation – chickpeas, faba beans and lentils*, J.R. Witcombe and W. Erskine (eds): 173–186. The Hague: Nijhoff/Junk.

Erskine, W. 1984. Evaluation and utilization of lentil germplasm in an international breeding program. In *Advances in agricultural biotechnology. Genetic resources and their exploitation – chickpeas, faba beans and lentils*, J.R. Witcombe and W. Erskine (eds) : 225–238. The Hague: Nijhoff/Junk.

Feine, L.B., R.R. Harwood, C.S. Kauffman and J.P. Senft 1979. Amaranth: gentle giant of the past and future. In *New agricultural crops*, G.A. Ritchie (ed.): 41–63. Boulder, Colorado: Westview Press.

Hogan, L. 1979. Jojoba. A new crop for arid lands. In *New agricultural crops*, G.A. Ritchie (ed.): 177–205. Boulder, Colorado: Westview Press.

ICARDA 1981. *ICARDA research highlights*. Aleppo, Syria: International Center for Agricultural Research in the Dry Areas.

ICARDA 1982. *Annual Report of the International Center for Agricultural Research in the Dry Areas*. Aleppo, Syria.

ICRISAT 1975–76. *Annual Report of the International Research Institute for the Semi-Arid Tropics*. Patancheru P.O., Andhra Pradesh 502 324, India.

ICRISAT 1982–83. *ibid.*

Jones, M.M., N.C. Turner and C.B. Osmond 1981. Mechanisms of drought resistance. In *The physiology and biochemistry of drought resistance in plants*, L.G. Paleg and D. Aspinall (eds): 15–37. Sydney: Academic Press.

Levitt, J. 1979. *Responses of plants to environmental stresses*. Vols. I and II. New York: Academic Press.

Mulroy, T.W. and P.W. Rundel 1977. Annual plants: adaptation to desert environments. *Bioscience* 27: 109–114.

Munns, R., C.J. Brady and E.W.R. Barlow 1979. Solute accumulation in the apex and leaves of wheat during water stress. *Aust. J. Plant Physiol.* 6: 379–389.

Nabhan, G.P. (ed.) 1983. *The desert tepary as a food resource*. Desert Plants 5.1. A journal symposium.

National Academy of Sciences 1975. Board on Science and Technology for International Development Report No. 16. *Underexploited tropical plants with promising economic value*. Washington, D.C.: National Academy of Sciences.

National Academy of Sciences 1979. Board on Science and Technology for International Development Report No. 25. *Tropical legumes: resources for the future*. Washington, D.C.: National Academy of Sciences.

Rachie, K.O. 1975. *The millets: importance, utilization and outlook.* India: ICRISAT publication.

Saxena, M.C. 1981. Agronomy of lentils. In *Lentils*, C. Webb and G. Hawtin (eds). Slough: Commonwealth Agricultural Bureaux.

Singh, K.B. and R.S. Malhotra 1984. Exploitation of chickpea genetic resources. In *Advances in agricultural biotechnology. Genetic resources and their exploitation – chickpeas, faba beans and lentils*, J.R. Witcombe and W. Erskine (eds): 123–130. The Hague: Nijhoff/Junk.

Summerfield, R.J. 1981. Adaptation to environments. In *Lentils*, C. Webb and G. Hawtin (eds). Slough: Commonwealth Agricultural Bureaux.

Summerfield, R.J., P.A. Huxley and R.M. Steele 1984. Cowpeas *Vigna unguiculata*. A review. *Field Crop Abstr.* 27: 301–312.

Weltzien, H.C. and J.P. Srivastava 1981. Stress factors and barley productivity and their implications in breeding strategies. In *Barley genetics IV*. Edinburgh: University Press.

Wilson, J.M. 1983. Interaction of chilling and water stress. In *Crop reactions to water and temperature stresses in humid temperate climates*, C.D. Raper and P.J. Kramer (eds): 133–148. Boulder, Colorado: Westview Press.

Wyn Jones, R.G. and R. Storey 1981. Betaines. In *The physiology and biochemistry of drought resistance in plants*, L.G. Paley and D. Aspinall (eds): 171–205. Sydney: Academic Press.

5 The nutritional composition of Australian Aboriginal food plants of the desert regions

J. C. Brand and V. Cherikoff

*Human Nutrition Unit and Commonwealth Institute of Health,
University of Sydney, Australia 2006*

Introduction

'A heavy, tasteless, evil-looking but nourishing cake' was Meggitt's (1957) description of a damper produced by a desert group of Australian Aboriginals. Over 100
years ago the English botanist and former Director of Kew, J.D. Hooker, writing of
Australian edible plants, suggested that many of them were 'eatable but not worth
eating' (Hooker 1859). To the modern Australian there may seem to be considerable justice in such a view. Nevertheless, the Australian flora together with the
fauna, supported the indigenous Aboriginal people. Before the European occupation of Australia, the Aboriginals lived as a hunting and food gathering society for
about 40 000 years.

During the last 200 years Aboriginal culture has undergone great change in
response to the domination of the immigrants and there is now a wide spectrum of
Aboriginal lifestyles. In desert regions which proved inhospitable to the immigrants,
change has occurred more slowly and there are now groups of Aboriginals living in a
traditionally-oriented manner. These 'homeland' groups are providing much of what
we know about Aboriginal foods in these regions. In other areas of the country,
information about earlier lifestyles comes primarily from fragments recorded by
those few outsiders who took a respectful interest in the Aboriginals around them
— explorers, missionaries, naturalists, government officials, pastoralists and others.

With two-thirds of its land mass receiving less than 50 cm annual rainfall and
more than half experiencing desert or semi-desert conditions, Australia is the most
arid continent. Drought with its associations of water shortage, searing heat, wind
and fires is possibly the worst calamity known to Australians. Australia has a
unique flora and fauna which is adapted to the prevailing climate and soil conditions and in many areas to the aridity. Aboriginal survival in the harsh desert
conditions was based on a strategy of small groups of people foraging over large
expanses of land, living in an unpredictable environment by exploiting a wide range
of resources as they became seasonally or locally available. Gould (1969) writes 'the
Western Desert people have managed to survive in the harshest physical environment on earth ever inhabited by man before the Industrial Revolution'. Although

many lists of edible plants have been published for various desert groups, there is only limited evidence on the composition of their daily diet because the lists fail to rank the foods by their importance (Sweeney 1947 is an important exception). Extensive lists of 60–100 known edible plant species reduce to around a dozen staples. While the staples vary a little from region to region, many are common throughout the desert. Because of irregular seasonality and rainfall, the dozen or so plant staples of a good year may be reduced to a core of three in a poor year (Peterson 1978). The vegetable staples include the seeds of various *Acacia* spp., of herbs and grasses such as *Eragrostis, Fimbristylis, Panicum* and *Portulaca* spp., the rootstocks of *Ipomoea costata, Vigna lanceolata* and *Dioscorea* sp. and the fruits of various Solanaceae and of *Ficus* and *Santalum* spp. Other species were certainly important and staples for varying periods.

This paper discusses the nutritional composition of the native food plants of the desert regions of Australia, the objective being to highlight the more 'nutritionally desirable' species. Background information on processing methods, seasonal availability and palatability will also be described. We have drawn on the nutrient data published by others as well as our own. Before 1980 there were only fragmentary data on the nutritional composition of Australian bushfoods but since then our own work and that of the Armed Forces (James 1983) have increased considerably the number of species analysed. Our study represents an ambitious attempt to document the composition of bushfoods while the range available is still capable of sustaining a traditional existence. The samples sent to our laboratory for analysis have been collected in the main by Aboriginals living on their homelands. They and their health workers as well as anthropologists, botanists, archeologists and the average bushwalker have increasingly demanded more nutrient information. We have assembled in the accompanying Table 5.1 all nutritional information available on 54 arid species including hitherto unpublished figures of our own. Our methods of analysis are described elsewhere (Brand *et al.* 1983).

Nutrient data

For any food a wide range in composition may be encountered. Much of the data presented in the tables represents the analysis of one single, often small, perhaps unrepresentative sample of food and as such its value is limited. It does, however, serve as a guide to the range in values that may occur. In a few instances different samples of a particular food item have been analysed by different workers and a more reliable estimate of the average nutrient composition can be calculated. The following discussion considers the foods by category – seeds, fruits, rootstocks, thirst quenchers – highlighting those foods which were staples plus those which may have made significant nutrient contributions at times. The samples have been analysed for macronutrients: water, protein, fat, total carbohydrate and ash. Total carbohydrate includes sugars, starch and dietary fibre. When dietary fibre has been

Table 5.1 Composition of foods per 100 g edible portion (raw unless otherwise indicated).
*previously unpublished analyses carried out by the authors; (T) = total carbohydrate including fibre

Foods	Description	Energy kJ	Water g	Protein g	Fat g	Carbo-hydrate g	Fibre g	Ash g	Na mg	K mg	Mg mg	Ca mg	Fe mg	Zn mg	Cu mg	Vita-min C mg	Source
Acacia aneura and Acacia kempeana mulga and witchetty bush	seeds	2220	4.3	23.3	37.0	25.5(T)	–	9.7	37	805	–	113	–	–	–	–	Peterson 1978
Acacia coriacea desert oak	green seed	627	56.8	23.7	3.3	6.4	8.2	1.6	3	362	75	137	3.2	1.0	0.4	–	*
	dry seed	1240	4.1	20.9	9.3	33.8	–	3.7	4	784	170	318	7.7	5.8	1.0	–	*
	dry seed	1491	17.0	23.8	7.7	48.1(T)	–	3.7	61	357	–	84	–	–	–	–	Peterson 1978
Acacia cowleana	seed	1507	15.6	22.2	10.1	44.6(T)	–	7.2	96	290	–	114	–	–	–	–	Peterson 1978
Acacia dictyophleba	seed	1519	11.2	26.8	6.3	49.0(T)	–	6.5	84	221	–	88	–	–	–	–	Peterson 1978
Acacia estrophiolata ironwood	gum exudate	1429	7.1	0.2	0	89.1	0	3.6	6	721	37	587	0.9	2.6	0.4	–	*
Acacia kempeana	seed	1631	5.6	22.9	10.2	51.0(T)	–	10.1	112	392	–	72	–	–	–	–	Peterson 1978

Table 5.1 – continued

Foods	Description	Energy kJ	Water g	Protein g	Fat g	Carbo-hydrate g	Fibre g	Ash g	Na mg	K mg	Mg mg	Ca mg	Fe mg	Zn mg	Cu mg	Vita-min C mg	Source
Acacia murrayana murray's wattle	seed	1107	5.4	18.1	5.8	37.6	28.9	4.2	4	800	218	213	6.6	3.7	0.7	–	*
Acacia tenuissima mulga	seed	1469	1.6	25.0	15.6	29.2	25.7	2.9	5	678	118	144	6.8	3.2	1.4	–	*
		1592	14.5	24.8	16.4	33.0(T)	–	11.1	89	372	–	80	–	–	–	–	Peterson 1978
Acacia victoriae bramble wattle	seed	1082	4.9	17.0	3.8	40.8	29.4	4.1	8	756	136	383	10.6	1.7	0.6	–	*
Albizia lophantha crested wattle	seed	1064	6.5	18.4	5.2	34.9	31.8	3.2	33	761	105	164	3.5	3.5	0.7	–	*
Amyema sp. mistletoe	fruit	533	74.3	2.2	3.6	–	–	–	–	–	–	–	–	–	–	20	James 1983
Boerhavia diffusa Tah vine	root	724	50.0	4.6	0.8	35.6(T)	–	4.2	–	1440	–	623	–	–	–	–	Dadswell 1934
Brachychiton diversifolius northern kurrajong	seed	304	82.5	1.2	1.8	–	–	–	–	–	–	–	(Thiamin 0 µg)			Tr	James 1983
Brachychiton gregorii desert kurrajong	root	177	81.1	0.5	1.0	8.2	7.6	1.6	2	275	47	175	5.0	0.4	0.1	–	*

Table 5.1 – continued

Foods	Description	Energy kJ	Water g	Protein g	Fat g	Carbohydrate g	Fibre g	Ash g	Na mg	K mg	Mg mg	Ca mg	Fe mg	Zn mg	Cu mg	Vitamin C mg	Source
Brachychiton paradoxum	seed	869	12.4	5.9	7.4	–	–	–	–	–	–	–	(Thiamin 0 µg)			6	James 1983
Brachychiton populneum Kurrajong	seed	1455	5.6	18.1	24.7	14.6	33.5	3.5	8	567	288	110	2.1	5.4	1.2	–	*
	exudate	–	14.3	0	0.4	–	–	–	–	–	–	–	–	–	–	–	*
Calandrinia balonensis	leaves	86	93.0	1.2	0.2	3.5	–	1.5	–	798	–	21	–	–	–	–	Dadswell 1934
broad-leaved parakeely	seed	1822	5.4	14.6	17.0	55.4	–	7.3	23	86	–	86	–	–	–	–	Peterson 1978
Capparis sp. wild orange	fruit	647	42.0	3.7	4.9	25.2	21.7	2.5	3	631	58	48	2.9	2.2	0.4	11	*
Capparis lasiantha native orange	fruit	343	54.1	10.1	4.9	18.9	10.2	1.8	3	42	60	58	1.2	0.9	0.4	6	*
Capparis mitchellii wild orange	fruit	500	57.9	7.3	1.5	20.0	11.4	1.9	1	580	63	25	0.7	0.1	0.3	49	*
Capparis spinosa coastal caper	fruit	263	79.6	4.6	3.6	3.2	7.2	1.8	18	383	39	28	0.9	0.4	0.1	23	James 1983
	fruit	869	64.2	7.2	7.4	–	–	–	–	–	–	–	(Thiamin 692 µg)			2	James 1983
Capparis umbonata native pomegranate	fruit	619	51.5	8.8	1.8	25.2	10.9	1.8	0.5	580	67	75	0.8	0.1	0.2	89	*

Table 5.1 – continued

Foods	Description	Energy kJ	Water g	Protein g	Fat g	Carbohydrate g	Fibre g	Ash g	Na mg	K mg	Mg mg	Ca mg	Fe mg	Zn mg	Cu mg	Vitamin C mg	Source
Carissa lanceolata Conkerberry	fruit	593	55.9	3.1	2.6	27.6	9.3	1.5	5	336	51	65	3.5	2.5	0.5	6	*
	fruit	601	70.3	1.9	2.1	–	–	–	–	–	–	–	(Thiamin 44 µg)	–	–	2	James 1983
Chenopodium rhadinostachyum green crumbweed	seed	1431	7.0	14.4	2.3	65.7(T)	–	10.5	10	148	56	–	–	–	–	–	Peterson 1978
	seed	–	–	20.1	2.9	–	–	–	–	–	–	–	–	–	–	–	Elphinstone 1971
Clerodendrum floribundum bleeding heart	root	918	52.3	1.6	1.8	–	–	–	–	–	–	–	(Thiamin 0 µg)	–	–	Tr	James 1983
Cyperus bulbosus bush onion, nut grass	corm	674	49.9	2.6	0.8	37.5	5.7	3.5	8.3	1000	286	47	7	0.3	1	–	Brand *et al.* 1983
	corm	698	59.9	2.9	Tr	–	–	–	–	–	–	–	(Thiamin 185 µg)	Tr	–	–	James 1983
	corm	1139	31.5	2.9	1.0	62.5(T)	–	1.8	86	381	–	63	–	–	–	–	Peterson 1978
	corm	–	–	5.5	1.8	–	–	–	–	–	–	–	–	–	–	–	Peterson 1978
Cyperus rotundus wild onion, nut grass	corm	–	54.4	–	0.7	–	–	–	–	–	–	–	–	–	–	Tr	James 1983
Cyperus sp. nut grass	stalk	200	86.2	1.4	0.1	–	–	–	–	–	–	–	(Thiamin 0 µg)	–	–	Tr	James 1983
Dioscorea bulbifera cheeky yam	root (cooked)	127	77.4	1.6	0.2	5.8	14.8	0.2	–	138	7	2	0.8	0.2	0.2	233	Brand *et al.* 1983

Table 5.1 – continued

Foods	Description	Energy kJ	Water g	Protein g	Fat g	Carbo-hydrate g	Fibre g	Ash g	Na mg	K mg	Mg mg	Ca mg	Fe mg	Zn mg	Cu mg	Vita-min C mg	Source
Eragrostis eriopoda woollybutt grass	seed	1333	5.2	17.4	2.3	58.2	13.2	3.7	5	237	156	235	31.0	5.0	1.3	–	*
	seed	1505	8.8	17.3	0.6	70.0(T)	–	2.3	36	144	–	40	–	–	–	–	Peterson 1978
	seed	–	–	12.9	2.4	–	–	–	–	–	–	–	–	–	–	–	Elphinstone 1971
Eremophila latrobei crimson turkey-bush	flower	377	77.4	2.0	1.5	–	–	–	–	–	–	–	(Thiamin 158 µg)			10	James 1983
Ficus platypoda	fruit	–	–	5.8	9.6	–	–	–	140	850	–	4000	–	–	–	–	Peterson 1978
Ficus racemosa cluster fig	fruit	150	87.9	0.8	0.3	–	–	–	–	–	–	–	–	–	–	Tr	James 1983
Fimbristylis sp. ? fringe-rush	seed	1713	3.8	13.9	20.9	40.8(T)	–	–	12.1	215	–	46	–	–	–	–	Peterson 1978
Ipomoea sp. sweet potato	root	959	40.0	1.9	0.8	51.9(T)	–	3.1	–	397	–	19	–	–	–	–	Dadswell 1934
Ipomoea costata wild potato	root	264	84.3	0.7	0.2	–	–	–	–	–	–	–	(Thiamin 108 µg)			3	James 1983
	root	530	66.8	2.5	0.4	–	–	–	–	–	–	–	(Thiamin 46 µg)			Tr	

Table 5.1 – continued

Foods	Description	Energy kJ	Water g	Protein g	Fat g	Carbo-hydrate g	Fibre g	Ash g	Na mg	K mg	Mg mg	Ca mg	Fe mg	Zn mg	Cu mg	Vita-min C mg	Source
Leichardtia australis bush banana, wild cucumber	fruit	221	77.7	3.4	2.0	4.8(T)	–	0.8	–	–	–	22	–	–	–	–	Dadswell 1934
cucumber	fruit	227	84.9	8.1	0.5	4.4	0.9	1.2	24	462	413	17	4.0	0.8	0.7	–	Brand *et al.* 1983
	fruit	360	75.2	2.6	0.5	18.6	2.1	1.0	4	245	31	22	0.9	0.7	0.2	2	*
	fruit	215	83.5	2.0	0.4	10.4	3.0	0.7	2	217	21	15	0.6	0.8	0.1	–	*
	fruit	506	71.7	4.1	0.7	–	–	–	–	–	–	–	(Thiamin 2935 µg)			4	James 1983
Leichardtia leptophylla bush banana	leaves	504	60.6	3.3	2.9	21.3	8.0	3.9	5	158	198	16	5.6	0.1	0.1	–	*
	fruit	261	84.1	9.8	0.4	5.0	–	3.9	15	325	224	8	1.8	0.6	0.4	9	Brand *et al.* 1983
Lepidium papillosum warty peppercress	leaves and stems	624	51.5	8.6	1.1	25.1(T)	–	4.0	–	1665	–	455	–	–	–	–	Dadswell 1934
Maireana villosa red berry	fruit	654	65.1	5.7	0.4	–	–	–	–	–	–	–	(Thiamin 83 µg)			Tr	James 1983
Panicum australiense Australian native millet	seed	1497	7.4	15.0	4.8	63.2(T)	–	9.4	11	90	–	–	16	–	–	–	Peterson 1978
	seed	1650	3.1	10.6	2.4	–	–	–	–	–	–	–	(Thiamin 0 µg)			3	James 1983
Pentatropis kempeana	fruit	193	85.8	1.1	0.5	9.7	2.2	0.7	2	192	21	21	0.4	0.8	0.1	20	*

Table 5.1 – continued

Foods	Description	Energy kJ	Water g	Protein g	Fat g	Carbohydrate g	Fibre g	Ash g	Na mg	K mg	Mg mg	Ca mg	Fe mg	Zn mg	Cu mg	Vitamin C mg	Source
Portulaca oleracea pigweed	seed	1794	8.1	19.6	15.6	51.9(T)	–	4.6	36	104	–	46	–	–	–	–	Peterson 1978
	seed	–	–	20.2	14.0	–	–	–	–	–	–	–	–	–	–	–	Elphinstone 1971
	damper cake	587	47.2	9.8	2.2	21.1	13.9	5.7	20	100	73	112	13	–	–	–	Brand *et al.* 1983
	leaves & stems	136	89	2.1	0.4	2.1	4.9(T)	–	–	709	–	112	–	–	–	–	Dadswell 1934
	plant	232	85.5	5.9	0.2	–	–	–	–	–	–	–	(Thiamin 131 µg) Tr					James 1983
	root	223	79	3.5	0.4	8.5(T)	–	–	–	1170	–	224	–	–	–	–	Dadswell 1934
Portulaca oleracea and *Portulaca interterranea*	seeds	874	5.6	18.5	1.1	32.4	30.4	12.0	12	616	329	181	275	6.5	2.4	–	*
Santalum acuminatum quandong	kernel	3000	1.6	15.9	57.6	3.1	20.8	1.0	–	–	–	–	–	–	–	–	G. Jones unpublished
	fruit	345	76.7	1.7	0.2	19.3	–	2.1	51	659	40	42	Tr	0.2	0.2	–	Brand *et al.* 1983
	fruit	410	67.7	2.8	Tr	22.6	4.5	2.4	71	781	34	59	1.1	0.4	0.2	–	*
Santalum lanceolatum bush plum, sandalwood	fruit	646	62.9	4.8	4.8	24.4	2.5	1.0	–	–	–	–	–	–	–	16	Brand *et al.* 1983
	fruit	741	62.2	2.0	5.9	–	–	–	–	–	–	–	(Thiamin 0 µg)			2	James 1983

Table 5.1 – continued

Foods	Description	Energy kJ	Water g	Protein g	Fat g	Carbohydrate g	Fibre g	Ash g	Na mg	K mg	Mg mg	Ca mg	Fe mg	Zn mg	Cu mg	Vitamin C mg	Source
Solanum sp.	fruit	669	69.8	4.5	0	–	–	–	–	–	–	–	–	–	–	0	James 1983
						(Thiamin 0 µg)											
Solanum centrale bush raisin	fruit	606	61.2	2.9	Tr	34.8(T)	–	1.1	–	–	–	–	–	–	–	–	Brand *et al.* 1983
	dried fruit	1273	2.0	8.6	1.1	67.9	15.5	4.9	4	1480	96	78	13.1	1.4	0.6	0	*
	partly dried fruit	1257	28.8	8.4	5.5	53.9(T)	–	3.3	119	195	–	46	–	–	–	–	Peterson 1978
Solanum chippendalei bush tomato	cooked fruit	339	78.6	1.1	0.7	18.4(T)	–	1.2	16	483	26	54	2.0	0.3	0.4	–	Brand *et al.* 1983
	fruit	294	78.2	1.8	0.6	15.1	3.2	1.1	21	503	23	38	1.5	0.5	0.5	49	Brand *et al.* 1983
	fruit	430	78.7	1.1	0.6	–	–	–	–	–	–	–	–	–	–	Tr	James 1983
	fruit	455	72.9	1.2	1.4	22.8(T)	–	1.7	18	639	–	98	–	–	–	29)	
	fruit	499	68.4	1.9	0.6	26.5(T)	–	2.6	21	795	–	130	–	–	–	34)	Peterson 1978
	fruit	357	77.8	2.5	0.3	18.0(T)	–	1.4	19	565	–	81	–	–	–	59)	
	fruit	305	81.7	1.4	0.5	15.5(T)	–	0.8	36	419	–	67	–	–	–	12)	
						(Thiamin 243 µg)											
Solanum ellipticum desert raisin	fruit	213	86	2.6	1.8	5.7(T)	–	1.3	–	355	–	35	–	–	–	–	Dadswell 1934
Stylobasium spathulatum nut bush	nut	1988	1.3	11.6	0.9	–	–	–	–	–	–	–	–	–	–	3	James 1983
						(Thiamin 0 µg)											

Table 5.1 — continued

Foods	Description	Energy kJ	Water g	Protein g	Fat g	Carbo-hydrate g	Fibre g	Ash g	Na mg	K mg	Mg mg	Ca mg	Fe mg	Zn mg	Cu mg	Vita-min C mg	Source
Vigna lanceolata	root	262	77.2	1.5	0.2	14.3	4.3	2.5	3	262	85	48	21.5	0.5	0.2	–	*
pencil yam,	root	362	76.0	2.8	0.3	17.5(T)	–	1.6	–	774	–	177	–	–	–	–	Dadswell 1934
maloga bean	root	19	98.7	3.5	–	–	–	–	–	–	–	–	(Thiamin 23 µg)		Tr		James 1983
	root	361	78.3	2.4	0.1	–	–	–	–	–	–	–	(Thiamin 0 µg)		1		James 1983

estimated separately, the figure has been subtracted from the total carbohydrate figure to give what is termed 'available carbohydrate'. The energy values will be an over-estimate in those foods where dietary fibre is unknown. Most samples have a figure for the minerals, Na, K and Ca and over half have information about other elements, Mg, Fe, Zn and Cu. Unfortunately vitamin analyses are the most incomplete with only a few figures for vitamin C and thiamin. We have also added our comments on palatability since people tend to select foods for their organoleptic qualities first and nutritional criteria become secondary.

Seeds

Collection of seeds was labour-intensive (except where ants' nests were raided) and usually done by the women. The seeds were brought back to camp in large quantities and prepared into a coarse meal by grinding between flat stones. In some cases the seeds were roasted first or water was added at the grinding stage. Winnowing in wooden food vessels might also be used depending on the species. The meal or flour would then be mixed with water into a dough or paste and damper cakes baked in the ashes of a fire. For some species, the raw paste was consumed.

The *Acacia* (wattle) seeds are strikingly nutrient rich with higher energy, protein and fat contents than crops such as wheat and rice and even higher than some meats. The wattle is Australia's national flower and they are generally unknown as food sources. The fruits and seeds are often poisonous or at least non-edible. Even domestic animals reject most of them (Cribb & Cribb 1974). Mulga (*A. aneura*) and witchetty bush *(A. kempeana)* have pods 2–4 cm long and carry several edible seeds about 3 mm in diameter. The high protein levels in the Acacias (range 17–25%) suggest breadmaking potential or possibly the food industry might find uses for the protein isolates. Their high fat contents (range 3.3–37.0%) suggest potential as oilseeds depending on fatty acid composition, stability and palatability. We found *A. coriacea* and *A. tenuissima* to have very palatable oils. At the immature green stage, *A. coriacea* seeds can be steamed from the pod and eaten directly. This stage has a protein content of almost 24%, a sweet, delicious flavour and a colourful appearance.

The grass and herbage seeds, *Fimbristylis* sp., *Calandrinia balonensis*, *Chenopodium rhadinostachyum*, *Eragrostis eriopoda*, *Panicum australiense* and *Portulaca* spp. are also nutrient dense food staples, particularly high in protein and often fat. *Portulaca oleracea* or pigweed (in Britain it is known as purslane) is one of those species which occur almost world-wide. Europeans recognize it as a garden weed but it was a prized food of many inland tribes of Aborigines. Dr Joseph Bancroft an eminent natural scientist and medical man of the 19th century is quoted as calling *P. oleracea* 'perhaps the most valuable of all specimens of native food under examination' (Cribb & Cribb 1974). The tiny, individual seeds which resemble black mineral sand are produced in such quantity that it is possible to collect an adequate amount without great effort. The entire plant is uprooted, the roots are removed to clear sand and dirt and the plants placed upside down on a sheet of bark, kangaroo skin or large sheet of paper, depending on one's ethnic background. The seeds are shed in a couple of days. Their flavour resembles linseed. The damper cake made from the ground meal contained 10% protein and appreciable quantities of dietary fibre and trace elements.

The seeds of the kurrajongs (*Brachychiton* spp.) appear to vary in nutrient density from species to species. *B. populneum* is very rich in protein (18.1%) and fat (24.7%) but *B. paradoxum* has the lowest level of protein (7.4%) of any of the mature seeds so far analysed.

Fruits

The desert tomato (*Solanum chippendalei*) was an important desert food available all year, particularly after rain (Hiddens 1980). They are green with a yellow tinge when ripe and filled with black seeds that are poisonous and not eaten. The fresh fruit contribute appreciable amounts of carbohydrate (20%) and a little protein (1–2%). They were probably one of the most consistent sources of vitamin C (range 12–49 mg/100 g, the recommended daily allowance for adults is 30 mg in many countries). James found his sample to be a rich source of thiamin or vitamin B_1 (0.24 mg/100 g, the Australian recommended daily allowance for adult men is 1.1 mg). The Aborigines of the Western Desert ate the fruit fresh but also dried the fruit in the sun and stored them indefinitely in tree caches (Gould 1969). Husks were generally dried impaled on a thin stick similar to a shish-ke-bab. The dried fruit were then mixed with a little water and then ground to a paste which was consolidated into a ball. Prior to eating the dried husks were dipped in water to make them palatable.

Another important Western Desert food was the desert raisin (*Solanum centrale* and *S. ellipticum*), which was sweet and palatable. They were eaten fresh or if collected in a dessicated state during summer, they were stone-ground, mixed with water to a seedy paste and then compacted into balls as large as 25 cm in diameter and weighting 1.5 kg. The outer surfaces dried to a crust and if kept dry the balls could be stored indefinitely (Gould 1969). The analysis of nutrients shows they were high carbohydrate foods but they would have contributed commendable amounts of protein (2.9–8.6%) and fat (1.1–5.5%) depending on the state of dessi-cation. There appears to be no vitamin C in the dried fruit.

Santalum fruits were staples in the summer and autumn. The trees are partial parasites. *S. acuminatum* (the quandong) fruits are bright red at maturity with a firm, fleshy layer surrounding the edible stone. The fruit 'rattle' when ripe. The flesh is eaten, although it is rather acid. It is a good source of carbohydrate (as high as a banana) and relatively high in protein compared with most fruit even when their lower moisture content is taken into account. It is a very popular food with both the Aborigines and the white settlers who made them into pies, jams and jellies. Both groups dried them which appeared to improve the flavour. The kernel is edible and oily (fat 58%) so much so that it will burn like a candle nut. CSIRO are presently studying the horticulture of the quandong with a view to commercial cultivation.

Santalum lanceolatum, the bush plum, is smaller than the quandong and deep blue-black in colour with a sweet, agreeable taste. It has relatively high levels of protein and fat for a fruit and contributes vitamin C (16 mg/100 g in one sample).

Native figs (*Ficus* spp.) were another staple fruit in spring and summer and they were favourites with the Aboriginals (Sweeney 1947). The sample of *F. platypoda* contained high levels of protein (5.8%) and fat (9.6%) and extraordinary levels of Ca (4000 mg/100 g); possibly this sample had been dried as *F. racemosa* contained more typical levels.

Australia is well endowed with *Capparis* spp. Capers, the pickled flower-bud of *C. spinosa,* are known throughout the world. Aborigines do not appear to have made use of the bud but the fruits of most species were eaten. Some fruits are 7 cm across. At maturity they are orange-yellow and contain numerous black seeds embedded in a sweet palatable pulp. The skins are not eaten. The five species analysed are reasonable sources of protein, fat and carbohydrate compared to most fruits even when their lower water content is taken into account. When available in summer—autumn they contributed vitamin C to the diet (2–89 mg/100 g) and one sample contained a very high amount of thiamin (0.7 mg/100 g).

The desert banana or cucumber (*Leichardtia leptophylla* and *L. australis*) is available throughout summer and autumn and has a wide distribution throughout the drier parts of Australia. The fruits are shaped like a pointed egg, 8 cm long with seeds that may or may not be eaten. Their taste has been compared to young peas, fresh and very crisp (Hiddens 1980). Aborigines prefer them roasted. In common with other bush fruits there is much more protein that one usually finds in a fruit (range 2.0–9.8%). They contribute a little vitamin C and one sample was found to contain extraordinary amounts of thiamin (2.9 mg/100 g). The leaves of *L. australis* are also eaten.

Mistletoe berries (*Amyema* sp.) being available for much of the year would have contributed vitamin C (20 mg/100 g) when other sources were scarce. The conker-berry (*Carissa lanceolata*) is useful food in the desert but the harvest only lasts for a few weeks (Sweeney 1947).

Rootstocks

Most people if asked to name the most important vegetable foods of the Aboriginals would put yams at the top of the list. Many species are thought to contain a poison which may be removed by slicing the yam, soaking them in running water for several hours, followed by cooking. Sweeney (1947) writes of the desert yam as 'the most remarkable of the native foods that nature has developed in the desert providing a nutritious food under hard conditions which can be harvested at any time of the year'. Meggitt (1957) later identified Sweeney's desert yam as *Dioscorea* sp., presumably *D. bulbifera*. The yams are found 50–100 cm underground and are harvested by the native women digging with yam sticks and using wooden food vessels as shovels. Tubers up to the size of a man's head are obtained. They can be eaten raw or cooked and have a slightly sweet taste (Sweeney 1947). Our analysis of a sample of *D. bulbifera* from tropical Arnhem Land showed that they contained a small amount of protein (1.6%) and carbohydrate (5.8%) but very high levels of vitamin C (233 mg/100 g after cooking).

The rootstocks of *Ipomoea costata* (wild potato) and *Vigna lanceolata* (pencil yam) were also important staples. They contain a little protein and carbohydrate (similar to a carrot), thiamin and vitamin C.

Nut or onion grass corms (*Cyperus* spp.) were eaten either raw or cooked and were available most of the year. They have a commendable amount of protein (2.6–5.5%) and appear to be starchy (total carbohydrate 62.5%).

Thirst quenchers

Plants with water storage as well as thirst quenching properties include kurrajong roots (*Brachychiton populneum*), the desert yam (*Dioscorea* sp.) and mulga apples (a large succulent gall produced by *Acacia aneura* trees and said to be 'very welcome to the thirsty traveller'. Kurrajong seeds (*B. populneum*) are also well known in Australia as an acceptable coffee substitute. They require a light roasting followed by pounding or grinding and brief boiling. Leafy vegetables like those of *Calandrinia balonensis, Portulaca oleracea* and *Lepidium papillosum* are high in moisture.

Some Aborigines also made mildly sweet, unfermented beverages by steeping flowers and other plant parts in water. *Hakea* spp., some gums, *Acacia* spp., mistletoe (*Loranthus* sp.) and *Leichardtia australis* were used for this purpose. *Canthium latifolium* berries are washed, imparting a sweetish flavour to the wash water which is eagerly drunk. Some Aborigines also made a drink from the mashed seed cap of *Acacia coriacea*.

Minerals

In general the Australian native plants are richer sources of minerals such as Ca and trace elements such as Fe, Zn and Cu than similar cultivated plants. The levels in some plants are astronomical, for example, *Portulaca oleracea/P. interterranea* seeds had 275 mg/100 g of Fe, more than one hundred times the recommended daily allowance in many countries. However high fibre levels in foods may decrease the absorption of minerals, necessitating a high intake. The contents of Na, K and Mg are usually within the expected range.

Conclusion

It appears that Australian native plants are richer sources of particular nutrients than one might expect on the basis of comparison with a similar cultivated plant. The seeds are almost all twice as high in protein as the common cereals and often many times higher in fat. The fruits have higher protein, fat and carbohydrate levels than cultivated fruits even after accounting for their lower moisture. The rootstocks have compositions similar to the potato or carrot, although mineral content may be higher. Some foods are good sources of vitamin C (*Capparis* sp. 89 mg/100 g; *Dioscorea bulbifera* 233 mg/100 g) and others are rich sources of thiamin (*Leichardtia australis* 2.9 mg/100 g). For most foods, the levels of these and other vitamins are unknown. In general, the native foods are higher in Ca and in the trace elements Fe, Cu and Zn than similar cultivated foods.

Acknowledgements

We are most grateful to Professor A.S. Truswell for advice and support and to Jeannie Devitt of the University of Queensland and the Aboriginal women of Utopia Station who collected most of the asterisked food specimens in the tables. The co-operation of other Aboriginal collectors and the dietitians of the Northern Territory Department of Health is also acknowledged. This study was funded by a research grant from the Australian Institute of Aboriginal Studies and the Sydney University Nutrition Research Foundation.

References

Brand, J.C., C. Rae, J. McDonnell, A. Lee, V. Cherikoff and A.S. Truswell 1983. The nutrional composition of Australian Aboriginal bushfoods 1. *Food Technol. Aust.* 35: 293–298.

Cribb, A.B. and J.W. Cribb 1974. *Wild food in Australia.* Sydney: Collins.

Dadswell, I.W. 1934. The chemical composition of some plants used by Australian Aborigines as food. *Aust. J. Exp. Biol. Med. Sci.* 12: 13–18.

Elphinstone, J.J. 1971. The health of Australian Aborigines with no previous associations with Europeans. *Med. J. Aust.* 2: 293–301.

Gould, R.A. 1969. Subsistence behaviour among the Western Desert Aborigines of Australia. *Oceania* 34: 253–274.

Hiddens, L.J. 1980. *Survive to live.* Thesis, James Cook University of North Queensland, Australia.

Hooker, J.D. 1859. *Flora Tasmaniae 1. Dicotyledons.* London: Reeve.

James, K.W. 1983. Analysis of indigenous Australian foods. *Food Technol. Aust.* 35: 342–343.

Meggitt, M.J. 1957. Notes on the vegetable foods of the Walbiri of Central Australia. *Oceania* 28: 143–145.

Peterson, N. 1978. Traditional patterns of subsistence in 1975. In *The nutrition of Aborigines in relation to the ecosystem of Central Australia,* B.S. Hetzel and H.J. Frith (eds): 25–35. Melbourne: CSIRO.

Sweeney, G. 1947. Food supplies of a desert tribe. *Oceania* 17: 289–299.

6 Khoisan food plants: taxa with potential for future economic exploitation

T. H. Arnold[1], M. J. Wells[2] and A. S. Wehmeyer[3]

[1],[2]Botanical Research Institute, Private Bag X101, Pretoria 0001, South Africa
[3]National Food Research Institute CSIR, PO Box 395, Pretoria 0001, South Africa

Introduction

The information presented in this paper is based on a broad literature review of knowledge of all plants used as sources of food and water by the Khoisan people, namely, the Hottentots or Khoi-Khoin and the San or Bushmen (Schapera 1963, Bruwer 1972), who inhabit the arid and semi-arid areas of Namibia, Botswana and southern Angola (mean annual rainfall 50–700 mm). The literature survey is supplemented by nutritional data, most of which has not been previously published.

Because the information gathered is too voluminous to be handled here, the bulk of it, comprising the baseline information, is to be published separately. The first part dealing with all 333 species used by the Khoisan as food plants will appear shortly in *Bothalia* the house journal of the Botanical Research Institute. This paper, part II, deals with some of the plants believed to have a potential for development as arid-area food resources.

SELECTION OF TAXA

Only the 66 species for which nutritional data was available were considered for inclusion in this paper*. These species were selected for chemical analysis, by the National Foods Research Institute of the CSIR, because they were considered to be the most important food plants. A further selection comprising 27 of the 66 species has been made on the basis of overall nutritional status, relative yield, previous cultivation record and Khoisan preference.

Nutritional status of plants

The nutrient composition of 66 Khoisan food plants, arranged alphabetically according to family, genus and species, is given in Table 6.1. Also included in this table is the average daily nutritional requirement per person (adapted from National

*The exception being *Cucumis anguria* included because it is already in cultivation.

Academy of Sciences 1979a). Where nutritional values per 100 g are greater than 20 per cent of the average daily requirement they have been emphasised by **bold** type. To put this nutritional data into perspective, equivalent data (Paul & Southgate 1978) have been included for eight cultivated plants (see Table 6.2, p. 73). Comparisons made between the nutrient composition of Khoisan food plants and cultivated plants are based on the edible parts in their natural (raw) state only, and do not take into account differences in moisture content.

Because a food plant has few or no nutritional constituents with values above the 20 per cent level this does not mean that it does not merit further consideration. *Coccinia sessilifolia, Cucumis kalahariensis* and *Vigna lobatifolia,* for example, have tubers with only 1–2 values above the 20 per cent daily requirement level. Despite this, they all have 7 or 8 constituents with values equal to or greater than those given for potatoes.

COMPARITIVE EVALUATION OF TAXA

The 27 taxa discussed in this paper are compared in Table 6.3, p. 74. They are arranged in seven groups according to life form and edible part and have been evaluated according to a six point scoring system applied to four major criteria. The individual scores for each criteria have been accumulatively multiplied to give a total score for each taxon (printed in **bold** type). The four criteria used in the evaluation are:

a. Nutritional composition : This is based on the number of chemical constituents which have a value greater than 20 per cent of the daily requirement for that constituent. Scoring was as follows : 0–1 constituent = 1, 2–3 = 2, 4–5 = 3, 6–7 = 4, 8–9 = 5, 10–12 = 6.
b. Domestication potential : This collectively takes into account : time to first crop, ease of harvesting, accessibility of edible part, distribution (width of habitat), ease of propagation, handling and storage qualities. *Adansonia digitata,* for example, was given a very low score due to its extremely slow rate of growth and because it is dependant on bats for pollination. Similarly *Acanthosicyos horrida* which has very specific habitat requirements also received a low score.
c. Relative yield : This takes into account two aspects : the size of the edible part and the number of edible parts per individual plant.
d. Desirability : This criterion reflects the desirability or potential acceptability of the plant part as a food and mainly takes into account palatability and the nature of the edible part. For example, the fruits of *Hyphaene ventricosa* are dry and very fibrous and are therefore relatively undesirable in their natural form.

Table 6.1 Nutrient composition of Khoisan food taxa. (Bold type indicates values greater than 20% of average daily requirement).

Name	Plant part	Mois-ture	Ash	Pro-tein	Fat	Fibre	Carbo-hydrate	Energy value kJ/100g	Ca	Mg	Fe	Na	K	Cu	Zn	P	Thia-min	Ribo-flavin	Nico-tinic Acid	Vit. C
				--- g/100g ---					--- mg/100g ---								1.1	1.3	1.4	60
Average daily requirement				55				13000	800	350	15	2200	4000	2.5	15	800	1.1	1.3	1.4	60
Sclerocarya birrea	flesh	85.0	0.9	0.5	0.4	1.2	12.0	225	20.1	25.3	0.5	2.24	317	0.07	0.1	11.5	0.03	0.02	0.27	**194**
Hyphaene ventricosa	nut	4.0	3.8	**28.3**	**57.3**	2.9	3.7	**2703**	118	**462**	**4.87**	3.81	601	**2.81**	**5.19**	**808**	**0.42**	0.12	0.72	–
Carralluma mammillaris	fruit	6.6	9.0	4.9	0.4	9.6	69.5	1265	103	**197**	2.04	–	**2560**	0.47	0.56	156	–	0.1	**4.62**	**19.7**
Ceropegia multiflora	leaves	88.4	1.3	0.7	0.2	1.6	7.8	150	**167**	63.8	1.42	112	174	0.38	0.44	19.1	0.1	0.04	1.85	**31.8**
Fockea angustifolia	Tuber	95.2	0.7	0.8	0.03	0.3	3.0	65	65.8	22.8	0.84	1.3	120	0.07	0.2	5.5	0.08	–	–	2.1
Microloma saggitatum	tuber	92.7	0.7	0.4	0.1	0.7	5.4	102	64.0	23.2	0.94	6.1	226	0.12	0.14	5.85	0.04	0.01	0.06	5.1
Pentarrhinum insipidum	fruit	74.3	2.0	4.3	0.5	3.3	15.6	353	127	**75.4**	2.27	11.1	649	0.46	1.07	1.02	**0.26**	0.18	0.89	**102**
	leaves	85.0	2.2	3.5	0.5	2.0	6.7	192	**370**	**78.7**	**8.75**	3.82	142	**0.43**	0.39	62.5	**0.21**	**0.26**	1.06	–
Adansonia digitata	leaves	88.1	1.3	2.3	0.2	1.5	6.6	157	71.8	**78.2**	0.79	6.3	323	0.37	0.15	46.6	–	–	–	**209**
	fruit	8.7	5.8	2.7	0.2	8.9	73.7	**1292**	**335**	167	2.65	11.2	**2409**	0.64	1.0	76.2	**0.62**	0.14	**2.73**	**209**
	kernel	8.1	5.9	**33.7**	**30.6**	16.9	4.8	**1803**	**273**	**640**	**6.55**	2.48	**1275**	**2.78**	**6.68**	5.12	**0.25**	0.14	1.0	6.9
Ehretia rigida	fruit	85.6	1.3	1.7	0.3	0.7	10.4	215	30.5	27.8	0.89	2.50	547	0.17	0.22	49.9	–	–	–	–
Commiphora pyracanthoides	tuber	81.0	2.6	–	–	–	–	–	285	53.0	0.8	7.5	635	0.2	0.4	–	–	–	6.9	6.6
Boscia albitrunca	root	68.1	1.8	6.5	0.2	3.6	19.8	449	63.6	**75.8**	0.38	37.2	504	0.08	0.78	14.4	0.02	0.03	0.24	6.6
Gynandropsis gynandra	leaves	85.0	3.6	5.1	0.6	1.3	4.4	181	**262**	**86.8**	**18.8**	33.6	410	0.46	0.76	12.0	0.1	0.22	1.49	–
Acanthosicyos horrida	flesh	84.0	1.6	1.4	0.3	1.0	11.7	231	21.4	19.0	0.5	14.1	654	0.3	0.6	22.4	0.01	0.02	0.75	–
	seed	5.3	3.4	**30.7**	5.7	1.3	2.3	**2709**	100	**363**	**4.0**	3.0	400	**3.9**	**5.5**	**8.11**	–	–	**2.17**	–
Acanthosicyos naudiniana	tuber	74.7	1.5	1.1	0.2	9.7	12.8	241	**273**	123	1.2	10.2	231	0.1	0.4	16.0	0.04	0.02	0.8	0.8
	fruit	90.6	1.0	1.3	0.04	2.1	4.8	111	21.2	23.6	0.52	1.53	270	0.12	0.24	25.1	0.09	0.03	**0.98**	**34.9**
Citrullus lanatus	flesh	94.2	1.1	0.4	0.04	1.2	3.1	60	31.6	16.0	0.3	2.1	267	0.27	0.09	5.1	0.06	0.01	0.34	9.2
	seed	5.7	1.9	**17.9**	**20.2**	**41.8**	12.5	**1274**	54.0	**209**	**6.55**	5.67	433	**1.14**	**3.4**	**474**	**0.59**	**0.27**	**2.45**	**19.3**
Coccinia adoensis	tuber	81.1	0.6	1.1	0.1	0.6	15.9	289	43.5	29.7	0.36	0.67	420	0.06	0.4	38.8	0.05	0.02	0.77	3.5
	fruit	91.7	0.8	1.8	1.0	2.1	2.6	112	28.4	28.1	0.61	0.52	269	0.11	0.17	4.1	–	–	–	3.5
Coccinia rehmannii	fruit	89.1	1.0	2.0	1.0	2.2	4.7	150	66.8	30.6	0.49	3.77	292	0.12	0.46	35.8	0.05	0.04	0.78	**24.5**
	tuber	89.1	1.3	1.2	0.1	1.6	6.7	137	**252**	37.2	0.36	4.5	154	0.19	0.73	11.7	0.02	0.02	0.31	6.7
Coccinia sessilifolia	fruit	82.3	1.4	2.1	0.2	1.3	12.7	256	37.9	2.2	0.2	4.9	685	0.2	–	24.0	0.19	0.13	–	**80.5**
	tuber	84.2	1.5	1.0	0.1	2.7	10.5	197	**351**	60.0	2.23	7.25	256	0.17	0.32	40.3	0.02	0.01	0.28	**12.8**
Cucumis africanus	leaves	92.2	1.6	1.3	0.3	1.2	3.4	90	**216**	175	**12.1**	11.3	109	0.17	0.31	11.1	0.2	0.11	0.34	5.4
	fruit	88.2	1.2	2.8	1.6	2.9	3.3	163	13.1	29.1	1.1	1.1	439	0.22	0.37	20.2	0.07	0.03	0.84	–
Cucumis kalahariensis	tuber	88.7	0.7	1.1	0.1	0.6	8.8	170	28.3	23.4	0.95	1.3	184	0.11	0.32	14.2	0.04	0.02	0.5	–
Cucumis metuliferus	fruit	91.0	0.9	1.1	0.7	1.1	5.2	134	11.9	22.3	0.53	2.08	319	0.11	0.25	25.5	–	0.02	0.55	–
Kedrostis africana	tuber	83.9	0.3	3.0	0.1	2.5	10.2	226	–	–	–	–	–	–	–	–	0.02	0.03	1.63	**18.6**
Momordica balsamina	fruit	89.4	1.6	2.0	0.1	1.8	5.1	123	35.9	41.2	2.61	3.25	533	0.2	1.0	35.8	0.04	0.06	0.55	0.5
	leaves	89.4	3.0	3.0	0.1	0.9	3.6	115	**340**	**87.1**	**12.7**	9.77	363	0.22	0.89	27.7	0.01	0.09	0.7	0.4

Table 6.1 continued. (Bold type indicates values greater than 20% of average daily requirement).

Name	Plant part	Mois-ture	Ash	Pro-tein	Fat	Fibre	Carbo-hydrate	Energy value kJ/100g	Ca	Mg	Fe	Na	K	Cu	Zn	P	Thia-min	Ribo-flavin	Nico-tinic Acid	Vit. C
		g/100g							mg/100g											
Cyperus fulgens	tuber	64.6	0.7	1.2	0.2	0.6	32.7	577	7.61	21.5	0.65	1.79	333	0.06	0.35	26.7	0.08	0.05	0.05	6.1
Cyperus rotundus	tuber	60.0	1.1	5.1	0.3	1.6	31.9	633	31.3	57.9	–	7.0	273	0.5	1.29	133	0.15	0.06	0.07	4.2
Dioscorea elephantipes	tuber	92.7	0.9	0.3	0.1	0.7	5.3	98	242	46.4	0.44	2.1	133	0.06	0.17	3.0	0.02	0.002	0.06	2.9
Diospyros mespiliformis	fruit	69.0	1.3	1.1	0.4	6.2	22.0	404	96.0	23.4	1.03	13.7	417	0.11	0.21	27.8	0.01	0.04	0.24	24.6
Ricinodendron rautanenii	flesh	8.6	5.2	7.8	0.5	2.9	**75.0**	**1410**	85.0	**214**	**2.54**	2.39	**2145**	**1.30**	**1.68**	74.3	**0.42**	0.13	**1.78**	**27.0**
	kernel	**4.2**	4.1	**26.3**	**58.1**	2.7	4.6	**2715**	223	**493**	**3.42**	3.35	674	**2.52**	**3.54**	869	**0.26**	0.22	0.27	2.2
Acacia albida	seed	6.5	3.9	24.8	2.2	6.8	55.8	1437	252	276	6.81	18.3	1125	1.57	2.55	390	0.9	0.17	2.04	–
Acacia erioloba	seed	8.0	4.2	25.8	4.6	11.4	46.0	1380	385	275	4.88	5.85	1100	1.0	3.48	314	0.77	0.15	1.34	–
Acacia karroo	pod	66.8	2.1	3.8	1.2	4.0	22.1	480	222	86.7	2.21	1.70	452	0.53	0.97	101	0.16	0.06	2.01	14.9
	gum	13.9	3.3	6.8	0.6	5.7	69.7	1308	963	111	16.6	36.0	183	1.01	0.27	33.3	0.02	0.01	0.04	–
Acacia tortilis	gr. pod	75.0	1.5	4.2	0.3	7.2	11.7	280	187	45.0	3.90	1.03	296	0.44	0.77	62.7	0.23	0.02	0.85	1.1
Bauhinia petersiana	seed	6.8	4.1	22.9	13.1	12.9	40.2	1554	237	220	3.87	1.24	1168	0.97	2.94	317	0.58	0.2	1.65	–
Schotia afra	seed	8.0	2.5	11.2	2.6	13.0	62.7	1340	168	119	15.6	2.54	974	1.4	2.19	174	–	–	–	–
Guibourtia coleosperma	seed	9.1	1.9	14.3	8.0	4.4	62.3	1589	323	163	4.69	20.3	390	0.87	2.7	198	0.07	–	–	–
Tylosema esculentum	seed	3.7	3.0	**32.9**	**37.8**	2.1	20.5	**2253**	183	295	3.87	22.6	780	1.38	**3.33**	463	**0.62**	**0.52**	**1.89**	2.3
Vigna lobatifolia	tuber	90.5	0.7	0.7	0.1	2.0	6.0	117	36.5	29.7	0.38	5.04	140	0.12	0.28	9.90	0.04	0.006	0.09	2.3
Pelargonium incrassatum	tuber	79.6	1.1	2.1	0.2	1.6	15.4	302	39.3	65.7	0.82	5.27	363	0.16	0.46	24.9	0.06	0.05	**0.94**	11.6
	tuber	59.8	1.3	3.5	0.2	1.6	35.6	665	237	59.0	0.7	12.4	280	0.29	0.8	78.8	0.13	0.03	0.67	68.3
Hydnora africana	fruit	69.9	2.2	1.7	2.0	5.9	18.3	412	8.49	27.1	0.93	80.1	738	**0.4**	0.6	88.9	0.12	0.09	0.76	13.8
Babiana dregei	corm	59.7	1.3	1.9	0.3	0.4	36.4	655	115	2.2	1.18	19.5	469	0.59	0.86	88.1	0.07	0.05	**0.86**	41.1
Hexaglottis longifolia	corm	58.6	0.6	1.9	0.2	0.4	38.3	683	27.4	17.5	0.7	19.5	163	0.14	0.47	45.1	0.11	0.02	0.67	92.2
Albuca altissima	infl.	93.5	0.4	0.3	0.1	0.2	5.5	101	20.7	14.9	0.45	18.5	114	0.09	0.11	8.8	0.01	0.004	0.7	7.4
Dipcadi longifolium	bulb	61.7	1.7	1.8	0.2	0.5	34.1	611	122	43.1	1.24	4.97	480	0.11	0.67	26.3	–	–	–	–
Dipcadi viride	bulb	86.3	0.7	0.8	0.1	0.3	11.8	214	75.1	13.9	0.45	1.81	13.0	0.1	0.1	6.2	0.05	0.01	–	5.1
Trachyandra falcata	infl.	84.9	2.1	2.2	0.5	2.8	7.5	182	165	56.7	**4.2**	99.3	385	0.27	0.65	57.4	0.05	0.01	0.57	96.2
Strychnos cocculoides	fruit	80.4	0.5	1.3	0.1	0.9	16.8	308	9.41	26.9	0.18	0.89	188	0.07	0.08	20.2	0.03	0.06	0.27	6.7
Strychnos pungens	fruit	72.1	1.0	1.1	0.8	6.2	18.9	367	29.3	38.1	0.62	2.0	478	0.25	0.34	27.1	0.05	**0.42**	**0.96**	10.7
Strychnos spinosa	fruit	78.8	1.8	2.7	0.1	1.4	15.2	305	45.8	43.6	0.75	4.55	328	**0.46**	0.12	22.6	**0.23**	0.1	**1.39**	10.6
Carpobrotus edulis	fruit	69.2	2.4	2.1	0.3	1.7	24.3	454	188	100	1.14	295	372	0.13	0.48	53.7	0.09	0.05	0.23	0.8
Ficus sycomorus	fruit	82.7	1.3	1.3	0.9	3.6	10.2	227	68.1	42.3	1.94	3.91	399	0.16	0.44	36.5	0.09	0.23	**2.2**	–
Ochna pulchra	fruit + seed	63.2	0.9	6.3	7.5	1.5	20.6	735	34.1	45.9	1.0	1.74	232	**2.65**	1.85	74.6	0.04	0.08	0.9	–
Ximenia americana	fruit	64.8	1.9	2.8	0.8	1.2	28.5	556	7.58	31.1	1.27	3.33	718	0.28	0.68	34.2	–	–	–	69.7
Ximenia caffra	fruit	77.4	1.3	1.9	1.5	0.9	17.0	374	8.17	19.0	0.49	1.25	558	0.17	0.29	35.4	0.03	3.02	0.48	68.2
Eulophia hereroensis	tuber	92.6	0.8	0.6	0.1	0.5	5.4	105	158	18.6	0.36	2.33	102	2.28	0.19	5.11	0.02	–	0.04	2.0
Oxalis flava	bulb	58.7	1.0	1.5	0.4	0.7	**37.7**	673	9.0	24.5	0.42	28.6	332	0.29	0.43	50.9	–	0.02	0.68	1.6

Table 6.1 continued. (Bold type indicates values greater than 20% of average daily requirement).

Name	Plant part	Mois-ture	Ash	Pro-tein	Fat	Fibre	Carbo-hydrate	Energy value kJ/100g	Ca	Mg	Fe	Na	K	Cu	Zn	P	Thia-min	Ribo-flavin	Nico-tinic Acid	Vit. C
		---- g/100g ----							---- mg/100g ----											
Raphionacme burkei	tuber	85.9	1.7	1.1	0.3	1.6	9.4	186	100	81.2	0.51	31.6	375	0.17	0.35	13.2	0.03	0.03	0.05	4.3
Raphionacme hirsuta	tuber	85.0	1.3	1.0	0.4	2.1	10.2	203	**163**	71.5	0.78	3.0	191	0.11	0.2	2.7	0.03	0.03	0.71	2.0
Talinium arnotii	leaves	93.1	0.2	1.6	0.2	0.7	4.2	105	126	**100**	1.54	0.92	655	0.14	0.54	22.6	–	–	–	4.9
Talinium crispatulum	leaves	91.0	1.4	3.4	0.1	0.6	3.5	120	**320**	**95.3**	–	–	368	0.1	0.26	25.9	0.05	–	–	8.0
Ziziphus macronata	fruit	56.4	3.2	5.6	0.4	2.0	**32.4**	653	80.9	54.9	0.89	2.82	57.1	**1.09**	0.6	32.0	0.07	0.07	0.9	86.4
Vanguaria infausta	fruit	64.4	1.4	1.4	0.1	4.7	28.0	498	24.9	39.1	1.07	**28.01**	521	0.25	0.24	36.6	0.04	0.03	0.61	4.7
Cyanella hyacinthoides	corm	55.5	0.7	3.3	0.3	1.2	**39.0**	722	66.8	24.4	1.25	9.78	176	0.34	0.77	58.3	0.13	0.03	0.98	33.3
Grewia flava	fruit	74.7	1.0	1.8	0.08	1.6	20.8	383	62.4	61.1	1.08	2.59	275	0.25	0.47	51.3	0.03	0.07	0.62	29.4
Grewia retinervis	whole fruit skin +	6.9	3.1	6.3	2.7	**38.1**	**42.9**	929	**624**	107	2.5	9.61	532	**3.03**	**2.33**	**128**	0.07	0.03	**1.63**	–
	flesh	6.9	3.2	4.9	0.3	15.9	68.8	1250	150	173	3.88	18.4	768	0.34	1.11	54.0	–	–	**1.31**	–
Grewia villosa	flesh	70.0	2.0	3.3	0.2	1.9	22.6	443	–	34.6	1.9	3.8	672	–	–	41.0	–	–	–	–
Terfezia boudieri	fr. body	80.1	1.6	4.1	3.5	2.6	8.1	338	9.23	91.1	**4.55**	3.24	290	0.6	0.56	1.42	0.13	0.14	**2.1**	–

Table 6.2 Nutrient composition of selected cultivated taxa. (Bold type indicates values greater than 20% of average daily requirement).

Name	Plant part	Mois-ture	Ash	Pro-tein	Fat	Fibre	Carbo-hydrate	Energy value kJ/100g	Ca	Mg	Fe	Na	K	Cu	Zn	P	Thia-min	Ribo-flavin	Nico-tinic Acid	Vit. C
		---- g/100g ----							---- mg/100g ----											
Potatoes	tuber	75.8	–	2.1	0.1	2.1	20.8	372	8.0	24.0	0.5	7.0	570	0.15	0.3	40	0.11	0.04	1.2	**20**
Carrots	root	89.9	–	0.7	–	2.9	5.4	98	48	12.0	0.6	95	220	0.08	0.4	21	0.06	0.05	0.6	**6.0**
Turnips	root	93.3	–	0.8	0.3	2.8	3.8	86	59	7.0	0.4	58	240	0.07	–	28	0.04	0.05	0.6	**25**
Onions	bulb	92.8	–	0.9	tr	1.3	5.2	99	31	8.0	0.3	10.0	140	0.08	–	30	0.03	0.05	0.2	**10**
Cucumbers	fruit	96.4	–	0.6	0.1	0.4	1.8	43	23	9.0	0.3	13.0	140	0.09	0.1	24	0.04	0.04	0.2	8
Soybeans	flour	7.0	–	**36.8**	23.5	11.9	23.5	1871	**210**	**240**	**6.9**	1.0	**1660**	–	–	**600**	**0.75**	**0.31**	2.0	tr
Brazil nuts	nut	8.5	–	**12.0**	61.5	9.0	4.1	2545	180	**410**	2.8	2.0	760	**1.1**	–	**590**	**1.0**	0.12	**1.6**	tr
Almonds	nut	4.7	–	**16.9**	53.5	14.3	4.3	2336	250	**260**	–	6.0	**860**	0.14	**3.1**	**440**	0.24	**0.92**	2.0	–

Table 6.3   Comparative evaluation of taxa (see text for explanation).

	Plant Name	Life form	Plant part	Nutritional composition	Domestication potential	Yield estimate	Desirability estimate	TOTAL	Recommended utilization	San preference rating
A	*Ricinodendron rautanenii*	tree	fruit	5	2	6	6	360	W(D)	5
	Sclerocarya birrea	tree	fruit	2	3	6	6	216	W(D)	5
	Hyphaene ventricosa	tree	fruit	5	3	5	2	150	W(D)	(3)5
	Strychnos cocculoides	tree	fruit	1	3	4	5	60	W(D)	4
	Grewia retinervis	shrub	fruit	3	3	2	3	54	W	(3)5
	Adansonia digitata	tree	fruit	5	1	3	3	45	W	(4)5
B	*Cucumis metuliferus*	herb	fruit	2	6	5	3	180	D	—
	Cucumis africanus	herb	fruit	2	5	4	3	120	D	—
	Cucumis anguria	herb	fruit	(1)	6	5	4	120	D	4
	Pentarrhinum insipidum	herb	fruit	2	5	3	3	90	D	4
	Acanthosicyos naudiniana	herb	fruit	2	3	5	3	90	D	(3)5
	Citrullus lanatus	herb	fruit	1	6	5	3	90	D	(4)5
	Coccinia sessilifolia	herb	fruit	1	5	4	3	60	D	3
	Cucumis kalahariensis	herb	fruit	1	4	4	3	48	D	5
	Coccinia rehmannii	herb	fruit	1	4	4	3	48	D	(4)5
	Acanthosicyos horrida	shrub	fruit	1	2	3	3	18	W(D)	—
C	*Ricinodendron rautanenii*	tree	nut	6	3	6	6	648	W(D)	5
	Sclerocarya birrea	tree	nut	6	2	6	6	432	W	5
	Guibourtia coleosperma	tree	nut	5	3	4	5	300	W	(2)5
	Adansonia digitata	tree	nut	6	1	3	3	54	W	(4)5
D	*Tylosema esculentum*	shrub	nut	6	6	5	5	900	D	5
	Citrullus lanatus	herb	seed	6	6	3	5	540	D	(4)5
	Bauhinia petersiana	shrub	seed	6	4	3	4	288	D	(3)5
	Schotia afra	shrub	seed	6	3	4	3	216	D	—
	Acanthosicyos horrida	shrub	seed	5	2	3	6	180	W(D)	—
E	*Trachyandra falcata*	herb	infl.	3	4	4	4	192	D	—
F	*Pentarrhinum insipidum*	herb	leaves	4	4	5	2	160	D	4
	Cucumis africanus	herb	leaves	3	4	5	2	120	D	—
G	*Pelargonium incrassatum*	herb	tuber	3	4	3	4	144	D	—
	Tylosema esculentum	shrub	tuber	1	5	6	4	120	D	5
	Coccinia sessilifolia	herb	tuber	1	5	6	4	120	D	3
	Coccinia rehmannii	herb	tuber	2	4	4	3	96	D	(4)5
	Babiana dregei	herb	corm	3	3	3	3	81	D	—
	Vigna lobatifolia	herb	tuber	1	5	4	4	80	D	4
	Cucumis kalahariensis	herb	tuber	1	5	4	4	80	D	5
	Fockea angustifolia	herb	tuber	1	4	4	4	64	D	5
	Cyanella hyacinthoides	herb	corm	2	3	2	3	36	D	—

W = utilization in the wild, D = domestication

RECOMMENDED UTILIZATION

Not all edible taxa are suitable for domestication, despite their possessing a number of positive characteristics. It takes only a single well developed negative characteristic, such as the inaccessible kernels in the nut of *Sclerocarya birrea,* to override its positive characteristics such as high yield and exceptional nutritional qualities, thus effectively lowering the suitability of this species for domestication for nut production.

The potential of such taxa is not completely lost as the option still exists to maximize their utilization in the wild, particularly in areas where they are abundant. This is not a new approach, having already been applied to taxa such as *Terfezia boudieri* (= *T. pfeilii*), *Acanthosicyos horrida* and *Sclerocarya birrea.*

A number of taxa listed in Table 6.3, particularly the trees, seem better suited to utilization in the wild than as domestic crops. However, the possibility of domestication should not be totally ignored, until ways of overcoming their apparent negative qualities have been studied.

SAN FOOD CLASSES

Table 6.3 shows the value placed on these plants by the San. The five classes used are: 5 — major food, 4 — minor food, 3 — supplimentary food, 2 — rare food and 1 — problematic food (Lee 1966, Tanaka 1976, Maguire 1978).

We would have liked to regard these classes as a 'control' to check against our own tentative efforts to evaluate Khoisan food plants. Unfortunately, the classes assigned vary from tribe to tribe and are obviously strongly influenced by the local availability of the plants, as observed by Maguire (1978).

Selected taxa

Sclerocarya birrea subsp. *caffra* (marula)

The female trees of this dioecious species, like *Ricinodendron rautanenii,* produce prolific quantities of fruit. According to Quin (1959) a single tree can yield between 21 000 and 91 000 fruits in a season. The fruits are 3—5 cm long and 2—3 cm in diameter, dropping from the tree when green and ripening on the ground (Wehmeyer 1971). The ripe fruits are slightly sour tasting with a characteristic aromatic, fruity, mango-like flavour (Maguire 1978), consisting of about 150 flavour components (Wehmeyer 1976).

The flesh has a high moisture content of *c.* 85 per cent (Table 6.1). It is rich in vitamin C, containing up to 200 mg/100 g of fruit, i.e. a better source of vitamin C than citrus fruits (Wehmeyer 1980). The pulp is commonly used by the indigenous adult peoples to make beer but is also suitable for making non-alcoholic fruit juices as well as jam and jelly (Wehmeyer 1976). Recently the fruits have also been used to produce a marula liqueur.

Within each stone are 2–3 embryos (12–17 $\times$ 6–8 mm). These have a hazel nut-like taste (Engeleter & Wehmeyer 1970) and are highly nutritious (Table 6.1), with a notably high content of protein and oil, most minerals, thiamin and nicotinic acid and a high energy value. Wehmeyer (1971) gives the two major fatty acids of the oil as oleic acid 70% and linoleic acid 8%. The embryos have a delicious flavour and are regarded by many indigenous peoples as the 'Food of Kings'. They are, however, small and extremely difficult to remove from the very hard, fibrous shell surrounding them (Wehmeyer 1976).

Hyphaene ventricosa (vegetable ivory palm, fan palm)

The edible fruits are 5–8 cm long (Maguire 1978). They have a *c.* 7 mm thick edible mealy layer immediately underlying the outer skin. This is fibrous and dry but sweetish with a pleasant flavour and can be chewed off the hard inner shell. The young fruits are boiled and eaten by the San (Story 1958). This fibrous part of the fruit has, due to its low moisture content, a high nutrient content (Table 6.1), especially with respect to the carbohydrate, sodium, potassium, nicotinic acid and its energy value. According to Dinter (1912), a single palm can yield 20–50 kg of fruit i.e. up to 2000 fruits per annum. These take 2–3 years to ripen (Maguire 1978).

The heart of the palm is also eaten by the San as a vegetable (Dinter 1912). The core of the young trunks is reported to be both crisp and delicately flavoured (Swart 1982)

Fockea angustifolia (khoa plant)

The tubers (resembling large turnips) have a mass of between 213–924 g and are sweet or slightly bitter depending on soil type (Archer 1982). The young tubers (*Fockea* sp. believed by Maguire to be *F. angustifolia*) are eaten raw by the San and have a faint but pleasant 'coconut ice' flavour (Story 1958). Story (1958) considered that it could possibly be bred into a plant worthy of cultivation.

The tubers have a high water content (Table 6.1) and, although not as nutritious as the potato, are either equal to or marginally more nutritious than the carrot and turnip (Table 6.2).

Pentarrhinum insipidum

The simple, heart-shaped, 3–9 cm long leaves (Maguire 1978) are eaten by most of the indigenous peoples of southern Africa. The !Kho San frequently pound the leaves together with those from other taxa, or with the tubers of several small Asclepiadaceae. The leaves contain significantly high amounts of calcium, iron and riboflavin (Table 6.1).

The fruits are solitary or paired follicles, 4—8 cm long. When young they are soft and may be eaten raw or cooked as a vegetable (Story 1958). They exude copious amounts of latex (which is harmless). They have a nutty, slightly peppery flavour and when boiled are reminiscent of asparagus (Fox & Norwood Young 1982). Watt and Breyer-Brandwijk (1962) state that the Hottentots eat the young pods which they describe as being fairly tasteless. Nutritionally they are rich in magnesium and copper (Table 6.1). According to Story (1958) the fruits can be stored for up to three weeks. The species grows easily from seed and may have some economic value.

Adansonia digitata (baobab)

The fruits are 12—15 cm long and 7—10 cm in diameter. Each contains numerous bean-sized seeds surrounded by a soft white, edible flesh. The dry flesh has a slightly tart, refreshing taste and is very nutritious, with particularly high values for carbohydrates, energy, calcium, potassium (very high), thiamin, nicotinic acid and vitamin C (very high).

The seeds (*c.* 10 × 5 mm) are eaten raw or are roasted and have a pleasant nutty flavour (Wehmeyer 1971). They are also very nutritious with high values for proteins, fats (oils), fibre, most minerals (Table 6.1). The fatty acid composition is palmitic acid 26.5%, stearic acid 4.4%, oleic acid 32.3% and linoleic acid 34.9% (Wehmeyer 1971).

Gynandropsis gynandra (= *Cleome gynandra*)

Like *Pentarrhinum insipidum* this plant is highly favoured by the indigenous peoples of southern Africa who prepare and eat it as a spinach. According to Quin (1959) this plant is so popular amongst the Pedi that cultivation is being considered. In Zimbabwe it is already being cultivated by the Ndebele and Shona people (Fox & Norwood Young 1982). According to these authors the leaves and young buds are washed and boiled with a little salt added. The leaves should be well cooked to get rid of the slightly bitter taste (Watt & Breyer-Brandwijk 1962).

The nutritional value of the leaves (Table 6.1) shows notably high values for protein, calcium, sodium and riboflavin.

Acanthosicyos horrida (narra, narra melon)

This plant is adapted to the extreme climatic conditions and loose substrata of the sand dunes of the Namib desert.

The Topnaar Hottentots still live for a considerable part of the year almost exclusively on the fruit (*c.* 20 cm in diameter) of the narra. They make a preserve from the flesh by boiling it to produce a pulp which, after straining the seeds, they allow to dry and solidify in the sun (Fox & Norwood Young 1982).

The Hottentots are also very partial to the seeds, which range in size from 14—15 mm long to 9—11 mm wide and 6—7 mm thick. They contain a considerable quantity of oil and were exported in large quantities to Cape Town, where they are sold as a substitute for almonds (Meeuse 1962).

The nutrient composition of this plant is given in Table 6.1. Although no exceptional values have been recorded for the flesh, the seeds were found to be rich in protein, magnesium, copper, zinc, phosphorus and nicotinic acid as well as having a notably high energy value.

Acanthosicyos naudiniana (herero cucumber)

The fruits vary in size from 6—12 cm long and 4—8 cm diameter and have a mass of *c.* 250 g (Meeuse 1962, Wehmeyer 1976). The skin is thick and is covered with blunt protuberances usually ending in a hard sharp apical spine. Beneath the skin is a translucent green, jelly-like flesh in which are embedded numerous flat seeds, 7.5—10 mm long and 4—6 mm broad (Meeuse 1962). The flesh contains *c.* 90 per cent moisture and is an important source of water for the San (Lee 1966, Maguire 1978). It can be eaten raw or roasted. The flesh of the roasted melons is delicious with a refreshing sweet-sour taste (Keith & Renew 1975), however, as in many cucurbit taxa, bitter fruit forms also exist (Story 1958).

Compared with the cucumber (Table 6.2), it has either similar or notably higher values for fibre, magnesium, potassium, nicotinic acid and vitamin C and energy (Table 6.1).

The seeds and skin are roasted and pounded by the San to make a meal (Story 1958). Their nutrient composition is not known.

Nutritionally, the tubers of this plant (Table 6.1) compare favourably with those of cucurbits such as *Coccinia sessilifolia* and *Cucumis kalahariensis,* but Maguire (1978) regards them as inedible.

Citrullus lanatus (tsamma, wild watermelon)

The fruits are globose or subglobose, 10—20 cm in diameter and have a very high water content (94 per cent). As a result they are a primary and often only source of water for the San during 8—9 months of the year when surface water is not available (Story 1958, Maguire 1978). Because of the high water content of the fruits their nutrient content (Table 6.1) is very low (Wehmeyer 1980).

The tsamma seeds are 7—12 mm long and laterally compressed. They are a special delicacy (Maccrone 1937). After roasting, the kernel and shell are ground into a course meal which has a very pleasant flavour.

The seeds are rich in protein, fats (oil), fibre (due to shell), various minerals (magnesium, iron and zinc), thiamin and nicotinic acid. The energy value is also notably high.

Coccinia rehmannii

The edible fruits are subglobose to ellipsoid and 3–4.5 cm long and 2–3.5 cm in diameter (Meeuse 1962). They are sickly sweet and flavourless (Fox & Norwood Young 1982). The fruits are more nutritious (Table 6.1) than the cucumber (Table 6.2) except for sodium and vitamin C. A notably high value was recorded for nicotinic acid.

The most important edible part is the tuber. These grow up to 40 cm long and 12 cm in diameter. They are collected when thick and juicy and are either roasted or boiled (Fox & Norwood Young 1982). The older tubers are considered to be rather tasteless (Watt & Breyer-Brandwijk 1962). They grow in hard sand at a depth of *c.* 70 cm and are therefore difficult to collect (Fox & Norwood Young 1982). The tubers are rich in calcium and nicotinic acid and are nutritionally comparable to the carrot and turnip.

Coccinia sessilifolia (red gherkin)

The tubers (tap roots) are up to 50 cm long. They are usually carrot-shaped and have a mass of up to 25 kg, with a firm, juicy, fibrous flesh (Maguire 1978, Keith & Renew 1975). Nutritionally, they compare favourably with the carrot, potato and turnip (Table 6.2) having notably higher carbohydrate, energy, calcium, magnesium, iron and phosphorus values (Table 6.1).

The fruits are 5–8.5 cm long and 2–3.5 cm in diameter and are either eaten green or cooked as a vegetable (Story 1958). The fruits have a firm texture and flavour similar to that of asparagus (Keith & Renew 1975). The fruits ripen from December to February and, according to ·Maguire (1978) perish fairly rapidly. Nutritionally, they show no outstanding qualities (Table 6.1) but are higher in most nutrients than the cucumber (Table 6.2).

Cucumis anguria var. *longipes* (West Indian gherkin)

This plant is considered by Meeuse (1962) to be the probable progenitor of the domesticated West Indian gherkin, *C. anguria* var. *anguria,* known only in a cultivated or semi-wild state in America. Meeuse (1958) states that the West Indian gherkin is a cultigen described from a non-bitter variant of this taxon, originally named *C. longipes* by Hook.f. If Meeuse is correct in linking this taxon to the cultivated *C. anguria,* then this plant has already proved itself in domestication. The usefulness of the African plants is therefore mainly as a source of germ plasm (possibly of more drought resistant forms).

Cucumis africanus

The leaves are deeply 3–5-lobed and 2–10 cm long and slightly less wide (Meeuse 1962), and are eaten by many tribes as a spinach. The leaves are rich in calcium, iron, nicotinic acid and vitamin C (Table 6.1).

The fruits, which grow up to 12 cm in length and 6 cm in diameter, are covered in stout, but soft prickles. The flesh is transparent green and jelly-like. Morphologically distinct bitter and non-bitter forms exist (Story 1958, Meeuse 1962). The bitter forms are slightly sour but not totally unpalatable (Story 1958). Nutritionally (Table 6.1), they are rich in nicotinic acid, and vitamin C and have an overall nutrient composition slightly better than that of the cucumber (Table 6.2).

Cucumis kalahariensis

The roots of this plant produce edible tuberous swellings. These are sausage-shaped, 5–20 cm long, 2–3 cm in diameter and covered with a thin light-brown bark which surrounds a white spongy inner tissue (Meeuse 1962). The tubers are juicy and can be eaten raw or cooked and have a pleasant turnip-like flavour (Story 1958). Nutritionally they have no outstanding qualities, (Table 6.1). Despite this, they are comparable with the carrot and turnip (Table 6.2), having higher values for magnesium and iron than these vegetables.

Cucumis metuliferus (jelly melon)

The flesh of the immature fruits is white and resembles the domestic cucumber both in appearance and taste. The flesh of the mature fruits is translucent green and jelly-like. The mature fruits, are palatable although insipid tasting, and bitter forms are known to exist (Story 1958, Meeuse 1962). The non-bitter types, reputed to be an excellent vegetable similar to the cucumber, are occasionally cultivated in South Africa (Meeuse 1962), although their cultivation is not restricted to this country (Story 1958).

The fruits are more nutritious than the cucumber, having notably higher values for most nutrient components with the exception of calcium and sodium (Table 6.1).

Ricinodendron rautanenii (mongongo, manketti)

The fruits are *c.* 3.5 cm long and 2.5 cm in diameter with a mass of *c.* 10 g (Wehmeyer 1980). They fall from the tree while green and ripen on the ground after several months (Lee 1966). The flesh is 2–3 cm thick and surrounds the seed. This comprises a very hard shell enclosing the inner kernel, which is difficult to remove without its being damaged (Wehmeyer 1980). Both the flesh and nut are edible and highly nutritious (Table 6.1).

The flesh is very sweet with up to 30% sucrose (Wehmeyer 1980). It contains small amounts of vitamin C and is a good source of magnesium and potassium. It is also notably high in its carbohydrate, energy and thiamin values. The nut is nutritionally rich, particularly in its protein and oil content and energy value. The major fatty acids of the oil are linoleic acid (42%) and oleic acid (18%) (Wehmeyer 1980). The oil cake remaining after the removal of the oil contains 60% protein. Besides the high values mentioned the nuts are also rich in most minerals and in their riboflavin content (Table 6.1). The mongongo compares favourably with the world's most nutritious foods, cultivated or otherwise (Lee 1966), both in energy value and in its vitamin and minerals.

According to Vahrmeijer (1976), between 1911 and 1914 about 2000 tons of these nuts were exported annually to Germany and in 1916 to England for the production of margarine. He also states that the mongongo can be propagated by means of seeds and truncheon cuttings. The !Kung San, however, report that the trees take up to 25 years before bearing fruit (Lee 1966).

Bauhinia petersiana subsp. *macrantha* (wild coffee bean)

The pod of this species contain 4–12 beans, each 10–15 mm in diameter and highly compressed (Maguire 1978). The beans are produced in reasonably large quantities and are considered very tasty (Fox & Norwood Young 1982). They are usually roasted in their pods and then removed and eaten without the testa (Maguire 1978). They have also been used as a coffee substitute (Palmer & Pitman 1972).

Like the Mongongo nut the beans of this legume are highly nutritious, with notably high values being recorded for proteins, carbohydrates, fibre, energy value, calcium, magnesium, potassium and riboflavin (Table 6.1). They also compare favourably with nuts such as almonds and Brazil nuts (Table 6.2) having more protein, carbohydrate and potassium.

Story (1958) comments that this plant might be useful in cultivation if improved by breeding. It grows well as a cultivated ornamental shrub as far south as the eastern Cape.

Schotia afra var. *angustifolia* (Hottentots bean)

The beans of this species were consumed by Stone Age Man and in the 1770s the Hottentots cooked them and pounded them into a meal or ate them while still green (Palmer & Pitman 1972). They are 10–15 cm long and 9–12 cm wide and are notably rich in fibre, carbohydrates, iron and potassium and have a high energy value (Table 6.1).

This plant is occasionally cultivated and germinates well, if erratically, and transplants easily. It is also reported to be tender to frost, is drought resistant and is usually slow growing (Palmer & Pitman 1972).

Guibourtia coleosperma (bastard mopane, chivi tree)

The pods of this tree are 1(−2) seeded. The seeds are bean-like and 1.2−2 cm long. They are roasted and then ground into a coarse meal and are excellent eating (Lee 1966, Maguire 1978). The beans have a fairly high nutritional value (Table 6.1) particularly with regard to their energy level and protein, carbohydrate and calcium content. According to Maguire (1978), with adequate water available the San can live solely on this plant, often for periods of up to three months.

Tylosema esculentum (marama or maramba bean)

The tubers vary in size depending on the age of the plant. Tubers weighing up to 250 and 300 kg have been excavated (Bergström & Skarpe 1981, Keith & Renew 1975, Bousquet 1982). The San, however, eat only the very young tubers (*c.* 1 kg). According to Wehmeyer (1971) these have a sweet flavour and make an excellent vegetable. Nutritionally, the tubers have no outstanding qualities (Table 6.1) but compare favourably with the carrot and the turnip (Table 6.2).

The greatest food potential of this plant is in the nuts. These are contained in a hard shell, 1.5−2 cm in diameter, from which they are fairly easily removed. The dried nuts have, after roasting, a pleasant, slightly coffee-like taste. Slightly bitter tasting forms are also known to exist. Besides being suitable as a dessert nut, the beans may also be potentially suitable as a coffee-substitute (Wehmeyer 1969).

The marama beans are highly nutritious (Table 6.1). They have a very high protein content, which compares favourably with that of soybeans (Table 6.2). The oil content is also high and on the basis of these constituents alone this plant rivals both the soybean and the peanut (National Academy of Sciences 1979b). The marama bean is also a good source of minerals, namely, calcium, magnesium and phosphorus as well as being rich in thiamin and nicotinic acid.

Attempts to cultivate this plant have as yet not been very successful (Clauss & Clauss 1979). According to Bousquet (1982) the plants do not flower within the first three years and the number of beans produced by different plants varies significantly.

Vigna lobatifolia (Sâ plant)

The Sâ plant has a branched root system which becomes swollen at intervals to produce tuber-like structure, 6−17 cm long and 3−5 cm in diameter. A single plant produces both small and large swellings separated along the root by distances of between 1−25 cm (Wehmeyer *et al.* 1969). These potato-like swellings can be eaten either raw or cooked and are said to taste similar to, if not better than, potatoes (Lee 1966). Maguire (1978) describes them as being crisp, and slightly sweetish in taste. Nutritionally the Sâ tubers are comparable with the potato, having less potassium and phosphorus (Table 6.1) and more calcium and magnesium.

Pelargonium incrassatum

According to Archer (1982), this plant together with *Cyanella hyacinthoides* are the most popular root storage organs collected and eaten by the Hottentots. The turnip-shaped subterranean tubers, are up to 4 cm long and *c.* 2 cm in diameter. They are eaten raw, cooked in milk or roasted. They are rich in carbohydrates, calcium and vitamin C (Table 6.1) and have a generally higher nutritional content than the potato, carrot and turnip (Table 6.2).

Babiana dregei (rock onion)

The corms of this species are a popular Hottentot food (Archer 1982). They are 3—4 cm in diameter and are covered by a matted, coarse fibrous tunic (Lewis 1959). They are best gathered and eaten during July and August as they tend to be dry and unpleasant later in the season (Archer 1982). The corms have moderately high carbohydrate and vitamin C values (Table 6.1). Their nutrient composition is, however, higher than that of the onion in nearly all respects (Table 6.2).

Trachyandra falcata (Hottentots cabbage)

This plant has edible inflorescences which grow to a height of 60 cm (Obermeyer 1962). They are considered a staple food of the Hottentots (Archer 1982). Schultz (1907) and Marloth (1915) refer to the young inflorescences (*Anthericum drepanophyllum*) being used as a vegetable. Nutritionally it has high values for calcium, iron, nicotinic acid and especially vitamin C (Table 6.1).

Strychnos cocculoides (yellow monkey-orange)

The fruits of this species are ± globose and 6.5—10 cm in diameter. The shell is woody and brittle, 3—4 mm thick containing a sweetish, pleasantly flavoured pulp in which numerous seeds are embedded (Maguire 1978).

Three species of *Strychnos* are eaten by the San, the others being *S. pungens* and *S. spinosa*. All are pleasant tasting with *S. cocculoides* considered to be the sweetest and most delicate tasting of the three, with *S. spinosa* rated second (Coates Palgrave 1977, Fox & Norwood Young 1982).

These three taxa are nutritionally comparable (Table 6.1) with the values for *S. spinosa* being marginally higher than those of the other two in most instances. The seeds are reputed to be poisonous, however, it is not certain whether this applies to all the species or not.

Cyanella hyacinthoides (raaptol)

The corms are an important food of the Hottentots. Without the enveloping, fibrous tunic they have a mass of *c.* 144 g (Archer 1982). They are usually roasted before being eaten and have a delicious nutty flavour (Metelerkamp & Sealy 1983). They are a good source of carbohydrate and vitamin C content (Table 6.1). The overall nutritional status of the corms is, in every respect, higher than that of the onion (Table 6.2).

Grewia retinervis (Kalahari raisin)

This is one of the many *Grewia* species of which the fruit is eaten. They are obovoid-subglobose and 6–8(–12) mm long. The thin layer of flesh is rather dry and moderately fibrous (Maguire 1978). They are described by Palmer and Pitman (1972) as sweet and pleasant tasting and are the staple food (raw or dried) of several other African tribes. Their desirability lies mainly in their abundance and ease of collection. The fruits ripen in winter and can be stored after drying (Story 1958). They are fairly nutritious (Table 6.1) with notably high values for fibre, carbohydrates, some minerals and nicotinic acid. They are considered better suited for increased utilization in the wild than for domestication.

Conclusion

The food plants eaten by the Khoisan people include a number of taxa with a high nutritional content. Some of these, for example, *Ricinodendron rautanenii* and *Sclerocarya birrea,* because of factors such as slow growth to production maturity, are considered mainly suitable for increased utilization in the wild. Others, such as *Cucumis metuliferus, Cucumis africanus, Tylosema esculentum, Pelargonium incrassatum* and *Coccinia sessilifolia* are seen as important candidates for domestication.

Four major areas requiring a greater research input have been identified. Firstly, there is a need to expand our knowledge of food plants at the most basic level, namely by compiling comprehensive lists of all known edible plants. Of the 333 Khoisan food plants listed in part I of this paper (Arnold & Wells in press), 124 (37%) were not included by Fox and Norwood Young in their recent (1982) book, in which they set out to list all the edible plants of southern Africa. This clearly indicates how incomplete our knowledge of indigenous food plants is. Secondly, our knowledge of the nutrient composition of edible food plants is poor, with nutritional data being available for only 20 per cent of the taxa eaten by the Khoisan people. Thirdly, we know little or nothing about the biology of most of these plants. Important aspects such as their breeding biology, germination, habitat requirements, rate of growth, number and size of edible parts produced, etc. are seldom studied.

Finally, there is a need for plants identified as having possible potential to be evaluated in the field in order to identify those taxa with the greatest potential for future exploitation, and in particular those taxa with the greatest potential for domestication. The most promising plants, with the greatest potential for improvement, should then be finally tested under cultivation.

References

Archer, F.M. 1982. 'n Voorstudie in verband met die eetbare plante van die Kamiesberge. *J.S. Afr. Bot.* 48(4): 433–449.

Bergström, R. and C. Skarpe 1981. The tuber of morama *(Tylosema esculentum)*. *Bots. Not. Rec.* 13: 156–158.

Bousquet, J. 1982. The Morama bean of the Kalahari Desert as a potential food crop, with a summary of current research in Texas. *Desert Plants* 2(4): 213–215.

Bruwer, J.P. 1972. Hottentots. In *Standard encyclopedia of South Africa,* D.S. Potgieter (ed.) 5: 606–610. Cape Town: Nasou.

Coates Palgrave, K. 1977. *Trees of Southern Africa.* Johannesburg: Struik Publishers.

Clauss, B. and R. Clauss 1979. *Some comments on possibilities of agricultural research on field foods in the central Kalahari.* (Contribution to ALDEP Conference) Gaborone, Botswana: Mimeo.

Dinter, K. 1912. *Die vegetabilische Veldkost Deutsch-Südwest-Afrikas.* Bautzen: Kommissionsverlag Ed. Rühls Buchhandlung.

Engelter, C. and A.S. Wehmeyer 1970. Fatty acid composition of oils of some edible seeds of wild plants. *J. Agr. Food Chem.* 18(1): 25–26.

Fox, F.W. and M.E. Norwood Young 1982. *Food from the veld – edible wild plants of Southern Africa.* Johannesburg: Delta Books.

Keith, M.E. and A. Renew 1975. Notes on some edible wild plants found in the Kalahari. *Koedoe* 18: 1–12.

Lee, R.B. 1966. *Subsistence ecology of* !Kung *Bushmen.* Ann Arbor, Michigan: University Microfilms.

Lewis, G.J. 1959. The genus *Babiana. J. S. Afr. Bot.* Suppl. 3: 1–149.

Maccrone, I.D. 1937. A note on the Tsama and its uses among the bushmen. *Bantu Studies* 11: 251–252.

Maguire, B. 1978. *The food plants of the* !Kung *Bushmen of north-eastern South West Africa.* Johannesburg: University of the Witwatersrand, M.Sc. Thesis.

Marloth, R. 1915. *The Flora of South Africa 4. Monocotyledones.* Cape Town: Dorter Bross.

Meeuse, A.D.J. 1958. The possible origin of *Cucumis anguria* L. *Blumea* Suppl. 7: 196–205.

Meeuse, A.D.J. 1962. The Cucurbitaceae of Southern Africa. *Bothalia* 8(1): 1–111.

Metelerkamp, W. and J. Sealy 1983. Some edible and medicinal plants of the Doorn Karoo. *Veld & Flora* 69: 4–8.

National Academy of Sciences 1979a. Recommended dietary allowances (revised 1980). *J. Am. Diet. Assoc.* 75(b): 623–625.

National Academy of Sciences 1979b. *Tropical legumes: resources for the future.* Marama Bean: 68–74. Washington, D.C.: National Academy of Sciences.

Obermeyer, A.A. 1962. Revision of the South African species of *Anthericum, Chlorophytum* and *Trachyandra. Bothalia* 7: 669–767.

Palmer, E. and N. Pitman 1972. *Trees of Southern Africa.* Cape Town: A.A. Balkema.

Paul, A.A. and D.A.T. Southgate 1978. *McCance and Widdowson's, the composition of foods.* London: Her Majesty's Stationery Office.

Quin, P.J. 1959. *Foods and feeding habits of the Pedi.* Johannesburg: Witwatersrand University Press.

Schapera, J. 1963. *The Khoisan peoples of South Africa. Bushmen and Hottentots.* London: Routledge and Kegan Paul.

Schultze, L. 1907. *Aus Namaland und Kalahari*. Jena: Gustav Fisher.

Story, R. 1958. *Some plants used by the bushmen in obtaining food and water*. Mem. Bot. Surv. S. Afr. 30.

Swart, W.J. 1982. Survival off the veld, 8 — Living from palm to mouth. *Farmer's Weekly* 72046: 66–67.

Tanaka, J. 1976. Subsistance ecology of central Kalahari San. In *Kalahari hunter-gatherers: studies of the !Kung San and their neighbours*, R.B. Lee and I. De Vore (eds): 99–119. Cambridge, Mass: Harvard University Press.

Tanaka, J. 1980. *The San hunter-gatherers of the Kalahari: a study in ecological anthropology*. Tokyo: Tokyo University Press.

Vahrmeijer, J. 1976. *Ricinodendron rautanenii* Schinz-manketti (Southern African Plants No. 4463.000–0010) Pretoria: Division of Agricultural Information. Department of Agriculture and Fisheries.

Watt, J.M. and M.G. Breyer-Brandwijk 1962. *Medicinal and poisonous plants of Southern and Eastern Africa*, 2nd ed. Edinburgh: Livingstone.

Wehmeyer, A.S. 1971. The nutritional value of some edible wild fruits and plants. In *Proteins and food supply in South Africa*, J.W. Claasens and H.J. Potgieter (eds): 89–94. (Proceedings of International Symposium, University of Orange Free State, Bloemfontein. April 1968). Cape Town: A.A. Balkema.

Wehmeyer, A.S. 1976. Food from the veld. *Scientiae* Oct.–Dec.: 2–11.

Wehmeyer, A.S. 1980. *Some Botswana veld foods which could possibly be used on a wider scale*. Pretoria: National Food Research Institute, CSIR, (unpublished).

Wehmeyer, A.S., R.B. Lee and M. Whiting 1969. The nutrient composition and dietary importance of some vegetable foods eaten by !Kung Bushmen. *S. Afr. Med. J.* 43: 1529–1532.

7 Food plants of prehistoric and predynastic Egypt

M. Nabil El Hadidi

The Herbarium, Faculty of Science, Cairo University, Giza, Egypt

Introduction

Early man penetrated to the Nile Valley in Egypt during the Lower Palaeolithic period some 250 000 years ago. The linear pattern of the Nile and the concentration of resources along its main course encouraged the establishment of numerous human settlements, which varied in size, character and density during the various periods or cultures. Most of these settlements availed themselves of the resources of the Nile; their proximity to the major desert wadis also indicate the utilization of the wadi fauna and flora. Their location close to the mesas provided access to raw materials commonly used in various industries, including the manufacture of tools and grinding stones.

According to Hassan (1980), there was a gradual increase in the number and the density of the occupation sites during the Late Pleistocene period between 22 000 and 14 000 years BP. The artifacts attributed to this period have been reported from many places along the main Nile, in Nubia, Upper Egypt and the Delta. There is a remarkable decrease in faunal remains with more reliance on fish and water fowl accompanied by a gradual shift towards the utilization of plants.

Industries of later episodes, e.g. Isnan Culture 13 000–12 500 years BP and later, included grinding stones and related diagnostic agricultural tools, denoting an economic shift to grain or seed consumption. The 12 000 years BP floods led to further drastic changes in the economy, which put an end to the extensive exploitation and dependence on wild grain or seed. The domestication of cereals by these early inhabitants of Egypt was an important part of the establishment of the first farming communities along the Nile.

The growth of agriculture led to the development of a unique economy during the Holocene period during which cultivation, herding, hunting and the exploitation of the Nile's aquatic resources are inseparable.

The available information about plant food resources is rather fragmentary, depending upon the number of excavated sites and the state of preservation of the archaeological remains. Personal experience on dietary traditions and food attitudes must be considered.

Two main stages can be recognised in the exploitation of plant food resources during the final stages of the Late Pleistocene, *c.* 20 000–12 000 years BP or earlier and Holocene, 12 000–4500 years BP (Table 7.1).

Table 7.1 Food economies of Palaeolithic and Neolithic man.

Period and duration	Stage	Artifacts
Middle Palaeolithic 20 000–16 000 BP	hunting and plant gathering	scanty plant remains
Late Palaeolithic 16 000–9500 BP	hunting and plant gathering	grinding stones, dentiolate tools, scanty plant remains
Terminal Palaeolithic 9500–6000 BP	hunting and plant gathering	grinding stones, dentiolate tools, scanty plant remains
Neolithic 6000–4500 BP	herding and cultivating	grinding stones, sickles, extensive plant remains

The hunting-gathering stage of the Late Palaeolithic

According to Roubet and El Hadidi (1981), the Late Pleistocene of Egypt extended from 20 000 years BP (or earlier) to 12 000 years BP and included several cultures belonging to the Middle and Late Palaeolithic periods (Hassan 1980).

Recovered organic remains and artifacts would suggest the existence of human settlements practising hunting and gathering. Very few plant remains are recorded from the sites belonging to this period (Roubet & El Hadidi 1981). These include fragments or charred remains of *Acacia, Tamarix* and *Salsola,* which are mainly sources of firewood and charcoal. No remains of food plants were recovered, although this may reflect the nature of the plant material eaten.

During these early times, man must have been practising some selection of plants for their food characteristics. The process of selection is deemed to have been multidisciplinary, depending on the application of the senses, particularly those of taste and smell, as well as observing the dietary habits of birds and bovids. This would have provided him with a wide range of wild plants within the environment.

It is envisaged that many of these food plants would have been annual herbs which nowadays are regarded as common weeds of cultivation. Such species are likely to be found on the flood plains exposed following the regression of the Nile waters at the end of each season's floods. Thus, early each spring, the green foliage of numerous annuals would provide fresh vegetables. The wild flora would provide such species of Compositae as *Cichorium endiva* subsp. *divaricatum* (= *C. pumilum*), *Sonchus oleraceus* and *Lactuca* spp., while the Cruciferae would provide wild

Raphanus sativus, Brassica nigra and *Eruca sativa.* It is suggested that other species eaten would include *Cynodon dactylon* (Gramineae), *Corchorus olitorius* (Tilia-ceae), *Malva parviflora* (Malvaceae), *Beta vulgaris* (Chenopodiaceae) and *Rumex dentatus* (Polygonaceae).

Some of these wild forms later became the progenitors of the present cultivars of raddish, mallow, etc. A wild form of *Lactuca,* endemic to Upper Egypt, is believed to be the progenitor of the cos cultivar which has been known in Egypt since Pharonic times and was early associated with Min, the god of vegetation and proc-reation (Darby *et al.* 1977).

The last days of spring would be the season for harvesting seeds and grain. A countless number of wild annual legumes would have provided ripe seeds of high nutritive value. Species of the following are worthy of mention: *Lotus, Cicer, Lens, Lupinus, Lathyrus, Trigonella, Pisum, Vicia, Vigna* and *Astragalus.*

Wild grasses as a source of grain were presumably utilized, although to a lesser extent. From a biogeographical viewpoint, species of the Panicoid genera *Pennise-tum* and *Sorghum,* were more likely than those from the Festucoid genera such as *Hordeum* and *Triticum.*

The author believes that the hunting and gathering communities were primarily dependent on leguminous seeds rather than grasses. The gathering of the mature grass grains by stripping the inflorescence is quite an elaborate process when compared collecting and popping legume pods; the yield of legume seed is also likely to exceed that of grass grains and for the same or less effort.

The symbiotic relation between legumes and grasses may have influenced early man in the use of grain. The observation of this relationship could have influenced the eventual development of crop rotations which were practised during the Predy-nastic period and continue to be practised in Egyptian agriculture.

The late spring must have witnessed the collection and consumption of other, non-leguminous seeds, namely those of oil and spice plants. Of these, the following are believed to be known to early man: Cruciferae – *Sinapsis alba, Lepidium sativum,* Compositae – *Carthamus tinctorius,* Umbelliferae – *Coriandrum sativum* and *Anethum graveolens.* These are still used today for the same purposes.

The summer floods and the following winter months were periods when collect-ing activities would have diminished due to lack of available species and would have been replaced by the ripening fruits on the trees and shrubs. Among these were the palms, *Phoenix dactylifera, Hyphaene thebaica* and *Medamia argun.* Other species with edible fruits are *Balanites aegyptiaca, Ziziphus spina-christi* and *Ficus sycomo-rus*; the pulpy cortical regions of the woody branches arising as the result of secondary growth are also edible.

Over-wintering storage organs can also provide food when other plant sources are unavailable. Bulbs of wild onions and garlic, *Allium* spp., as well as the tubers of *Cyperus esculentus* are known food sources from the Neolithic period (Täckholm & Drar 1950) and apparently earlier. The rhizomes of the water lilies *Nymphaea lotus* and *N. coerulea* were and are still valued for their high nutritive content.

In summary, early man during the Late Pleistocene in Egypt, would have gathered a wide range of plant material for food, depending on the availability during different seasons of the year. The gathering of food following a repeated annual sequence is considered to have resulted in the development of food-producing communities in the following period.

The herding-cultivation stage of the Holocene

By the end of the Pleistocene, the communities along the Nile were apparently sufficiently successful to allow large human populations to build up (Wendorf 1968). It has been suggested that high population densities at this time must have encouraged increasingly higher levels of exploitation which may have approached the beginnings of animal and plant domestication. The diverse and abundant resources along the Nile were perhaps among the reasons why the earlier inhabitants never needed to resort to food procurement practices that were as labour-intensive as agriculture (Clark 1971).

During these early days, the people were able to support themselves with a wide selection of plant foods, including grains of wild grasses, seeds of various legumes, etc. According to Clark (1971), the teeth of the people occupying terminal Upper Palaeolithic sites were worn down in a way that suggested a large proportion of their diet came from gritty, stone-ground grain.

By the beginning of the Holocene, including Terminal Palaeolithic and Neolithic periods, 9500—4500 years BP, the populations along the Nile were no longer homogenous or as conservative as previously. Connections with other populations and cultures, primarily from the south, i.e. tropical Africa, but mainly from the east, southwestern Asia, led to the development of new communities which were not only depending on gathering but also on the cultivation of selected grain or seed plants.

Cultural equipment at Early Egyptian agricultural sites were in part African, with bifacial elements and microlithic forms, as well as Asiatic elements, with adzes, sickle blades, etc. This mixture of African and Asiatic elements strongly suggests that the cultivators were not all migrants from Asia. It is more probable that small groups of people moved from southwestern Asia to North Africa and Europe bringing their domesticated plants and animals with them and that the Egyptian peoples adopted the agricultural system and some of the associated tools and technology (Stemler 1980).

Clark (1971) has suggested that the very deep levels of animal and plant exploitation pre-adapted the early Egyptians to an agricultural way of life. Clark argues that agriculture would not have been accepted so readily, as it apparently was, if it had not been preceded by experience gained in the exploitation of wild grains and the cultivation of indigenous Egyptian plants.

It is possible that the changes in climatic conditions following 10 000 years BP resulted in decreased populations of native grasses. This could have been the first stimulus to practice primitive forms of agriculture, with the sowing of seed along the exposed banks of the Nile as the flood waters receded. The sown seed presumably being selected for maximum yield.

The earlier experience of seasonal gathering may also have been preparation for sowing seeds at certain intervals, an initial stage in the development of crop rotations.

Other techniques that must have evolved during this period were those of water economy and water requirements of crop plants. This seems to be related to the changes in the Nile hydrological regime, particularly during the last millenia.

The earliest well-documented evidence of agriculture in Africa is from Egypt. Grain bins containing carbonized barley and emmer wheat have been found in settlements at Faiyum, Marimde and Badari; dating from *c.* 6000 years BP (Stemler 1980). Few remains of legumes were recovered, which can perhaps be attributed to the poor conditions for preservation.

It is generally accepted that Neolithic and Predynastic populations had shifted gradually to the cultivation of a wide range of food plants that must have been selected and domesticated during the Palaeolithic period. With the introduction of the southwestern Asian domesticates, early Egyptians gave up some of their local cultigens in favour of more productive crops of alien origins.

Most of the food plants that were known in Egypt since Predynastic period are still known today. The accounts of Täckholm (1951), Darby *et al.* (1977), El Hadidi (1982) and Wetterström (in press) are among the more important references.

In summary, the changes in environmental conditions during the Holocene led to a gradual shift from food-gathering to food-producing societies. Climatic changes plus changes in the hydrology of the Nile resulted in decreased populations of wild grasses with a consequental increased reliance on cultigens. The inhabitants discovered that a constant supply of grain, i.e. crop production, might be assured by sowing seeds in the silt left by the receding flood waters of the Nile. Pluvial conditions also made possible a free exchange of people and domesticates between southwestern Asia and Africa, accompanied by an exchange of experience in domestication and agriculture.

The adaption of agriculture during the Neolithic—Predynastic period, 6000—4500 years BP led to the development of a unique economy in which cultivation and herding, as well as hunting and the exploitation of the Nile resources were inseparable.

References

Clark, J.D. 1971. A re-examination of the evidence for agricultural origins in the Nile valley. *Proc. Prehistoric Soc.* 37: 34—79.

Darby, W.J., P. Ghalioungui and L. Grivetti 1977. *Food: the gift of Osirus, 2.* London, New York, San Francisco: Academic Press.

El Hadidi, M.N. 1982. The Predynastic of Hierakonpolis: an interim report. Chapter 4. The Predynastic flora of the Hierakonpolis region. *Egyptian Stud. Ass. Publ.* No. 1: 102–115.

Hassan, F. 1980. Prehistoric settlements along the main Nile. In *The Sahara and the Nile, quaternary environments and prehistoric occupation in northern Africa,* M.A.J. Williams and H. Faure (eds): 421–450. Rotterdam: Balkema.

Roubet, C. and M.N. El Hadidi 1981. 20,000 ans d'environnement préhistorique dans la Vallée du Nil et le Désert Égyptien. *Bull. Centenaire (Suppl.) Bull. Inst. Française Arch. Orient* 81: 445–470.

Stemler, A.B.L. 1980. Origins of plant domestication in the Sahara and the Nile valley. In *The Sahara and the Nile, quaternary environments and prehistoric occupations in northern Africa.* M.A.J. Williams and H. Faure (eds): 503–526. Rotterdam: Balkema.

Täckholm, V. 1951. *Faraos blomster.* Stockholm: Natur & Kultur.

Täckholm, V. and M. Drar 1950. *Flora of Egypt,* Vol. 2. Bull. Fac. Sci. Fouad 1 Univ.

Wendorf, F. (ed.) 1968. *Prehistory of Nubia,* Vol. 2. Dallas: Southern Methodist University Press.

Wetterström, W. in press. Palaeobotanical studies at Predynastic sites in the Nagada–Khattara region. In *Predynastic studies in the Nagada–Khattara region of Upper Egypt,* F. Hassan (ed.). New York: Academic Press.

Wetterström, W. in press. Crops and agricultural practices at El Hibeh, a late Pharonic and Ptolemaic site in Egypt. In *Archaeological investigations at El Hibeh.* Malibu, Ca.: Undena Press.

8 Place and role of trees and shrubs in dry areas

B. Ben Salem and Christel Palmberg

Forest Resources Division, Forestry Department, FAO, Rome, Italy

Introduction

More than 50 per cent of the land surface of the developing countries is situated in the arid and semi-arid zones. In many of these countries, in which more than 80 per cent of the population lives off agriculture and/or animal husbandry, a tragic and dangerous imbalance is developing between requirements and available supply of wood and other products and services traditionally provided by the natural, woody vegetation (FAO 1981b, 1982a & 1984a).

Dwindling vegetation cover will adversely affect all facets of rural life, in which trees and shrubs generally serve not only as fuel but also as shade and shelter for man, animals and crops. Other uses are for building materials, food and fodder, as well as a source of a range of products such as gums, resins, tannins and medicines. In the long term, depletion of the natural vegetation will increase ecological fragility and contribute to gradual degradation of the resource base as well as the natural resources themselves (Eckholm 1975, Palmberg 1981a & b).

The history of arid zones is replete with records on drought, crop failures and widespread human suffering. Should these trends continue, entire development programmes of countries in these zones and the very survival of their populations will be in jeopardy.

Past development efforts, particularly within the agricultural sector, have largely focussed on the more productive lands; there are growing reasons for re-examining such policies and increasing efforts to introduce appropriate land management and development systems into low rainfall areas. Trees and shrubs have a major role to play in this development (FAO 1984a).

The present paper examines the influence on the environment of annual plants grown in the arid and semi-arid regions and pinpoints some of their characteristics which may contribute to ecological fragility and the marginalization of agriculture; it discusses the place and role of perennial woody vegetation in an equilibrated rural economy and discusses priorities for action to more fully utilize, on a sustainable basis, the potential of trees and shrubs in coordinated land management programmes.

Characteristics of dry zone plants: annuals versus perennials

Both theory and practice show that, in areas with a long dry season, perennial plants make, in the long run, the best use of soil and climate (Ben Salem 1980). In addition to their capacity for self-protection through a range of physiological and morphological mechanisms, they also afford protection for other plants growing under their shelter through biological and micro-climatic improvement of the environment.

The main groups of perennials found in arid and semi-arid lands are: succulents, palms and woody species. All of these have their given role and importance. This paper will concentrate mainly on discussing the role of woody perennial species.

One of the most common and unrecognized abuses of arid and semi-arid eco-systems involves the cultivation of annual plants in monocultures without adequate conservation measures. Annual plants leave the soil unprotected over lengthy periods of time and, if used unwisely, rapidly degrade and impoverish its relatively fertile, upper layers where their shallow root systems extend. Annuals are vulnerable to climatic fluctuations and year-to-year variation in, e.g. rainfall, being dependent on specific climatic conditions (e.g. the on-set of rains) during their establishment and development (Ben Salem & Eren 1982).

It has been argued that annual dry zone plants are often not, in fact, xerophytes: they develop from sprouting seed, carry through their cycle of development, and remain dormant in the form of heavily coated seeds during the drier part of the year. During their growth and maturation, their water needs are often high, even in absolute terms (McGinnies 1981).

Perennial plants cover the soil over long periods of time and stabilize it with frequently deep and extensive root systems. Their growth does not depend on rain starting within a few days or weeks of a certain date and, in addition, most of the species growing in arid and semi-arid zones have developed physiological and morphological characteristics which efficiently assist them in withstanding extended periods of drought. They are thus biologically buffered against conditions generally considered adverse to plant growth.

As a range of woody species of different genera and varying growth habits can provide specified goods and services, it may not be obligatory to establish large-scale cultures consisting of single species, even if artificial planting and establishment of plantations are used; diversification and the planting of mosaics of different species is often desirable both from the biological and the socio-economic points of view, and possible when woody perennials are used.

Whether native or introduced, perennial woody plants generally provide a range of products and services in addition to wood, such as fruits (*Butyrospermum paradoxum* subsp. *parkii, Adansonia digitata, Ceratonia siliqua, Tamarindus indica, Ziziphus mauritiana*); fodder (*Acacia* spp., *Prosopis* spp., *Parkinsonia* (= *Cercidium*) spp.); essential oils (*Argania spinosa, Pistacia* spp., *Rosmarinus officinalis*); resins (*Pinus halepensis*); latex (*Parthenium argentatum*); tannin (*Acacia* spp.); gum

(*Acacia senegal, A. laeta, Commiphora* spp., *Sterculia* spp.); and pharmaceutical products (*Cadaba farinosa, Combretum glutinosum, Khaya senegalensis, Catunaregam nilotica**). They also stabilize the soil and, in some cases, improve it.

These types of perennials, of which the above are but examples, play a vital role in the daily life of most rural communities. Even if fluctuations in climate temporarily adversely affect, e.g. growth (wood production) and/or the development of fruit and seeds, this does not render the plant useless from the point of view of environmental protection, nor does it generally signify a complete 'crop failure' for the population in the area. Once conditions again change and become more favourable, the plants are likely to resume their full potential.

Even when under heavy use, woody vegetation can be managed on a sustained basis, through the regulation of livestock browsing and pruning, pollarding and harvesting of products on a rotational basis, with due consideration of the biology of the species, Traditional practices, which foster the wise utilization of tree and shrub resources and do not destroy them, have been developed over the ages in many regions (McKell 1980).

Present land management systems

Agricultural systems in dry areas can be classified, roughly, under two concepts: traditional and conventional agriculture.

TRADITIONAL AGRICULTURE

In areas with a prolonged dry season and outside irrigable lands, traditional agriculture is characterized by grazing and crop-growing practised cyclically, without any clear allocation of separate areas for either land use over a given period of time; thus, all intermediate stages between them can be found simultaneously and side by side. The specific use of any unit of land at any one time will depend on the conditions of the soil, climatic fluctuations and alternatives available to the peasant farmer. Crops are grown as soon as trees and shrubs have restored the soil; once the soil has been consumed by erosion, leaching and loss of humus under intensive crop growing, it is abandoned until once again restored under a fallow of woody perennials. Such fallow areas are grazed, and provide recourse during difficult periods to stock-raising.

In this agricultural system, in which crops and pastures alternate, the main activities of the farmer centre around annual plants. Owing to the unreliability of the climate and the intrinsic characteristics of annual plants which leave the soil largely unprotected, the lands are prone to degradation. The very agricultural system is thus marginal, and yields and life itself precarious.

**Catunaregam nilotica* syn. *Lachnosiphonium niloticum, Xeromphis nilotica* and *Randia nilotica* (G.E.W.).

CONVENTIONAL AGRICULTURE

In the same climatic zones described above, another type of agriculture is also practised; this agricultural system, adopted from temperate regions, may be called conventional agriculture.

In conventional agriculture, which differs distinctly from the traditional one, the system tends to adapt the soil to prevailing needs rather than to adapt crops and cropping systems to the soil and climate. The flocks and herds of domestic animals have at their disposal land specially set aside for them and sometimes artificially maintained to meet the requirements of stockraising; similarly, agriculture is practised on separate areas, specially set aside for this purpose.

Organized management is the outstanding feature of this type of agriculture. Thanks to greater means at its disposal, it occupies the same piece of land more lastingly than does the traditional type of agriculture. However, despite this more intensive, directional management and acknowledged respect for basic rules of soil conservation, conventional agriculture uses essentially the same plant species as traditional agriculture; when the climate becomes arid, the range of cultivated plants is narrowed down to the traditional annual crops of barley, maize, sorghum and millet.

The means, the rules, the power of which conventional agriculture disposes, increase its potential for agricultural production, but also its potential for causing eriosion and degradation of the soils.

Place and role of woody vegetation

As evidenced above, arid and semi-arid areas present severe problems of land use and stabilization of agricultural production. Large areas have been severely degraded as a result of deforestation, unwise cultivation of annual crops and overgrazing, which failed to take into consideration the vulnerability of the ecosystems and their slow restoration once disturbed.

The use of trees and shrubs combined with the cultivation of annual plants has an important role to play as one of the primary means of arriving at a balanced rural economy, while representing at the same time perhaps the only way of avoiding degradation of the environment and destruction of the resource base. As elements essential to the overall stability of agriculture in these zones, trees and shrubs thus cannot be relegated to a secondary role, but must form an inseparable component of rural land use management (Ben Salem 1980, FAO 1980c & 1984a).

Trees and shrubs can be integrated in the rural landscape in three ways which are reviewed below: permanent cover of woody vegetation, trees and shrubs as part of rotation cropping and trees and shrubs in symbiosis with agriculture.

PERMANENT COVER OF WOODY VEGETATION

The woody vegetation of dry zones constitutes nature's choice, and maximizes adaptation and survival under harsh environmental conditions. By appropriate management, it will be possible to improve the yield of such vegetation, be it for livestock production, wood or fibre, resin, pharmaceuticals or other products and services. Advances in management will be dependent on applied research to provide answers to a number of practical questions on autecology and synecology of natural woody vegetation and on physiology, phenology and inherent variation of individual species, gradually leading to improved techniques suitable for a range of conditions and socio-economic requirements.

Many of the cultural and management practices (regeneration, weeding, thinning, exploitation) commonly employed by foresters, deserve special attention for their adaptation and application to dry areas and to management systems aimed at increased productivity of natural woody vegetation and a diversification of available end products.

TREES AND SHRUBS AS PART OF ROTATION CROPPING

Trees and shrubs have been used traditionally as part of rotation cropping in many regions of the dry zones. This 'bush fallow' system has been the mainstay of nearly all stable biological cropping systems in the tropics. The culture of gum arabic in Sudan is an example of such a system. Apart from restoring soil fertility and protecting it against wind erosion, *Acacia senegal* trees grown during the fallow period provide substantial income from gum arabic in addition to fuelwood, fencing material and fodder.

Considering both the technological and economic limitations in the use of fertilizers in arid areas, the bush fallow is likely to remain the most widely-used method to improve soil fertility and agricultural production for some time to come. However, there is considerable scope for improvement in this system through research into species to be used, optimal for different sites and conditions; adequate fallow periods; cultural and management questions, including utilization and processing of both wood and non-wood products, etc.

TREES AND SHRUBS IN SYMBIOSIS WITH AGRICULTURE

In the two systems described above, trees and shrubs are separated from agricultural crop growing in either space (permanent woody cover), or time (bush fallow). However, trees and shrubs can also be grown, with advantage, in 'symbiosis' with agricultural crops, co-existing with them for mutual benefit. In such a system, trees and shrubs are planted in rows (shelterbelts) to protect crops against wind erosion and desiccation, or intermingled with crops to reconstitute and enrich the soil and, simultaneously, protect the crops.

Numerous examples can be quoted of both systems, which constitute well-balanced production models based on the principle of sustained long-term management for a range of products and services, as well as environmental stabilization.

Examples of species used for multi-layered and multi-purpose windbreaks include, e.g. *Azadirachta indica/Parkinsonia aculeata* in P.D.R. Yemen and *Eucalyptus microtheca/Acacia mellifera* on black cotton soils of the Sudan. Models of trees/shrubs intermingled with crops and often traditionally protected by peasants include the *Acacia albida*/millet system in West and East Africa; the *Prosopis cineraria*/millet system in India and Pakistan; the *Leucaena leucocephala*/maize system in Central and South America, etc. Such systems were generally developed independently in various and totally different socio-economic conditions, as a response to actual needs brought about by rapidly diminishing fertility of the soil and dangers of desertification.

Integrated land use

In the light of the above options, a possible management strategy for integrated land use in arid lands can be outlined.

While permanent woody vegetation will be limited to non-agricultural lands and managed for a range of goods and services in multi-purpose production systems, the more gentle slopes, peneplains and plains will be converted, over most of their area, into a network of plots managed under a bush fallow system, or using agricultural crops and woody perennials in symbiosis.

Such a comprehensive and mutually complementary system, if properly planned and managed, will ensure that land is used more efficiently, production per unit land area is maximized and diversified to meet a wide range of needs, and destruction of the resources and degradation of the resource base are minimized.

Priorities for action

Although during recent years there has been a gradual shift of emphasis in national, bilateral and internationally assisted projects towards meeting the needs of rural communities in hardship areas, including arid and semi-arid zones, and this is gradually leading to increased understanding of desirable land use and cropping systems (see, e.g. FAO 1981d,e, 1983c & 1984a, Anon.1981b & 1982, Le Houérou 1980, McKell *et al.* 1972), there is little coordination of such activities. The consequent duplication of effort and the loss of time, energy and scarce financial and manpower resources, are considerable.

For the incorporation and full, long-term utilization of trees and shrubs in development, a multi-purpose orientation is needed by land-use planners, botanists, foresters, range managers and agriculturists alike. Such an orientation has, to date, often been overshadowed by a tendency to compartmentalize problems and to heavily slant discussions and recommendations passed at meetings and seminars towards meeting one main end-use criterion, e.g. the production of browse or wood for energy; shelterbelts; land amelioration through the use of research into nitrogen-fixing species, etc. Such a tendency is even more pronounced between different disciplines, and contradictory of mutually incompatible recommendations by different professional groups are not uncommon.

In view of the above, there is an urgent need to create a network of coordinated, balanced, multi-disciplinary programmes, aimed at developing sustainable land use practices for arid and semi-arid areas, in which the full potential of trees and shrubs is realized as a means towards improving the livelihood of people and maintaining a stable ecological balance. Training in the management of natural woody vegetation, which has been largely overlooked and neglected as compared to training efforts devoted to a better understanding and utilization of other natural resources, must play an important role as part of such a network, together with improved dissemination of information and an efficient exchange of experiences and results.

Although some basic knowledge exists on a number of arid zone woody perennials, considerable gaps can still be identified in questions related to intraspecific variation, phenology, management, cultural requirements and utilization. In many cases, even the most basic taxonomic and genecological information is lacking, thus impeding any efforts to conserve and wisely utilize many species of present importance to local communities. A concerted effort to solve such technical problems is urgently needed and should be carried out within the framework of the network described above.

Priorities for immediate action to support the establishment and early implementation of action programmes in such a network can be summarized as follows:

(a) Reviewing and evaluating present knowledge on the use of woody vegetation (trees, shrubs, bushes) in arid and semi-arid zones, with a view to identifying, for each environment, the potentially most suitable species and species combinations to serve the needs of rural communities; identifying gaps in technical knowledge to be filled so as to increase the contribution of woody vegetation to the development of rural economy.

(b) Disseminating information on available knowledge and experiences and promoting the application of such knowledge into practice. This will include making widely available information presently found in local journals, bulletins and documents which are not easily accessible; and translation into appropriate languages of information and publications of practical interest at regional and global levels.

(c) Strengthening national capabilities in research and development aimed at improved management and utilization of the woody vegetation in arid and semi-arid areas, through increased training and cooperation between research institutes.

(d) Establishing a network of pilot and demonstration areas, with a view to demonstrating the value of known techniques within the framework of integrated agro-silvo-pastoral development activities in rural areas in arid and semi-arid areas.

Activities of FAO's forestry department in the field of arid zone forestry

Recognizing the specific role of tree and shrub resources as discussed above, FAO's Forestry Department has initiated a number of activities to assist member nations in the use and development of trees and shrubs for the benefit of rural communities in arid and semi-arid zones, as well as to stimulate regional and international co-operation to this end (see, e.g. FAO 1982b, 1983c & 1984a). Coordination between different disciplines is assured through a number of Inter-Departmental Working Groups, covering such subject complexes as environment and energy, shifting cultivation, etc.

Among the programmes underway the following, considered of particular relevance to the subject under discussion, can be mentioned:

(a) Exploration, collection, evaluation and conservation of genetic resources, aimed at better utilization of arboreal species in arid and semi-arid areas (FAO 1975, 1977 & 1980a, Palmberg 1981a,b, 1983).

(b) Planning, monitoring and evaluation of community forestry programmes (FAO 1978b, 1981b).

(c) Evaluation and monitoring of woody resources and analyses of supply and demand trends (FAO 1981c, 1982a, 1983a & 1983m).

(d) Wood energy programme coordination and liaison within the UN-system, in the framework of the Nairobi Programme of Action on New and Renewable Resources of Energy (FAO 1980d, Anon. 1981a).

(e) Elaboration of monographs and manuals on: Lesser-known Multipurpose Tree Species (in preparation); Forest Food and Fruit Trees (FAO 1983b, subsequent volumes in preparation); Medicinal Plants (in preparation); Forestry in Arid and Semi-Arid Zones (FAO/UNEP, in preparation); and Management Systems for Natural Woody Vegetation in Arid and Semi-Arid Areas for Energy Production (in preparation).

(f) Establishment of research priorities (FAO 1980b, FAO/WB 1981).

(g) Assistance in institution building (FAO 1983e), dissemination of information and preparation of training materials (FAO 1973–83, 1976, 1978a, 1983d, 1983f–1 & 1984b).

(h) Technical meetings and training courses in, e.g. the role of forestry in combating desertification, sand dune stabilization, conservation of genetic resources, tree improvement and seed handling, wood energy, etc. (FAO 1980e, 1981a, 1982c, 1983n & 1984c).

References

Anon. 1981a. *Report of the United Nation's Conference on New and Renewable Resources of Energy. Nairobi 10–21 August 1981.* New York: UN.

Anon. 1981b. *Primera Reunión Nacional sobre Ecología, Manejo y Domesticación de las Plantas Utiles del Desierto.* Instituto Nacional de Investigaciones Forestales, México D.F. Publicación Especial No. 31.

Anon. 1982. *Congreso Nacional sobre Essencias Nativas.* Revista do Instituto Florestal 16A (Special Issue).

Ben Salem, B. 1980. Arid-zone forestry: where there are no forests and everything depends on trees. *Unasylva* 32(128): 16–18.

Ben Salem, B. and T.M. Eren 1982. Forestry in a sandy world. *Unasylva* 34(135): 8–12.

Eckholm, E.P. 1975. *The other energy crisis: firewood.* Washington D.C.: Worldwatch Institute.

FAO 1975. **The methodology of conservation of forest genetic resources: report on a pilot study.* FO:Misc/75/8. Rome: FAO.

FAO 1976. *Conservation in arid and semi-arid zones.* FAO Conservation Guide No. 3. Rome: FAO.

FAO 1973–1983. *Forest Genetic Resources Information* numbers 1–12 (Newsletter published by FAO's Forestry Department in two issues per biennium). Rome: FAO.

FAO 1977. *Report on the 4th Session of the FAO Panel of Experts on Forest Gene Resources. Canberra, Australia 9–11 March 1977.* Rome: FAO.

FAO 1978a. *Forestry for rural communities.* Rome: Forestry Department, FAO.

FAO 1978b. *Forestry for local community development.* FAO Forestry Paper No. 7. Rome: FAO.

FAO 1980a. *Genetic resources of tree species in arid and semi-arid areas: a survey for the improvement of rural living in Latin America, Africa, India and South-West Asia.* Rome: FAO.

FAO 1980b. *Needs for forestry research in the tropics and what international action can do to meet them.* FO:Misc 80/3, February 1980. Rome: FAO.

FAO 1980c. *Forests and trees in the rural economy of arid regions.* Secretariat Note, Committee on Forest Development in the Tropics. FO:FDT/80/4, April 1980. Rome: FAO.

FAO 1980d. *United Nations Conference on New and Renewable Sources of Energy: Report on the 2nd Meeting of the Technical Panel on Fuelwood and Charcoal.* Rome: FAO.

FAO 1980e. *Mejora genética de arboles forestales.* FAO Forestry Paper No. 20. Rome: FAO.

FAO 1981a. *Report on the FAO/UNEP Expert Consultation on in situ Conservation of Forest Genetic Resources. Rome 2–4 December 1980.* Rome: FAO.

FAO 1981b. *Forestry and rural development.* FAO Forestry Paper No. 26. Rome: FAO.

FAO 1981c. *Map of the fuelwood situation in the developing countries.* Rome: FAO.

FAO 1981d. *Wood Energy I. Special Issue.* Unasylva 33(131).

FAO 1981e. *Wood Energy II. Special Issue.* Unasylva 33 (133).

FAO 1982a. *Tropical forest resources.* FAO Forestry Paper No. 30. Rome: FAO.

FAO 1982b. The FAO Forestry Department's Programme of Work and Budget for 1982–83. *Unasylva* 34(135): 26–37.

FAO 1982c. *FAO/DANIDA Intensive Training Course on Sand Dune Stabilization. Somalia, June 1982.* Forestry Department, Forest Resources Division. Rome: FAO. (Unpublished; in English and Somali).

FAO 1983a. *Fuelwood supplies in the developing countries.* FAO Forestry Paper No. 42. Rome: FAO.

*All FAO publications are generally available in English, French and Spanish (some also in Arabic and Chinese).

FAO 1983b. *Food and fruit-bearing forest species. I. Examples from Eastern Africa.* FAO Forestry Paper No. 44/1. Rome: FAO.

FAO 1983c. *The Director-General's Programme of Work and Budget for 1984–85.* Rome: FAO.

FAO 1983d. *Wood for energy.* Forestry Topics Report No. 1. Rome: Forestry Department, FAO.

FAO 1983e. *Forest administration for development. Special Issue.* Unasylva 35(142).

FAO 1983f. *Handbook on taxonomy of Acacia species* (Based on the work of J.P.M. Brenan). Rome: FAO.

FAO 1983g. *Handbook on taxonomy of Prosopis species* (Based on the work of P.F. Ffolliott and J.L. Thames). Rome: FAO.

FAO 1983h. *Handbook on seeds of dry-zone acacias* (Based on the work of J.C. Doran *et al.*) Rome: FAO.

FAO 1983i. *Handbook on collection, handling, storage and pre-treatment of Prosopis seeds in Latin America* (Based on the work of P.F. Ffolliott and J.L. Thames). Rome: FAO.

FAO 1983j. *Handbook on seed insects of Acacia species* (Based on the work of B.J. Southgate). Rome: FAO.

FAO 1983k. *Handbook on seed insects of Prosopis species* (Based on the work of C.D. Johnson). Rome: FAO.

FAO 1983l. *Simple technologies for charcoal making.* FAO Forestry Paper No. 41. Rome: FAO.

FAO 1983m. *Wood fuel surveys.* FAO/SIDA Forestry for Local Community Development Programme. GCP/INT/365/SWE. Rome: FAO.

FAO 1983n. *FAO/DANIDA Training Course on Desertification Control and sand Dune Stabilization. Sudan, September 1983.* Forestry Department, Forest Resources Division. Rome: FAO. (Unpublished; in Arabic only).

FAO/UNEP 1983. *Notes on trees and shrubs in arid and semi-arid regions.* EMASAR, Phase II. Rome: FAO.

FAO 1984a. *Forestry beyond 2000. Prospects and problems in arid and semi-arid zones.* Secretariat Note. Committee on Forestry, COFO 84/5, May 1984. Rome: FAO.

FAO 1984b. *The fuelwood crisis and population: Africa.* Forestry Department. Rome: FAO.

FAO 1984c. *FAO/DANIDA Training Course on Sand Dune Stabilization, Shelter-belts and Afforestation in Dry Zones. India, March 1980.* Rome: FAO. (in press).

FAO/World Bank 1981. *Forestry research needs in developing countries: time for a reappraisal? 17th IUFRO Congress, Kyoto, Japan (6–17 Sept 1981).*

Le Houérou, H.N. 1980. *Browse in Africa. The current state of knowledge. Papers presented at the International Symposium on Browse in Africa, Addis Ababa 8–12 April 1980, and other submissions.* Addis Ababa: International Livestock Centre for Africa.

McGinnes, W.G. 1981. *Discovering the desert. Legacy of the Carnegie Biological Desert Laboratory.* Tucson: The University of Arizona Press.

McKell, C.M. 1980. Multiple use of fodder trees and shrubs: a world wide perspective. In *Browse in Africa. The current state of knowledge. Papers presented at the International Symposium on Browse in Africa, and other submissions.* H.N. Le Houérou (ed.): 141–149. Addis Ababa: International Livestock Centre for Africa.

McKell, C.M., J.P. Blaisdell and J.R. Goodin 1972. *Wildland shrubs: their biology and utilization. An International Symposium, Utah State University, Logan, Utah, July 1971.* Logan: USDA, Forest Service General Technical Report INT–1.

Palmberg, C. 1981. *Genetic resources of fuelwood tree species for the improvement of rural living.* Paper presented at the FAO/UNEP/IBPGR International Conference on Crop Genetic Resources, Rome 6–10 April 1981. Rome: FAO.

Palmberg, C. 1981b. A vital fuelwood genepool is in danger. *Unasylva* 33(133): 22–30.

Palmberg, C. 1983. FAO Project on genetic resources of arid and semi-arid zone arboreal species for the improvement of rural living. A report on progress. *Forest Genetic Resources Information* No. 12: 32–35. Rome: FAO.

9 *Prosopis tamarugo* in the Chilean Atacama – ecophysiological and reforestation aspects

H. Stienen

Universität Hamburg, Hamburg 80, W. Germany

Introduction

More than in the past it is nowadays a dream of the rapidly increasing mankind either to make the desert permanently arable or to give it a natural vegetation cover. The best way to achieve this would be a 'forest' which reproduces itself quickly and provides usable wood, edible fruits, forage for animals that are important for meat-production as well as living space for other species and finally to give a place for relaxation and refreshment for the people. A 'forest' that gives the region economic and ecological stability will guarantee, by careful use, the survival of its inhabitants for a very long time.

In the North-Chilean Atacama desert a little bit of this dream once was and is now being realized again. The tree which combines the above mentioned characteristics is *Prosopis tamarugo* (tamarugo). Yet its reforestation today is only because scientists have convinced the government that this step is also of great economic advantage and can be used for the production of fodder for large flocks of sheep.

Geography and climate

The 'Pampa del Tamarugal', the only area of natural distribution of *Prosopis tamarugo,* is part of the Atacama desert on the west coast of the Latin American continent (Fig. 9.1a). The 'Pampa del Tamarugal' embraces approximately 12 000 km² and lies in the southern tropics between 19°07′ and 21°03′S. Travelling east from the Pacific coast to the Andes a 1000 m high range rises from the coast followed by an eastward sloping plain ascending from 300 to 1200 m, bounded by the Andean foot hills. The 'Pampa del Tamarugal' lies in this plain or central valley, with an extension of approximately 60 km between the coastal range and the main range (Fig. 9.1b).

The arid character of the land is largely due to the influence of the cold Humboldt Current flowing northwards along the Pacific coast affecting the humidity. The current is too cold to produce large quantities of water vapour into the

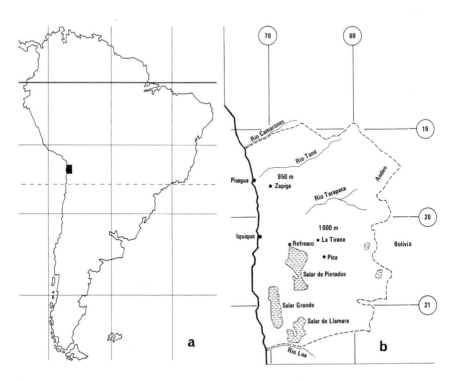

Figure 9.1a   Map of Latin America showing the central Atacama region.
9.1b Map of the 'Pampa del Tamarugal' and North Chilean Atacama. Both drawn by H. Schäfer.

atmosphere, so that the winds reaching the mainland are exceptionally cold and dry for these latitudes. Moreover, the air flow does not find enough condensation nuclei for rain-formation. Only fog 'camanchaca' can be observed during the cold nights. The relative humidity of the air can be above 80 per cent for some hours, but rainfall, however, never occurs in what is one of the driest regions in the world. The average temperature during the day lies between 13°C in the winter and 21°C in the summer. Overall temperature fluctuations are high. In winter temperatures can reach almost freezing point (Cadahia 1970). Geomorphologically the plain represents a tectonical depression which has filled with post Tertiary deposits eroded from the Andes. The 'Pampa' appears like an 'ocean' of sand, dry mud and salt, in parts devoid of vegetation. In large depression areas 'salares' the tamarugo can be found. Large parts of the depressions are covered with a salt crust 10–60 cm thick (FAO 1970), consisting of Ca-, Mg-, K-, and Na-chlorides and sulphides (CORFO 1970). The soils are slightly alkaline. Due to the subterranean water supply from the Andes approximately 10 m (1.5–40 m) below the surface, water is draining to the sea at the rate of at least 3 m³ per day. The quality of the water ranges from very salty (5000 ppm) to sweet (500 ppm). The soil contains only a relatively small amount of organic matter and has a well developed capillary structure. The salt crust and deeper salt veins create a suction force which draws the underground water to the surface, which then evaporates either into the open desert or into the tamarugo canopy (Castillo 1960).

Figure 9.2 A Tamarugo tree more than a thousand years old (Stienen 1983).

Botanical notes

The endemic *Prosopis tamarugo* is a tree of up to 20 m in height (Fig. 9.2); it has a large, much branched, spreading crown with the branching commencing a few meters from the ground. Some specimens may be up to 3000 years old (personal communication from C. Munoz via C. Klein-Koch). After a short period of rapid initial growth, increases in height and diameter diminish. After a few decades it grows slower and adjusts to seasonal growth conditions. The leaves are bipinnate with 10–15 pairs of linear leaflets. It possesses typical sessile, yellow, mimosoid flowers. The first flowering begins when the trees are 8 years old. The tamarugo usually has a second winter flowering period called 'desvareo'. Its legumes vary from dark yellow to light brown; they are cylindrical, sickle-shape and turgent in the early stage, forming a sickle 2–3.5 cm in diameter. The seeds are brown, ovate, 3–4 mm long, with hard testa and cuticle (Burkart 1976). Three other, non-endemic *Prosopis* species can be found interspersed between the tamarugo stands. These are *P. strombulifera, P. burkartii* and *P. chiliensis.* They may be distinguished from *P. tamarugo* by the characters given in Table 9.1.

Table 9.1 Major characteristics of *Prosopis* species in the 'Pampa del Tamarugal' (from Ffolliott & Thames 1983).

	P. tamarugo	*P. strombulifera*	*P. burkartii*	*P. chiliensis*
Distribution	endemic N. Chile	S. Peru, N. Chile W. Argentina	endemic 'Pampa del Tamarugal'; and adjacent valleys, N. Chile	S–central Peru, N–central Chile
Habit	tree to 18 m	stoloniferous shrub to 3 m	stoloniferous shrub to 1.5 m	tree to 10 m or more
Pinnae Leaflets	1 pair 10–15 pairs	1 pair 3–8 pairs	1 pair 7–12 pairs	1–3 pairs 10–29 pairs
Inflorescence	spicate 3–5 6.5 cm long	globose 1.5 cm in diameter	shortly spicate 3 × 1.5 cm	spicate 7–12 cm long
Legume	cylindrical, falcate, dark	cylindrical, tightly corkscrew-like lemon yellow	no data	flat, straight falcate
Seeds	ovate, 3–4.4 mm long, brown	ovate, 4.5– 5.4 mm long	no data	ovate, flattened, 6–7 mm long

The wood of *P. chiliensis* (algarrobo) is of great importance in some regions and it has been successfully used for reforestation in certain parts of Latin America; it also provides fodder for domestic animals. Reforestation is now being attempted in the Chilean Atacama using *Prosopis tamarugo.*

In the 'forest' and especially in certain areas outside the stands there are plant communities composed of a small number of species (Stienen 1981), some of which are of economic importance, as for example, the 2 m high, bushy composite *Tessaria absinthioides* which provides fuel.

The anthropo-ecological structure and its changes

Until the reforestation during recent decades the ecological history of the 'Pampa del Tamarugal' since the first human settlement can be divided into two phases:

1. The settlement history of the present Chilean part of the Atacama is described by Billinghurst (1868) as 'intensive' and as 'extensive and sporadic' by Nunez (1962).

More than 10 000 years ago indian tribes from the transandean zones, today called Argentina, came and settled in the strip of land along the Pacific coast, the humid Andean valleys, oases and some parts of the desert. In contrast to the settlers of the valleys and oases, the people who had chosen the tamarugo woodlands remained gatherers for a longer period of time, living on the tamarugo fruits and other herbs. For hunting they went to the nearby Andes. Later they kept Tylopoda such as llamas and their domestic forms in the 'forests'.

The tamarugo seed germinate well after pre-digestion of the testa and thus the relation of these animals with the trees can perhaps be regarded as co-evolutionary. The animals ate the fruits and twigs of the trees and thus supplied the first humans in the 'Pampa' with hides, meat and other essentials. The wood was used for burning and charcoal production. However it was hardly ever used for the construction of houses because such dwellings were usually made with blocks of dry earth or even blocks broken from the salt crust; roofs were both unknown and unnecessary in the area because it never rained. Beside the use of various wild herbs the indians later irrigated the eastern (upper) parts of the Pampa, which were almost free of salt. There they grew alfalfa (*Medicago sativa*), beans (*Phaseolus vulgaris* var. *aborigineus*), quinoa (*Chenopodium quinoa*), pumpkins (*Lagenaria siceraria*), etc. Old irrigation systems can still be found buried under the sand. The fruits of tamarugo with their high protein contents had presumably been used for flour.

Until the 'conquista' by the Spaniards in 1535, the primitive desert indians lived in total independence, contrary to the highland indians who were largely controlled by the Incas. The Likan-antai, as they called themselves, had a stabilizing influence on the 'Pampa del Tamarugal' ecosystem because the population was always kept at a low level.

2. After the 'conquista', under the regime of the viceroy of Peru, the region was rapidly integrated into the world of commerce; everywhere new settlements were established, ports built, and the oases and valleys were intensively cultivated. Sheep were now led to pasture in the tamarugo stands. The number of indians rapidly declined and nowadays only small numbers of half-castes remain.

The number of new settlers doubled every 10–20 years. In 1800 the port Iquique had about 100 inhabitants, which a century later had increased to nearly 40 000 (Antonioletti 1966). Timber was now needed to build the Spanish-style houses of Iquique, Arica and Antafagasta; these still exist today. In addition to the favoured, easily workable wood of algarrobo, the knotty stems of the tamarugo soon had to be used. Because of the cutting of trees, the humidity within the stands was reduced and the reproduction of the trees declined. With the beginning of the exploitation of nitrates and copper in the region, the tamarugo 'forests' were systematically felled to provide firewood for the smelting furnaces and this increased even more after the discovery of silver in the region, by which time even the roots were being torn from the impoverished soil for fuel. After only 100 years the ecological balance of the region was destroyed and after 300 years the former woodlands became a desert. By the beginning of the 20th century only solitary trees were left in their natural habitat; *Prosopis tamarugo* as a species was close to extinction.

This, one of the most heavily populated desert regions on earth, has after several periods of economic growth, periodically encountered new problems. These include the collapse of the nitrate exploitation at the beginning of this century because of cheaper synthetic products and followed 50 years later by the decline of the fishing industry due to overfishing of the Humboldt Current. So the decision of the government to reforest the desert with tamarugo was based on the need to make the region economically viable and provide food for its people (Stienen 1983).

In the 1930s a successful reforestation experiment with tamarugo was carried out (Habit *et al.* 1980). Allesch (1974) stated that more than 600 000 ha are suitable for reforestation with tamarugo, whereas Rolando (1974) restricts his estimate to only 200 000 ha. The trees are pollinated by the solitary bee *Centris mixta*. This task nowadays is partly taken over by the introduced *Apis mellifera* which, moreover, yields honey.

A tree in order to meet the requirements of reforestation has to have several characteristics, such as adaptability to the environment to guarantee high productivity and healthy trees; high yields as well as wood quality and other useful attributes, e.g. for animal nutrition. *Prosopis tamarugo* fulfills all these demands as shown below.

Qualities of tamarugo

The heartwood of *Propopis tamarugo* is extremely heavy, with a density of about 1.00 gm/cm³. The wood is dark brown and variegated due to the increment zones and yellow-brown deposits in the pores. The extremely durable heartwood occupies more that 80 per cent of the stem volume. The wood hardly weathers and in this region lasts unprotected for several hundred years. Even at depths of 40 m, completely preserved stems thousands of years old were found in the moist soil (Contreras 1978). In fact tamarugo appears to have better timber properties than those of *Prosopis chilensis*, which is currently being used for construction because of its more favourable dimensions. The wood of tamarugo has a large proportion of fibres, is diffuse-porous and has paratracheal to banded parenchyma. It sometimes contains rhomboid silica bodies which blunt cutting tools. The wood is prone to splitting when dried rapidly but serves well for heavy construction, tool and knife handles, tubes and barrels, sleepers, furniture, turnery and, because of its hardness, for parquet floors. It yields high quality charcoal as well as firewood of high caloric value, an aspect which should not be underestimated in developing countries (Maydell 1982).

After 25–40 years the trees may reach 10 m in height; any increase after this is very slow. After 150 years the trunk attains a diameter of *c.* 1 m. However even 1 000 year old specimens may not exceed 1.5 m. With increasing age the treee diverts more and more energy from photosynthesis to the production of leaves and fruits. With a planting distance of 10 m the crowns of the young trees touch each other after 25 years (Habit *et al.* 1980). In mature trees the diameter of the crown may reach c. 20 m. Often the low crown touches the ground. The trees produce 3.1 kg fruits/m² (dry weight) projected crown area, altogether 3.4 kg/m²/year (dry weight) biomass. The relation of fruit to leaf production is about 50:50 (Oyarzun 1967). Thirty-five year old stands already produce 15 000 kg/ha/year biomass dry weight (Cadahia 1979).

The fruits of tamarugo, with 55–65 per cent digestibility, are an excellent feed for sheep and cattle. The leaves are rich in carbohydrates and proteins and are of better feeding value than hay. For good animal production a certain amount of fruits with the fodder is necessary (Lanino 1966, Lamagdalaine 1972). Estimates vary greatly as to how many sheep could continually subsist on one hectare; 10 is the number given by the Chilean Instituto Forestal 1964 and 3–5 in a study by Latrille and Garcia (1968). Experiments have show that sheep living free in the forests are selective in their feeding and consume large quantities of fruits. Unless the numbers are strictly controlled damage to the plantations could occur. The sheep produce 3.1–5.4 kg/wool/fleece, when their number is controlled. The tamarugo biomass is rich in macro- and micronutrients. Only the addition of vitamin A is sometimes necessary; supplementing the diet with mineral salts increase the growth rate. Supplementing the leaves and fruits of tamarugo with alfalfa did not increase the yield of the wool but did give a 20 per cent increase in meat production. Growth deficits were observed when the flocks had to travel long distances to reach water.

The productivity of tamarugo is impaired by various insects. The legumes in particular suffer losses of up to 30 per cent. Mainly Lepidoptera, such as the fruit-moth *Crytophlebia carpophagoides* and *Leptotetes trigemmatus*, are involved together with the endemic Coleoptera, *Scutobruchus gastoi,* which completes its life cycle in the fallen legumes. Other insects have been reported by Klein-Koch and Campos (1978), see also Johnson (1983) for further information. In the past they had been exterminated using, unfortunately, very high doses of Endosulfane.

The many positive and versatile characteristics of *Prosopis tamarugo* show that in the 'Pampa del Tamarugal' this can help solve many of the more pressing economic problems and it could possible be used with advantage in other parts of the world with similar conditions.

Ecophysiology of tamarugo

Prosopis tamarugo owes its high drought resistance, high growth increment and productivity on poor, salty substrates to special morphological and physiological characteristics. One of these is the ability to make use of nitrogen-fixing bacteria in form of symbiotic root nodules. Information about the physiology of tamarugo, particularly about the water up-take conditions are available from studies by Sudzuki (1969) and her students Botti (1970) and Acevedo (1977). The following summarises the existing information.

Tamarugo has a double root system, a tap-root, which may divide, reaches a depth of several meters and supports the tree. Only 50–100 cm below the surface a second, richly ramified system of fine roots develops, extending laterally to a diameter equalling that of the crown (Fig. 9.3). Neither system reaches any water-conducting horizons. Many experiments with polyethylene bags wrapped around the roots have proved that tamarugo does not need to take up water via roots and that intensive watering may even damage the tree. The plants grow well without watering when the relative humidity of the surrounding air reaches 80 per cent or more for some hours daily, as is often the case during night in the 'Pampa del Tamarugal'. From repeated experiments using tritium markers, in the tamarugo stands as well as in the laboratory, we know that water vapour is taken up by the leaves. In reality this type of humidity originates not only from the 'camanchaca' but also from soil transpiration (see above) which provides an additional 210 m³ of water/month/ha to the crowns of the trees. The leaves possess a water conducting parenchyma which take up the water through the stomata. Part of the water taken up eventually reaches the ramified subsurface root system directly under the salt crust. Here the roots deliver water to the dry rhizosphere to dissolve the nutrient salts within the immediate surroundings, which in turn, are taken up by the roots and transported to the active tissues. The water flow to the roots is presumed to be actively transported through the phloem, while the transport from the roots occurs regularly by the water and ions moving through the xylem to the crown. Pastenes (1972) found that young tamarugo plants which are subjected to more fog than normally, respond by increasing leaf size. This also happens when there are frequent short periods of high humidity. Thus the trees take optimum advantage of the available water supply. This capability is intensified during periods of excessive dryness of the soil. Additional water taken up by the tree is stored in the rhizosphere. For short periods the soil humidity can exceed that of the air. The flow of water to the soil can be passive because the soil has a potential of -52 atmos.

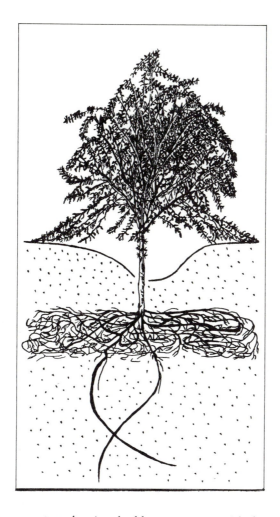

Figure 9.3 Tamarugo tree showing double root system with deep central roots to anchor the plant and fine upper roots to secrete water and take up mineral salts (Stienen 1981 after Sudzuki 1969).

while the roots have a potential of only -32 atmos. The water potential of the leaves varies from -15 to -47 atmos. in the summer, and from -24 to -80 atmos. in the winter. During the experimental periods, conditions which would have favoured a passive diffusion into the leaf were never encountered. One has to conclude that this process is also an active one. The stomata of tamarugo are mainly open during the night but never entirely; on reaching a certain critical humidity an opening of the stomata can be observed during the day time. At the present time nothing is known about the photosynthetical adaption to these conditions. Where the soils reach higher humidities, the stomata will also open during the daytime when the relative humidity of the air underneath the crowns reaches 95 per cent. The water flow to the root lasts 4–8 hours; during the rest of the day this process is largely reversed, when taking into account the net water balance. Areas that offer such minimum ground water depths are limited also in the 'Pampa del Tamarugal'. Consequently reforestation is restricted to such areas.

Reforestation

Taking into consideration the described ecophysiological adaptions of *Prosopis tamarugo* in certain sectors of the Atacama desert, and regarding its excellent properties as a timber and fodder tree, the Chilean authorities finally decided in 1966 on the large scale reforestation of several thousands of hectares with this tree. The spiritual father of such an adventure was Billinghurst (1886) who described, almost 80 years ago, the prehispanic life in the Andean valleys, oases and in the 'Pampa del Tamarugal', and who in his works on the water budget of the Atacama gave hints of the possibility of reforestations in the region.

By knowing certain growth patterns of tamarugo one could start from the hypothesis that young trees, once reintroduced, would develop satisfactorily. Therefore emphasis in the trials have concentrated on early growth and survival after transplanting of seedlings raised in glass-houses. During this early phase the project has been supported by FAO and carried out by the Chilean Corporación de Fomento de la Produccion (CORFO) co-ordinating several groups from universities within Chile.

Eventually a successful method was found how to overcome a number of problems (Habit *et al.* 1980). The legumes selected, using phenotypic characteristics of the mother trees, are ground and the seeds extracted and treated with 0.2% Aldrine. They are then immersed for seven minutes in sulphuric acid to dissolve the cuticle, such treatment being analogous to pre-digestion by animals. Sowing is done in a substrate of 2:1 'Pampa'-soil: sheep manure, which has been previously sterilised with methyl bromide to prevent fungal infection. Plastic bags are used as plant containers. The seed is then watered at increasingly longer intervals. The quickly germinating seedlings remain for 3–5 months in the glass-house before planting begins in 1 km² plots with 20 or 15 m between each seedlings. A main excavation of 80 cm diameter and 40 cm depth is dug or bored through the salt crust either mechanically or by hand. At the bottom another small pit is made (Fig. 9.4), into which the seedling is planted with pure sheep manure. Over this lower hole a transparent plastic sheet is placed in order to maintain a water saturated atmosphere. Initially the plants are watered every 10 days, later every 20 or 30 days; the quantities required differ very much from stand to stand. Sometimes plastic bags are filled with water and placed in the hole and the contents allowed to

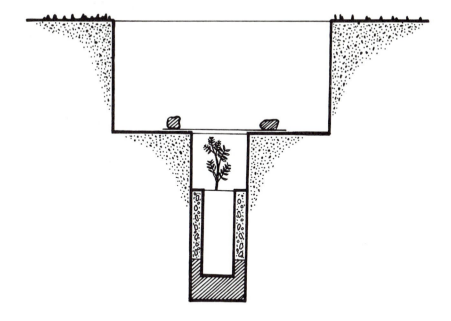

Figure 9.4 Planting hole for young Tamarugo tree. The deeper hole is covered with a transparent plastic sheet to help to create a water-saturated atmosphere (drawn after Habit *et al.* 1981).

trickle through a tiny opening. The plastic covers are removed after 90 days. The plants grow very rapidly and after a certain time, pruning every 5–6 years is needed in order to create sufficient space to allow stock to move between the trees.

Over the years different breeds of sheep, cattle and goats were tried in the newly planted areas, e.g. the sheep breeds Corriedale, Early Maturing French Merino, Australian Merino, Romney Marsh, Suffolk Down, Karakul and the Angora goat (Fig. 9.5, p. 114). The Australian Merinos and the Karakul performed best, while Romney Marsh proved to be the least suitable. Cattle at first were very successful but in the end did not live up to expectations. Apparently they disliked the available food, which might be attributable to the mineral salts in the plants. The promoters of the project intended to continuously graze 2.8 million sheep, which, from the ecological point of view, seems an excessive number. This new endeavour is expected to give work to the unemployed and stop migration from northern Chile (Bähr 1976). After a decade without much concern for the project the government agency CORFO started anew, and while preparing this paper an 'International Round Table on *Prosopis tamarugo,* Desert Fodder Tree'was actually taking place in Arica at the Universidad de Tarapacá, where the successes so far, problems, and new experiments were discussed.

Figure 9.5 Sheep eating fallen pods and leaves under old Tamarugo trees (Steinen 1983).

Small irrigated plantations of fruit trees and bushes as well as vegetables at the border of the 'Pampa' are considered a useful addition to the reforestation activities carried out under this programme. Plans to establish large citrus plantations under continuous irrigation will have to be rejected, because nothing is known about the huge quantities of water required to realise such an undertaking. Nobody knows where this water might come from because too little is known about the water resources underneath the 'Pampa'. One fears that such large-scale planting of exotic species will require so much water that the finely woven network of the ecosystem in the 'Pampa del Tamarugal' (Fig. 9.6) will be jeopardized due to a probable lowering of the ground-water level. This will prevent sufficient water reaching the atmosphere under the crowns of the trees through the soil capillary system. All reforestations would be at a risk. So far such plans have been successfully prevented by prudent scientists convincing the government of their inadvisability.

A transfer of the experience gained with *Prosopis tamarugo* to other parts of the world is considered somewhat limited, and should only be realised after careful long-term investigations. One of the major problems is to find desert regions with similar soil and climatic conditions to the 'Pampa'. These can perhaps be found in small sections of the coastal deserts of the USA, Africa or Australia.

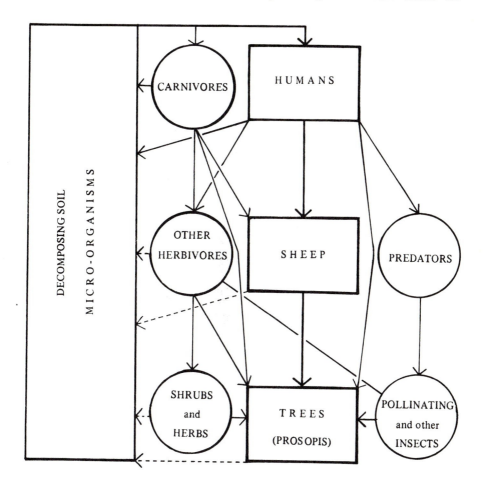

Figure 9.6 Trophical chain in the 'Pampa del Tamarugal' (Stienen 1983 after Klein-Koch & Campos 1978).

Acknowledgement

The author wishes to thank Senor M.A. Herman Silver R., Centro de Estudios Zonas Aridas, Universidad de Chile, La Serena, for his valuable information and spontaneous help in preparing this paper which was introduced at the last moment to fill a sudden gap in the conference proceedings.

References

Acevedo, E. 1977. *Relaciones suelo-planta en estudios relacionados con Tamarugo (Prosopis tamarugo Phil.)*, unpublished.

Allesch, R. 1974. Experencias agrícolas y ganadero-forestales en el norte árido chileno (Pampa del Tamarugal). *Rev. Geogr. Valparaiso* (5)1: 11–32.

Antonioletti, R. 1966. Las funciones de la ciudad de Iquique. *Inf. Geogr. Chile* 16(1): 133–149.

Bähr, J. 1976. Migration in Lateinamerika mit besonderer Berücksichtigung des Chilenischen Norte Grande. *Erdkundl. Wiss.* 43: 34–52.

Billinghurst, G.E. 1868. *Estudios sobre la geografia de Tarapacá.* Santiago: El Progresso.

Botti, C. 1970. *Relaciones hídricas del Tamarugo (Prosopis tamarugo Phil.) en la localidad de Canchones.* Santiago: Universidad de Chile. Thesis.

Burkart, A. 1976. A monograph of the genus *Prosopis* II. *J. Arn. Arb.* 57: 450–525.

Cadahia, D. 1970. *Informe sobre el plan ganadero Pampa del Tamarugal. Estudio FAO for BID.* (Internal report). CORFO, Chile. (Mimeo.).

Castillo, U. 1960. El agua subterránea en el norte de la Pampa del Tamarugal. *Inst. Invest. Geol. Santiago Bol.* 5.

Contreras, D. 1978. *Estudio actual de conocimiento del Tamarugo.* Santiago: Inf. CORFO.

CORFO 1971. *Programa forestal Pampa del Tamarugal.* Inst. CORFO Norte.

Ffolliott, P.F. and J.L. Thames 1983. *Handbook on taxonomy of Prosopis in Mexico, Peru and Chile.* Rome: FAO.

FAO 1970. *Informe sobre el proyecto de plantaciones de Tamarugo y explotación ganadera en el Norte Grande, Chile. 1/70 Chi. I b, Cooperative Programme FAO/BID.* Santiago: FAO.

Habit, M.A.D., D. Contreras, and R.H. Gonzalez 1980. *Prosopis tamarugo: Arbusto forrajero para zonas áridas.* Santiago: FAO. Reprinted 1981 *Prosopis tamarugo: Fodder tree for arid zones.* Rome: FAO Plant Production and Protection Paper No. 26. (Mimeo.).

Instituto Forestal 1964. *Posibilidades de reforestación de la Pampa del Tamarugal.* Santiago: Inf. CORFO.

Johnson, C.D. 1983. *Handbook on seed insects of Prosopis species.* Rome: FAO.

Klein-Koch, C. and L. Campos 1978. Biocenosis del Tamarugo (*Prosopis tamarugo* Philippi) con especial referencia de los artrópodos fitófagos y sus enemigos naturales. *Z. Angew. Entomol.* 85(1): 86–108.

Lamagdelaine, L. 1972. Programa forestal ganadero Pampa del Tamarugal y programa Altiplano de Tarapacá. CORFO, Chile. *Centre. Doc. Dep. Cienc. Soc.* 2: 1–49. (Mimeo.).

Lanino, R.I. 1966. *Comparación en tres razas ovinas alimentadas con Tamarugo (Prosopis tamarugo Phil.).* Santiago: Universidad de Chile. Thesis.

Latrille, L. and X. Garcia 1968. Evaluación nutritiva del Tamarugo (*Prosopis tamarugo* Phil.) como forraje para rumiantes. Univ. de Chile. *Dep. Prod. Anim., Mem. An.*: 257–273.

Maydell, J. von 1982. Der Beitrag der Forstwirtschaft zur ländlichen Entwicklung. *Hdb d. Landw. u. Fern.* 1: 229–242.

Nunez, L. 1962. Contactos culturales prehispanicos entre la costa y la subcordillera Andina. *Bol. Univ. Chile* 31: 42–47.

Oyarzun, S. 1967. *Algunas variaciones lanimétricas con ovejas merino encastadas en dos épocas diferentes Canchones, Pampa del Tamarugal.* Santiago: Universidad de Chile. (Mimeo.).

Pastenes, G.J. 1972. *Efecto de aplicación foliar de humedad en plántulas de Prosopis tamarugo Phil.* Univ. de Chile, Fac. Fil. Educ., Dep. Biol., Antafagasta.

Rolando, N. 1974. *Evaluación económica del rubro ovino en la Pampa del Tamarugal y su posibilidad de desarrollo.* Santiago: Universidad de Chile. Thesis.

Stienen, H. 1981. Die Tamarugal-Pampa Nordchiles Botanische Aspekte. *Nat. u. Mus.* 111(9): 289–298.

Stienen, H. 1983. Anthropogene Veränderungen einer Naturlandschaft in der Atacama-die Tamarugal-Pampa Nordchiles. *Nat. u. Mus.* 113(1): 1–11.

Sudzuki, H.F. 1969. *Absorción foliar de humedad atmosférica en Tamarugo (Prosopis tamarugo Phil.).* Minist. Agric., Bol. Tecn. 30.

10 Forage and fuel plants in the arid zone of North Africa, the Near and Middle East

H. N. Le Houérou

CEPE/Louis Emberger, BP 5051, Montpellier-Cedex 34033, France, and formerly Caesar Kleberg Wildlife Research Institute, Texas A & I University, Kingsville, Texas 78363, USA

Introduction

The present paper is concerned with the Mediterranean isoclimatic zone of Africa north of the Sahara and of the Tropic of Cancer, and with the Near and Middle East, and excluding the tropical zones, i.e. having summer-rain climates. The countries included are

North Africa: Morocco, Algeria, Tunisia, Libya, Egypt.

Near East: Cyprus, Turkey, Israel, Jordan, Syria, Lebanon, Saudi Arabia, Iraq, Kuwait, Bahrain, Oman, Qatar, United Arab Emirates, North and South Yemen.

Middle East: Iran, W Afghanistan, SW Pakistan.

This area covers some 12 million km² (133% of the USA) which are distributed as shown in Table 10.1.

Table 10.1 Land distribution according to bioclimatic zones, in 10^3 km². P = mean annual rainfall in mm.

Area	Total land surface	A Desert $P\langle 100$		B Arid $P\ 100-400$		C Semi-arid $P\ 400-600$		D Sub-humid & Humid $P\rangle 600$		E A+B+C $P\langle 600$	
	km²	km²	%	km²	%	km²	%	km²	%	km²	%
North Africa	5751	4864	85	563	10	180	3	144	2	5607	97
Near East	3705	2816	76	589	16	120	3	180	5	3525	95
Middle East	3100	518	17	2132	68	120	4	330	11	2770	89
Total	12 556	8198	65	3284	26	420	4	654	5	11 902	95

The flora of these territories comprises probably more than 20 000 species of flowering plants. North Africa harbours some 5000 species (Le Houérou 1974b, 1984b), the Sahara 2000, Turkey and Iran about 7000 each. Perhaps as many as half of these are to be found in the arid and semi-arid zones, below the 600 mm iso-hyet. A small country like Tunisia has no less than 1300 species in the 140 000 km² of its arid and semi-arid zone (Le Houérou 1959).

It is obviously out of the question to review in the present paper all the useful species of potential economic value as fodder, fuel or fibre; I shall therefore restrict myself to species or groups of particular interest which are either little known or deserve further investigation.

The list of species thus selected is necessarily somewhat arbitrary and reflects not only author's field experience but also, to some extent, his unavoidable biases.

For the sake of exposition, the species examined will be classified as forage producers and fuel producers, although some may produce both commodities. Forage species may be native or exotic, perennial or annual and belong to Poaceae (or Gramineae), Leguminoseae and other families such as Rosaceae, Crucifereae, etc. Browse species are both fodder and fuel and belong to various families; those of particular importance are the Chenopodiaceae, Leguminoseae (Mimosoideae, Caesalpinioideae and Faboideae) where the arid zone is concerned. Fuel species may have many other uses besides the production of energy: soil conservation, sand binding, honey production, chemicals, fibres, reclamation of salt affected land, etc.

Bioclimatic adaptation is described using two very simple criteria: (a) mean annual rainfall P which is correlated with many other climatic factors influencing water availability for plant growth and (b) mean minimum daily temperature m of the coldest month, which is correlated with the number of freezing days and inversely correlated with the length of growing season for a given amount of annual rain. This classification is shown in the matrix of Table 10.2.

Table 10.2 Mediterranean bioclimatic types (Le Houérou 1969b, 1973).

Winter temperature sub types	Per humid $P > 1200$	Humid P 800–1200	Sub humid P 600–800	Semi-arid P 400–600	Arid P 100–400	Hyper arid $P < 100$
Very warm $m > 9$	PH/VW	H/VW	SH/VW	SA/VW	A/VW	HA/VW
Warm $m > 7–9$	PH/W	H/W	SH/W	SA/W	A/W	HA/W
Mild $m > 5–7$	PH/M	H/M	SH/M	SA/M	A/M	HA/M
Temperate $m > 3–5$	PH/T	H/T	SH/T	SA/T	A/T	HA/T
Cool $m > 1–3$	PH/C	H/C	SH/C	SA/C	A/C	HA/C
Cold $m > -2–1$	PH/CC	H/CC	SH/CC	SA/CC	A/CC	HA/CC
Very cold $m > -5--2$	PH/VC	H/VC	SH/VC	SA/VC	A/VC	HA/VC
High mountain $m < -5$	–	H/HM	SH/HM	SA/HM	A/HM	–

This classification has been thoroughly discussed elsewhere (Le Houérou 1969b, 1971 & 1977, Daget 1971, 1977).

Forage species

Four synthetic reports have been published; two are concerned with northern Africa: (Villax 1962, Le Houérou 1974a) and two deal with the Near and Middle East (Pabot 1965, Kernick 1978). The following draws, to a large extent, on these.

PERENNIAL GRASSES

Perennial fodder grasses are numerous throughout the region; some are rather easy to 'domesticate', e.g. *Agropyron, Festuca, Bromus,* etc. and thus to be used in re-seeding programmes. Others are more difficult, either because of poor germination, e.g. *Hyparrhenia,* or because of the presence of a physical impediment to mechanical handling of the seeds, such as the long or branched awns in *Stipa, Aristida, Stipagrostis,* etc., which become entangled and prevent their use in a drill or other mechanical sowing device. Others have seeds which are difficult to harvest due to uneven ripening and/or shedding, e.g. *Hyparrhenia* and *Phalaris,* or have germination inhibitors as in *Cenchrus.*

Agropyron and *Elymus*
Most species of *Agropyron* and *Elymus* (which includes many species formerly included in *Agropyron*) are particularly well adapted within our region to climates with cold winters. The most successful reseeding programmes so far, in the arid zone of the region, actually happen to use these genera in cold winter areas, e.g. *A. cristatum, A. desertorum, E. tauri* (= *A. tauri*), *E. hispidus* subsp. *hispidus* (= *A. intermedium*) at Homand and Kari in Iran and *A. cristatum* and *E. hispidus* subsp. *hispidus* at Midelt in Morocco. *E. elongatus* (= *A. elongatum*), however, gave encouraging results in the mild winter semi-arid zone of Tunisia (Corriols 1965).

The best fodder among the native species are *E. elongatus* in North Africa and the Near East, *A. cristatum, E. tauri, E. hispidus* subsp. *hispidus* and subsp. *barbatus* (= *A. trichophorum, A. aucheri*) in the semi-arid mountains of Turkey, Syria, Iraq and Iran, where they are common and seemingly with many different ecotypes. Species of *Agropyron* and *Elymus* have been successfully used in small scale pasture development in Iraq, Iran, Tunisia, Algeria and Morocco under annual rainfalls of 250–300 mm and above, with average yields of 1000–2000 kg DM/ha/yr (Corriols 1965, Pabot 1967, Fallon 1974, Kernick 1978).

Arrhenatherum

There are three main native species, *A. elatius, A. kotschyi* and *A. palaestinum.* The first species is native on deep, well drained, medium textured soils in the semi-arid mountains of North Africa and the Near East, while the two last are only found in the Middle East. *A. elatius* has several subspecies and varieties, many of which seem to have a good agronomic potential under semi-arid and sub-humid climates with mild to cold winters.

Bromus

B. cappadocicus, B. inermis, B. tomentosus and *B. tomentellus* are native to the Near East. *B. inermis* has given excellent results (2500–3000 kg DM/ha/yr) in semi-arid zone cultivation trials in Morocco and Tunisia (Villax 1963, Corriols 1965), while *B. tomentellus* was reported as an outstanding forage crop (2000 kg DM/ha/yr) at Homand in the cold winter arid zone of Iran under 150–300 mm of rain (Pabot 1967, Kernick 1978).

Cenchrus ciliaris

This is a tropical species which is also native, but fairly rare, on light to sandy soils in the warm to mild winter arid and semi-arid Mediterranean zone of North Africa, Near and Middle East, under rainfalls of 100–600 mm/yr. There are many different populations and ecotypes, showing various habits, productivity and cold tolerance. The development of local strains is made difficult by the presence in the glumes of a hydrosoluble germination inhibitor. Some degree of apomixis has been shown to occur (Delisle 1964); inhibitor-free cultivars have been developed in Australia and the USA.

Cynodon dactylon

This is a very common grass in pastures and fallows of the warm and mild winter zones above the 100 mm isohyet, particularly in areas receiving some summer rains and preferably on sandy soils, but sometimes on vertisols. There are many populations and ecotypes from the equator to 53°N and from sea-level to 4000 m elevation. Some are fairly tolerant to salinity such as var. *hirsutissimus* in North Africa and var. *villosus* in the Near East.

Dactylis

Dactylis glomerata subsp. *hispanica* is common throughout the Mediterranean arid zone on sandy shallow soils under rainfalls as low as 150–200 mm particularly in southern Tunisia, northern Libya and northern Egypt. Temperate zone cultivars cannot survive the Mediterranean summer droughts and, in addition, have proven to be sensitive to fungal diseases. Yields of 3000–4000 kg DM/ha/yr have been obtained in semi-arid climates of Algeria, Tunisia and Morocco (Kernick 1978); while 1500–3000 kg DM/ha/yr could be obtained in the cold winter arid zone of Iran (Kernick 1978) and 3000–8000 kg in the sub-humid mild-winter zone of Iran (Dumancic 1975).

Digitaria commutata subsp. *nodosa*

Digitaria commutata subsp. *nodosa* is a tall bunch grass whose spikes may reach a height of 150 cm or more. It is found, albeit rare, on the sandy and sandy shallow soils of central and southern Tunisia, southern Algeria and southern Morocco under Mediterranean arid climates with warm/mild winters, i.e. rainfall 150–300 mm and *m* above 5°C. Its ecology is very similar to *Cenchrus ciliaris,* including cold sensitivity. But *Digitaria* seems far more productive than *Cenchrus,* judging from cultivation trials made at Tunis by the writer in 1963–67. Although it is extremely resistant to heavy grazing, the species has become very rare due to overexploitation. The species is also known from Egypt, Arabia, Iran, Afghanistan, Pakistan and India where it also seems to have become rare. Seed production is good with a germination rate of the order of 60 per cent. This is a high potential forage species for arid and semi-arid Mediterranean zones with mild to warm winters (*m* ⟩ 5) and perhaps for the dry tropics, which has been unduly neglected so far.

Festuca arundinacea (= *F. elatior* subsp. *arundinacea*)

Mediterranean populations of tall fescue present ecological and physiological characteristics greatly different from the temperate climate populations or cultivars (Robson 1967, Robson & Jewiss 1968), e.g. lower zero growth temperatures, summer dormancy, drought tolerance, etc. No less than ten botanical varieties have been described in the Mediterranean basin. There are many different ecotypes adapted to a large array of soil conditions and water regimes as shown by the writer (unpublished) in pot experiments over 20 populations from Tunisia. Agronomic trials in various climatic conditions always showed that Mediterranean populations were far superior to cultivars imported from abroad, e.g. Europe, USA, New Zealand (Lapeyronie 1965, Corriols 1965, Maignan 1971, Le Houérou 1974a).

F. arundinacea is found in waterlogged soils generally of fine texture and sometimes fairly saline (10–20 ms/cm in the saturation extract).* Production, in dry farming, under 300–600 mm of rainfall, reaches 4000–12 000 kg DM/ha/yr, yields of 15 000–20 000 kg DM/ha/yr have been obtained in irrigated farming from October to May (the plant remains summer dormant, even under irrigation). Seed production is easy and yields up to 1000 kg/ha/yr have been obtained with supplementary irrigation (Le Houérou 1974a).

Hordeum

Several native species of barley are present in the semi-arid zone of North Africa and the Near East. Some have a forage value, e.g. the perennials, *H. bulbosum, H. violaceum, Psathyrostachys fragilis* (= *H. fragile*) and the annual *H. spontaneum.* However, these species do not seem to present any outstanding characteristics neither in terms of ecological adaption, productivity and feed value.

* 1 ms/cm = 1 ds/m = 1 mmho/cm #0.06% NaC1 #0.01 mo/1 NaC1.

Lasiurus

Lasiurus hirsutus (= *Elyonurus hirsutus, Rottboellia hirsuta*) is a native in the wadis of those parts of the Sahara and deserts of the Middle East having mild to warm winters, where it is found in association with *Panicum turgidum* and *Pennisetum divisum* (= *P. dichotomum*). Unlike its close relative *L. scindicus* from India and Pakistan, *L. hirsutus* has not been used in revegetation programmes, although its potential would seem quite similar to *L. scindicus,* which yields 1000–3000 kg DM/ha/yr of good fodder in the Rajasthan under 200–300 mm of annual rainfall (Gupta & Saxena 1970).

Lolium

Mediterranean ecotypes of *Lolium perenne* greatly differ from temperate climate cultivars, particularly in their drought tolerance and summer dormancy. The Mediterranean ecotypes, however, show a low herbage productivity (Le Houérou 1965) as compared to other perennial fodder grasses (*Festuca, Oryzopsis, Phalaris*); but they are very tolerant to trampling and may therefore be used in establishing drought tolerant, least maintenance, swards, turfs or lawns.

Oryzopsis

The genus *Oryzopsis* has two main species of forage value: *O. miliacea* and *O. holci-formis* and three species of seemingly lower potential. These are *O. thomasii, O. coerulescens* and *O. molinioides.*

O. miliacea is found in arid and semi-arid zones with annual rains as low as 100 mm (Le Houérou 1959) with cold to warm winters ($m \rangle$ -1°C). *O. holciformis* is rather a semi-arid zone species growing under annual rainfalls of 300–900 mm in areas with warm to cold winters. Within North Africa, *O. holciformis* is only found in Cyrenaica where it had been confused with *O. coerulescens* (Le Houérou 1969b). *O. miliacea* is clearly more drought resistant but also more difficult to establish than *O. holciformis.*

Both species have been successfully sown and established in various parts of the arid zone. *O. miliacea* may grow on rather shallow calcareous soils or sandy soils while *O. holciformis* demands deeper medium textured soils (Le Houérou 1974a). Fodder yields in sown pastures may vary from 2000 to 10 000 kg DM/ha/yr under rainfalls of 250–600 mm. Under semi-arid conditions on good deep soils the latter may produce up to 15 000 kg DM/ha/yr, i.e. 50 per cent more than the former (Corriols 1965).

O. thomasii is similar to *O. miliacea* but grows on acid soils under rainfall above 600 mm. *O. coerulescens* is native in semi-arid open pine forests and shrublands. Its production potential is inferior to that of *O. miliacea* and *O. holciformis,* while *O. molinioides* is a high mountain species in the cedar forests of the Near and Middle East.

Panicum

Panicum turgidum is a common species in the sandy wadis of the Sahara and Near Eastern desert in association with *Pennisetum divisum, Cyperus conglomeratus, Lasiurus hirsutus* and *Acacia tortilis* subsp. *raddiana.* It produces a forage of moderate to poor quality, but has a value for revegetating purposes and sand binding.

Panicum antidotale is native in arid zones and deserts of the Near and Middle East having mild to warm winters. It has been used successfully in pasture reseeding on a semi-large scale in Rajasthan and on a small scale in North Africa. It is a long lasting species of good fodder value and extremely drought tolerant.

Paspalum vaginatum

This is a rather rare species in the region, but has been used elsewhere for its salinity tolerance (Malcolm & Clarke 1973).

Phalaris

The genus *Phalaris* has four important perennial species of fodder value to the region. They are: *P. aquatica (= P. tuberosa, P. bulbosa* and *P. nodosa), P. truncata, P. coerulescens* and *P. arundinacea.*

P. aquatica is a semi-arid to sub-humid zone species native on calcareous clay soils receiving 500–800 mm annual rainfall. Five botanical varieties are recognised in North Africa. It has been extensively cultivated, particularly var. *stenoptera* in Australia, USA and South America.

P. truncata is an arid and semi-arid zone, short-lived, perennial species common on calcareous clay soils under 300–600 mm of rainfall, where it often grows in association with *Hedysarum coronarium* in North Africa and southern Italy. *P. truncata* has three botanical varieties in North Africa. There are also many ecotypes of various eco-climatic adaption and productivity potential. It has been the subject of semi-large scale cultivation in Tunisia during the 1960's.

The forage value of both *P. aquatica* and *P. truncata* is excellent, with digestibility coefficients of over 75 per cent organic matter and over 70 per cent dry matter. Forage yields of 3000–9000 kg DM/ha/yr have been recorded for both species in various countries of North Africa and Italy under semi-arid bioclimatic conditions.

P. coerulescens is a humid zone species growing on wet soils with over 600 mm rainfall, very similar to tall fescue in ecological requirements but clearly less productive. Mixtures of *P. truncata, P. coerulescens* and *Hedysarum coronarium* have attained yields of up to 20 000 kg DM/ha/yr in the sub-humid zone of Tunisia (Le Houérou 1965).

P. arundinacea is a marshy species of the humid zone and is of limited forage value in our region.

Puccinellia

Puccinellia distans from North Africa and mainly its closely related species *P. ciliata* from the Near East are fodder grasses highly tolerant to salinity under high water table conditions. *P. ciliata* has been used in Western Australia to revegetate water-logged and highly saline areas where it produces good yields under semi-arid bioclimatic conditions with mild to temperate winters (Malcolm 1969).

Secale

Secale montanum is a perennial bunchgrass from the semi-arid mountains of Turkey and Iran, with two botanical varieties and a number of ecotypes (Pabot 1967). A hybrid with *S. cereale* known as *S.* × *ceremont* has been developed in the USA. Sown pastures of *S. montanum* were successfully established in the 1960's in the arid zone of Iran under 200 mm of annual rainfall (Pabot 1967, Kernick 1978) with an average yield of 1500 kg DM/ha/yr over a period of 4 years, i.e. an excellent rain use efficiency (RUE) of 7.5 kg DM/ha/mm/yr (excellent natural pastures usually having a RUE of 4–6).

Sporobolus ioclados (= S. marginatus)

This is a remnant perennial fodder grass of seemingly potential value from central Tunisia: Kairouam, Cherichera, Sbeitla, which has become very rare.

Stipa

Stipa parviflora, S. lagascae and *S. barbata* are excellent native fodder grasses in the arid rangelands of North Africa and the Near East. Unfortunately, their germination rates are very low and multiplication is thus very difficult, the more so as vegetative propagation is also unsatisfactory. The same remarks apply to a number of species of the genus *Stipagrostis.*

PERENNIAL AND SEMI-PERENNIAL FODDER LEGUMES

Astragalus

The genus *Astragalus* has several hundred species in the Near East; most of them are unpalatable or of little fodder value. Pabot, however, drew attention to a few drought tolerant perennial species of fodder value, which were the subject of promising small scale trials at the Homand Research Station, Iran in the early 1960's (Pabot 1967). These are: *A. loboghorus (= A. siliquosus), A. chaborasicus, A. teheranicus* and *A. aduncus (= A. kotchyanus).*

Coronilla

Local populations of *Coronilla varia* were subject to cultivation at Kharaj in Iran (250 mm of annual rainfall), according to Pabot (1967), and gave very encouraging results in the early 1960's.

Hedysarum

Hedysarum coronarium is a biennial fodder legume native on well drained, calcareous clay soils in the semi-arid and sub-humid bioclimates of the western Mediterranean with temperate to warm winters (North Africa, Spain, Italy). Erect leafy, high yielding cultivars have been selected in Italy. Types with various degrees of cold tolerance, habit, leafiness, perenniation, feed value and proportion of hard seeds may be selected from the native populations. *H. flexuosum* is a closely related species from Morocco, Algeria and Spain.

H. carnosum is an arid zone species growing between the 100–350 mm isohyets in eastern Algeria and Tunisia, with temperate to warm winters ($m > 3$) on gypsoferous and saline clay soils, with electric conductivities up to 30–40 ms/cm in the saturation extract. It has been successfully cultivated on farm scale demonstration plots and has shown a forage production potential of 5000–8000 kg DM/ha/yr. Native stands in southern Tunisia may produce 2000–3000 kg DM/ha/yr in good years (Le Houérou 1959). There are, however, problems of hard seeds and regrowth after cutting that should be solved before this species may come into routine agricultural practice in the reclamation of salt affected lands, for which it has a great potential.

Other species of *Hedysarum* from the clay calcareous hills in the Aleppo pine forests of North Africa may be of interest in forage research. *H. humile, H. naudinianum* and *H. perralderianum* are deep-rooted, strongly perennial, cold tolerant species, which because of grazing pressure, have become rare in the clay soils of semi-arid *Pinus halepensis* forests of Algeria. They may too be of interest in breeding programmes.

The stoloniferous *H. pallidum* is not palatable and the annuals, *H. spinosissimum* and *H. glomeratum (= H. capitatum)* do not seem to have a high production potential.

Lotus

Mediterranean populations of *Lotus corniculatus* subsp. *decumbens* are found in association with *Festuca arundinacea, Trifolium fragiferum, Tetragonolobus maritimus, Elymus elongatus* and *Agropyropsis lolium* in the native meadows on waterlogged vertisols, which are sometimes more or less saline. To my knowledge, these populations have not been tested under cultivation.

The *Lotus creticus* complex is represented in North Africa by *L. creticus (= L. creticus* subsp. *commutatus), L. collinus (= L. creticus* subsp. *collinus)* and *L. cytisioides (= L. creticus* subsp. *cytisioides)*. The first two species are found on sandy to medium textured soils in the arid and semi-arid zones under annual rainfalls of 150–600 mm. The last species together with *L. creticus* are sea-shore psammophytes of low palatability. *Lotus creticus* and *L. collinus* were tested in pasture reseeding in Spain and Tunisia in the early 1960's; they produced very good results yielding 6000–8000 kg DM/ha/yr under 300–400 mm of annual rain (Le Houérou 1965, 1974a). They could be successfully used in mixtures with *Oryzopsis miliacea, O. holciformis* or *Agropyron* on shallow sandy to medium textured soils in the arid and semi-arid zone.

Lotus maroccanus has characteristics very similar to *L. creticus* and *L. collinus* and could be used as sown pastures in the sandy coastal plains of Morocco and Portugal (Villax 1963).

Medicago

Apart from the well known case of *Medicago sativa* subsp. *sativa* and subsp. *falcata* and their hybrid (*M.* × *varia*) there are a few, little-known, perennial species of use as fodder, with the exception of *M. marina* which has no fodder value. *M. suffruticosa* var. *leiocarpa* is a high mountain species from the *Cedrus atlantica* forests of Morocco and the *Abies pinsapo* forests of southern Spain.

M. gaetula (= *M. tunetana*) is a strongly rhizomatous and stoloniferous species growing on calcareous clay soils and marls in eastern Algeria and western Tunisia (Batna, Thala). It could be of great value in breeding programmes where tolerance to clay, to cold, to grazing and a stoloniferous habit are desired. Multiplication by rhizomes is very easy. Cultivation experiments by the present writer in the early 1960's at Sidi Thabet in the mild-winter, semi-arid zone of Tunisia gave rather poor results, perhaps due to a lack of cold.

Melilotus

There are a number of semi-perennial sweet clovers with a fodder potential within the region, such as *Melilotus alba* and *M. italica*, which are cultivated in northern Morocco either for their salt tolerance, e.g. *M. alba*, or for their high yields, e.g. *M. italica*.

Onobrychis

Onobrychis viciifolia (= *O. sativa*) is a drought-tolerant fodder species adapted to shallow calcareous soils. Over 50 000 ha are cultivated in Anatolia in the semi-arid and arid cold winter zone.

O. argentea subsp. *africana* is a native of the arid and semi-arid *Pinus halepensis* forest of North Africa. It is adapted to shallow calcareous soils and cold winters. Cultivation tests by the writer at Tadmit (Algeria) in the late 1960's gave encouraging results.

There are also a number of *Onobrychis* spp. of potential fodder value in the semi-arid mountains of the Middle East (Pabot 1967), particularly: *O. gaubae, O. persica, O. melanotricha, O. scrobiculata,* and *O. pinnata,* in addition to a number of undescribed species.

Tetragonolobus

Tetragonolobus maritimus (= *T. siliquosus*) is a strongly rooted, perennial fodder legume associated with *Lotus corniculatus, Trifolium fragiferum* and *Trifolium subterraneum* var. *yanninicum* in the *Festuca arundinacea* meadows, apparently fairly tolerant to salinity and waterlogging. To my knowledge, it has not been tested under cultivation.

Trifolium

Among the perennial clovers, Mediterranean populations are characterized by their tolerance to water stress and their summer dormancy. Of particular interest in this respect are the North African and Near Eastern populations of *Trifolium pratense, T. repens* and *T. fragiferum.* The latter is particularly tolerant to waterlogging and salinity (10–15 ms/cm of EC in the saturation extract). It has an ecology close to that of *Festuca arundinacea* with which it is often associated in natural meadows along the margins of swamps.

Trigonella

According to Pabot (1967) there are in Iran and Afghanistan a few perennial trigonels of fodder value, seemingly with many ecotypes. *Trigonella elliptica, T. teheranica, T. cachermiriana,* and *T. griffithsii* and possibly a few undescribed species. The subject has been hardly explored.

Vicia

A perennial vetch, *Vicia onobrychioides* from Morocco, is a valuable fodder in the cold winter, semi-arid to humid climates.

ANNUAL LEGUMES

The magnificent work carried out by the Australian agronomists on Mediterranean legumes, particularly on *Trifolium subterraneum sens. lat.,* is justifiably famous and sets an example. There are at present perhaps more than 40 cultivars of subterraneum clovers developed over the past 50 years in southern Australia and western Australia together with half a dozen species of annual *Medicago* with their many cultivars. But this is only a minute fraction of the potential. There are many more possibilities within these species and in other genera as well, particularly *Vicia: V. monantha, V. benghalensis (= V. atropurpurea), V. villosa* including subsp. *varia (= V. dasycarpa), V. sativa,* etc., *Trigonella arabica, Lathyrus, Melilotus* and *Trifolium,* not to mention the classical fodder species *T. alexandrinum* and *T. resupinatum.*

FORAGE SPECIES FROM OTHER FAMILIES

ROSACEAE

Sanguisorba minor has been successfully used in pasture reseeding on shallow calcareous soils in the arid and semi-arid Mediterranean zone (200–500 mm). It is an extremely hardy species which reseeds itself profusely and is therefore very easy to establish; yields of 2000–5000 kg DM/ha/yr may be expected under 300–500 mm annual rainfall (Le Houérou 1974a).

FODDER SHRUBS

A synthesis study was published by the writer (Le Houérou 1980) to which the reader should refer for further details.

ASCLEPIADACEAE

Periploca laevigata, P. aphylla and *P. gracea* are browse species. *P. laevigata* is native to the arid and semi-arid zones of North Africa and the Near East under rainfalls of 80–500 mm mild to warm winter conditions (*m* ⟩ 5). *Periploca* has been success-fully planted in Libya in areas under 120–150 mm mean annual rainfall. Germina-tion is excellent, establishment of nursery grown seedling is easy, but growth is rather slow. Establishment takes two to three years, but once it is established the shrub is virtually indestructible due to its extreme tolerance to browse and drought; it may remain leafless for several years and break into leaf again when favourable conditions return. It is found on shallow rocky deserts and wadis, with under 60–100 mm annual rainfall, in Tunisia (Sidi Toui, Remada, Dehibat) and Libya (Nalut, Djosh, Tiji, Bou Grain, Bouerat, etc.). This shrub is heavily browsed by sheep, camels, goats and gazelle. A total production of 600 kg DM browse/ha (0.3 kg/shrub) was measured by the writer from one year's regrowth with 115 mm of rainfall, at Wishtata, Libya, in 1983.

CACTACEAE

Opuntia ficus-indica var. *inermis* was brought to Spain during Columbus's second expedition, the spineless cacti spread throughout the Mediterranean basin during the 16th to 18th centuries. They were cultivated for their fruits and as an anti-scorbutic for sailors (Diguet 1928). Spineless cacti are now cultivated over some 200 000 hectares in the arid and semi-arid zone of North Africa (Monjauze & Le Houérou 1965). Yields may be very high under appropriate technology with a RUE up to 25 kg DM/ha/yr mm compared with 5 for the best native arid zone pastures and up to 10 for semi-arid zone sown pastures (Le Houérou & Barghati 1982). The species has become naturalized but the spineless variety is only found in inaccessible places, such as cliffs, steep slopes, etc. where it can escape elimina-tion by browsing animals. The spiny form var. *amyclaea* is common over the arid semi-arid and sub-humid zone with temperate to warm winters. (*m* ⟩ 3, *P* ⟩ 200).

CHENOPODIACEAE

Atriplex
Among the native Chenopodiaceae only four species deserve mentioning: *Atriplex halimus, A. mollis, A. glauca* and *A. leucoclada.*

A. halimus is represented by two subspecies: the type, var. *halimus* in the semi-arid to humid zone and var. *schweinfurthii* in the arid and semi-arid zone (Franclet & Le Houérou 1971). *A. halimus* var. *schweinfurthii* has been successfully planted over several tens of thousands hectares in Libya and Tunisia under rainfalls as low as 120–130 mm. The native population and plantations are genetically extremely heterogeneous in terms of productivity potential, leaf: stems, ratios and palatability to ruminants (Le Houérou *et al.* 1982, Le Houérou 1984b).

A. mollis is a subsaharan phreatophyte of potential fodder value in the reclamation of waterlogged saline area. *A. glauca*, a prostrate dwarf shrub, is very easy to establish by direct broadcasting in the arid zone (100–400 mm). The same remark applies to *A. leucoclada*, a semi-perennial, sprawling, bush-like plant growing in extremely arid conditions in the Near East. For further information on *Atriplex* see Franclet & Le Houérou (1971) and Le Houérou (1984a).

Among the *c.* 20 exotic species of *Atriplex* introduced to the region, particular mention must be made of *A. nummularia* and *A. canescens.* *A. nummularia* was introduced from Australia in the late 19th century and has proved a very useful fodder shrub in the arid and semi-arid zone of North Africa and Near East. It is now grown on several tens of thousands hectares over some ten countries of the region. Both productivity and palatability are higher than in most other species of *Atriplex*; but this is a fragile species requiring careful management and cannot therefore be recommended in all types of situations; this species was recently split into three subspecies (Parr-Smith 1982) on morphological and caryological grounds.

A. canescens is genetically very heterogeneous with diploid to dodecaploid populations extending from central Mexico to southern Canada. Some populations are adapted to sandy soils (subsp. *linearis* and var. *gigas*), others to clay soils (subsp. *canescens*); moreover it will readily hybridize with other species (Stutz & Sanderson 1979). *A. canescens* has been successfully planted over some 30 000 ha of arid range-lands in Iran (Nemati 1976).

Haloxylon

Haloxylon persicum*, the white saxaoul, is a psammophyte of the Near and Middle Eastern deserts from Jordan to southern USSR. It has been established over several thousand hectares of dunes at Karman, Yadz, Bam, Qom, etc. in Iran under rainfalls as low as 60–80 mm. It is found in areas with very warm winters $(m \rangle 9)$ (Akaba, Eilath) as well as in very cold winter zones in Iran (Meched) and Turkmenistan (Kara Koum). Large genetic variations in terms of ecological adaptation and frost tolerance or cold requirement are to be expected. Seed viability does not exceed six months, which may be a serious limitation for the use of this species outside its native area of distribution.

*Some authorities include *H. persicum* in *H. ammodendron.* The genus is in need of revision (G.E.W.).

Haloxylon ammodendron (= *H. aphyllum*), the black saxaoul, contrary to the white, is adapted to silty/loamy soils and is fairly tolerant to salinity; its fodder value is subject to discussion, again by contrast with white saxaoul. It was successfully planted in Libya in the early 1960's, albeit on unsuitable drifting sands in Hashian; it still survives there after 20 years, in spite of heavy browsing.

ELEAGNACEAE

Eleagnus angustifolia, the Russian olive, is sometimes planted as fodder hedges or windbreaks and may be useful because of its tolerance to both frost and salinity.

LEGUMINOSEAE *sens. lat.*

Acacia
Native Acacias, which require mild to warm winters ($m > 5°C$), are of limited interest so far, as they are difficult to establish from seedlings, in the Mediterranean region. This is particularly the case of *Acacia tortilis* subsp. *raddiana*, a remnant of the wadi vegetation of the warm winter parts of the Sahara and mid-eastern deserts.

Many exotic *Acacia* spp. have been introduced to the region, particularly Australian wattles (Phyllodineae), for the purpose of sea-shore sand-dune fixation. The most successful, which also happens to be a good browse species, has been *A. saligna* introduced into Libya as early as 1916 (Leone 1924) and consistently planted ever since in sand stabilization programmes in Morocco, Algeria, Tunisia, Libya, Egypt, Israel, etc. The total area planted may reach half a million hectares including zones with less than 200 mm of rain, such as the area around Syrte in Libya. Fodder production, palatability, intake and feed value have been experimentally studied (Dumancic & Le Houérou 1980, Le Houérou & Barghati 1982, Le Houérou *et al.* 1982, Le Houérou *et al.* 1983). Consumption, over long periods, of up to 1.6 kg DM/ha/day have been recorded in sheep; the feeding value is of the order of 4 MJ of digestible energy per kg DM with a crude protein content of 10–20%.

Acacia victoriae from Australia has also been successfully used in Israel and Libya under 150–200 mm rainfall. There are spiny and spineless individuals. Production and feed value is comparable to *A. saligna.*

Adenocarpus
Adenocarpus telonensis and *A. decorticans* including var. *decorticans* and var. *speciosus* are considered as excellent browse species on the siliceous soils of the semi-arid to humid mountains of Morocco and Algeria, but have become rare.

Cassia sturtii
This species was introduced from Australia where it is considered unpalatable. It has been planted successfully in the northern Negev, Israel (Forti 1971) and more recently in Libya (Le Houérou *et al.* 1982).

Ceratonia

Ceratonia siliqua is a common browse species in the mild to warm winters arid, semi-arid and sub-humid bioclimates (*m* 〉 5; *P* 〉 200). It has been subject to large scale plantations in Sicily and Cyprus for the forage value and pharmaceutical properties of its pods. The genus was long considered monospecific until a second species was recently discovered in northern Somalia and the mountains of Oman: *C. oreothauma* (Hillcoat *et al.* 1980).

Colutea

Colutea arborescens is a semi-arid and sub-humid associate of the oak and pine forests of North Africa and Near East. It is highly appreciated by herbivores but, to my knowledge, has rarely been planted.

Colutea istria from the eastern Mediterranean and the Near East has shown very good adaptation and vigorous growth in the arid zone of Libya with under 150 mm of annual rainfall. Its strong ability to suckering is a very positive factor in establishment and survival; but the fact that this species tends to shed its leaves when under water stress is an unfavourable characteristic from the browsing viewpoint. Leaves and twigs are highly appreciated by herbivores.

Coronilla

The genus *Coronilla* includes a number of vestigial, fodder shrubs in the semi-arid open forests and garrigues of *Quercus ilex, Q. coccifera, Pinus halepensis* and *Tetraclinis articulata* in North Africa. These are *C. emerus* subsp. *emerus* and subsp. *emeroides (= C. emeroides), C. valentina* subsp. *valentina,* subsp. *glauca (= C. glauca)* and subsp. *pentaphylla (= C. pentaphylla)* which are more or less related species, with a rather confused taxonomy. Semi-large scale trials for three decades at Ksar-Tyr, Montarnaud in Tunisia (1940–1970), in semi-arid, mild-winter conditions showed that *C. valentina* is similar to *Medicago arborea* in terms of production and feed value but somewhat more drought-tolerant.

Cytisus

Cytisus albidus (= *Chamaecytisus mollis*) is regarded as one of, if not 'the' best browse species in Morocco (Shoenenberger, pers. comm. 1980) where it is an endemic species in the atlantic, coastal, semi-arid and arid sandy plains matorrals. Direct sowing has been used in the improvement of forest rangelands at altitudes up to 1000 m of elevation in the western High Atlas (Shoenenberger, *ibid.*)

Cytisus arboreus (= *Sarothamnus arboreus*) is a semi-arid to humid zone species on neutral to acid soils, more or less associated with the cork oak *Quercus suber* forests and matorrals, keenly sought for by livestock. It can be easily multiplied by seeds. *C. palmensis* (= *C. proliferus* var. *palmensis*), a 'macaronesian' species from the Canary Islands, Cape Verde Islands and Morocco has been successfully used as fodder hedges on neutral to acid soils under semi-arid to humid climates, including in New Zealand. *Cytisus villosus* (= *C. triflorus*) is a high rainfall browse species in the *Quercus faginea* and *Q. suber* forests.

Gleditsia triacanthos

The American honey locust has been successfully planted on various kinds of soils in the semi-arid to humid Mediterranean bioclimates. It is frost tolerant. There are many ecotypes and populations, thorny or thornless, with various production potential and large variations in the sugar content of the pods (less than 10% to 38%). Pods from individual trees, within any population, may be relished or ignored by livestock; immature pods produce a kind of syrup which is highly appreciated by honey bees, hence the vernacular name.

Medicago arborea

Tree medic or tree lucerne is native to the Balearic Islands, Sicily, the Aegean Isles and Asia Minor; it has been cultivated as an ornamental hedge or a fodder shrub all over the region. *Medicago arborea* is typically a species from semi-arid bioclimates with mild to warm winters (P 300–600; $m \rangle$ 3). Under these conditions forage yields may be quite high, both in quantity and quality (2000–6000 kg DM/ha/yr with 10–20% crude protein). In the arid zone ($P \langle$ 300), under severe water stress, tree medic tends to shed its leaves during the second half of the summer and thereby loses much of its interest as a fodder reserve.

Hedysarum

Hedysarum membranaceum is a vestigial, high-mountain shrub in the semi-arid zone of Morocco with cold winters, highly appreciated by livestock. *H. argentatum,* recently discovered in the dunes of southern Morocco (Tarfaya), is a greyish, up to 4 m tall, ephedroid, psammophytic shrub, keenly sought for by livestock. It is an endangered species of great potential value in the reclamation of desert sand dunes under mild to warm winter conditions.

Parkinsonia aculeata

This tree was introduced from North American deserts. *Parkinsonia* has proved well adapted to the Mediterranean arid zone with cool to warm winters ($P \rangle$ 100 mm; $m \rangle 1°C$). Growth is rapid but forage production is low; the life span of the plantations is rather short (15–20 years). Under 150 mm annual rainfall one may expect 600–800 kg DM/ha/yr of browse from a density of 600 trees/ha (Le Houérou 1980, Le Houérou *et al.* 1982).

Prosopis

There seems to be a great confusion in the literature concerning *Prosopis* spp. introduced to the Mediterranean region; most introductions refer to *P. juliflora,* which in a number of cases has been confused with *P. chilensis, P. glandulosa,* and perhaps other species. There are, however, a number of mature individual trees in Tunisia, Libya, Israel and perhaps other countries which are well adapted and productive. Trials with species of known origin have shown that *P. juliflora* is very sensitive to cold while *P. glandulosa* and *P. chilensis* are much less so. Drought tolerance is good down to 150 mm of annual precipitation on sandy soils. The *Prosopis* spp. seem of interest in the coastal arid belts, particularly for honey production, pods and fuel.

Prosopis cineraria (= *P. spicigera*) is native to the Persian Gulf in southern Iran and Pakistan in the very warm winter arid zone and has been the subject of intensive studies in Rajasthan (Mann & Saxena 1980).

P. tamarugo from Chile is a phreatophyte linked to particular ecological conditions; very few, if any, successful establishments have been reported outside its natural area of occurrence in northern Chile.

MORACEAE

Morus alba and *M. nigra* were introduced from the Far East in the Middle Ages and are cultivated as ornamental trees and sometimes as a source of browse, producing a highly digestible, protein-rich fodder (20–25%).

OLEACEAE

The olive tree *Olea europaea* and its wild forms (= *O. europaea* forma *oleaster, O. europaea* var. *silvestris*) is perhaps the most important browse species in the region since the leaves and twigs from the pruning of the 150 million cultivated olive trees present in the region produce about 1.5 million tons of browse (10 kg DM/tree/yr) in addition to the perhaps as many millions of wild olives that are intensely browsed all over North Africa and the Near East ($P \rangle 200$ mm; $m \rangle 2°$C).

POLYGONACEAE

Calligonum comosum and *C. azel* have been successfully planted as sand binding fodder shrubs in their native sandy deserts of Tunisia, Libya and Iran; and *C. polygonoides* in the desert zones of southern USSR.

SAPOTACEAE

Argania spinosa (= *A. sideroxylon*) occupies an area of some 600 000 ha of bushland and parkland in southwestern Morocco, where it plays an important economic role as browse and as the producer of an edible oil from the kernel. Attempts to establish the argan tree outside its native habitat have so far met with little success, perhaps due to the disturbance of the main tap-root in nursery-grown seedlings.

VITACEAE

American vines introduced as rootstock for the cultivation of grapes (*Vitis vinifera*) are sometimes used for grazing; because of their prostrate habit they may also be efficiently used in soil conservation programmes. The main species are *Vitis vulpina* (= *V. riparia*), *Vitis berlandieri* and *Vitis rupestris* and their numerous hybrids.

Fuel species

NATIVE SPECIES

Native species are rarely planted for fuel or timber in the arid zone because of their slow growth, with the exception of *Pinus halepensis, Cupressus sempervirens, C. atlantica* and *C. dupreziana.* Arid zone native species such as *Tetraclinis articulata, Juniperus phoenicea, Argania spinosa, Acacia tortilis* subsp. *raddiana, Pistacia atlantica, Olea europaea, O. laperrinei* and *Ceratonia siliqua* are virtually never included in reafforestation programmes. It is true that their growth, under controlled conditions in arid zone arboreta, is usually of the order of 20–30 cm/yr in height (Le Houérou 1981); nobody is interested in having a 2 m high tree after having waited for 20 years. Their survival rate, however, is usually very high (more than 90% after 20 years).

Pinus halepensis however, has been successfully planted in areas with only 150–200 mm annual rainfall and, when soil conditions are good, may produce 10 m high trees in 25–30 years. There are, as ever, very large differences in growth and ecological adaptation according to the provenance of the plant material used.

Local strains of *Cupressus sempervirens* forma *numidica* and *C. atlantica* are semi-arid and forest remnants from Tunisia, Libya and Morocco which have been sometimes successfully planted in areas receiving more than 300 mm rainfall. *C. dupreziana*, a relict from the Central Saharan mountains of Tassili N'Ajjers, is a riparian species whose performance in reafforestation projects remains to be assessed.

Tamarix aphylla is a phreatophyte from the Sahara and Near Eastern deserts which has been planted, usually as an ornamental or a windbreak, in the arid zones around the world. It is a fast growing tree, extremely tolerant to drought and salinity, albeit usually discarded from arid zone reafforestation projects. There are large variations in habit, growth, and frost tolerance. A straight stemmed, fastigiate type, cv. erecta, is grown in Israel; it could be more widely used.

Tamarix stricta is a native from the Persian Gulf (Iran, Pakistan). It has been grown with great success in reforestation projects in southwestern Iran under rainfalls of 150–200 mm, where it grows much faster than all exotics, including some 35 species of arid zone *Eucalyptus* spp. (Bhimaya 1974). Five year old plantations near Ahvaz, Iran reached a height of seven meters in 1975 with straight stems (Le Houérou 1975). The quality of its wood has been tested and found suitable for service timber (crates, boxes, etc.). Average yields from these plantations of 60 kg dry wood/tree at the age of five years, were reported by Mahdizadeh and Bhimaya (1974). With a final density of 600 productive trees/ha this corresponds to an average annual timber yield of 7200 kg DM/ha/yr for the first five years of plantation. The average rainfall being 180 mm, the RUE (7200 ÷ 180) is 40, which is one of the highest figures recorded in the literature. Similar figures have however, been given for *Haloxylon persicum* in southern USSR (Rodin 1979).

Calligonum arich is a small tree, 6–8 m high, from the Great Eastern Sand Sea on the Algerian–Tunisian border, near El Borma. Only a few trees are left in the Sand Sea, as several thousand were destroyed in 1941–43 for making charcoal during the second World War. This species has been successfully used in southern Tunisia in dune fixation programmes; it germinates well and is easy to propagate from seedlings. It possesses the ability to grow adventitious roots from the stems and branches when they become buried in sand; in the large dunes such roots may be found up to 50 m from their mother tree.

Populus alba and *P. euphratica* are phreatophytes in the semi-arid, arid and desert zones of North Africa and Near East. They have a similar cold tolerance $(m \rangle$ -2°C.) and salinity tolerance (15 ms/cm); they therefore may be used for fuel plantations using poor quality water that would be unfit for irrigation cropping or in areas which are too cold for other fast growing species such as *Acacia* spp. or *Eucalyptus* spp.

Haloxylon persicum and *H. ammodendron*: see p. 129–130.

EXOTIC SPECIES

Exotic fuel species for the arid zone belong essentially to two genera: *Eucalyptus* and *Acacia.* Unfortunately all arid zone species from these genera are cold sensitive and cannot be used in the high plains of North Africa or in the Irano-Turranian zone of the Near and Middle East, where winters are too cold for them $(m \langle 3)$.

Acacia

Among the *Acacia* species one should mention the Australian phyllodineous species *A. saligna, A. cyclops, A. salicina* and *A. ligulata,* each of them having its particular merits. *A. saligna* is a successful producer on deep sandy soils above 250 mm annual precipitation $(m \rangle 5$°C.); it has been planted in sand dune fixation projects over perhaps half a million hectares in North Africa and Israel since the 1920's in Libya, Tunisa, Algeria, Morocco and Egypt (Leone 1924). The fodder value is not negligible, as mentioned on p. 130.

A. cyclops shows a drought tolerance at least equal to *A. saligna* but is much more tolerant to sea-spray and is thus preferred to the latter in sea-shore sand dune fixation; the fodder value is mediocre.

A. salicina is a late comer, introduced in the early 1960's, which produces excellent results under 150–300 mm annual rainfall even on very poor, shallow, gypseous soils in Tunisia (Oued Gabes), Libya (Bir Ayyad) and Israel (N. Negev). Its strong suckering from roots makes it an aggressive colonizer in spite of its poor seed production. It has no fodder value (Le Houérou *et al.* 1983).

A. ligulata is perhaps the most drought tolerant exotic so far undergoing trials in the region. In Tunisia and Libya (Medenine, Tatahouine, Bir Ayyad) it is growing on shallow sandy soils with only 120–130 mm annual precipitation; it is a good sand binding shrub of no fodder value (Le Houérou *et al,* 1983).

A. pycnantha, the South Australian 'golden wattle' has a valuable ornamental possibility on sandy soils with more than 250 mm rainfall. *A. karroo* from South Africa (sometimes mistakenly named *A. eburnea* and *A. horrida*), is a valuable species for defensive thorn hedges on deep sandy soils with more than 200 mm rainfall; it is often used as an effective protective hedge around the citrus groves in North Africa and the Near East. Other arid zone *Acacia* spp. have been tried in North Africa with mixed results: *A. sowdenii, A. peuce, A. pendula, A. kempeana, A. cambagei* and *A. aneura,* the mulga. Surprisingly enough the latter never produced convincing results in the Mediterranean arid zone.

Eucalyptus

Perhaps as many as 150 species of arid zone *Eucalyptus* have been tested in the region over the past 50 years. The result is a short list of less than 10 species (Table 10.3).

Table 10.3 Arid zone *Eucalyptus* species.

Species	Rainfall mm	Soils	Remarks
E. astringens	⟩ 150	tolerant to mild salinity and gypsum or deep sandy soils	
E. brockwayi		tolerant to shallow soils	drought tolerant
E. camaldulensis	some provenances may be successful with ⟩ 250	some provenances may be successful on deep sandy soils; many provenances sensitive to $CaCO_3$	large genetic variation within strains
E. gomphocephala	200–400	tolerant to $CaCO_3$	sensitive to the longhorn beetle *Phoracantha*
E. microtheca	successful ⟩ 150	tolerant of poor soils	drought tolerant
E. occidentalis	⟩ 150	tolerant of mild salinity in depressions	drought tolerant
E. oleosa	⟩ 150	tolerant to mild salinity and gypsum or deep sandy soils	
E. salmonophloia	⟩ 200	sandy soils	
E. toquata		tolerant shallow calcareous soils	

Casuarina

Several species have been tried with mixed success in the arid and semi-arid zone with mild winters. The best performers have been *Casuarina stricta* and *C. cunninghamiana*.

FUEL PRODUCTIVITY

Native stands of fuel species are often located on poor shallow soils and rocky hills; productivity under these conditions is very low (Le Houérou 1969b). The best native stands of Aleppo pine in semi-arid climates produce some 1000 kg dry wood/ha/yr; the average figure would be less than half as much (below 500 mm mean rainfall the best Aleppo pine forests do not produce more than 1.5 m³ of timber/ha/yr). In general, the RUE is of the order of 2 kg DM/ha/yr/mm in the native woodlands of the arid zone. Some plantations of native species, such as *Tamarix*, Aleppo pine, *Populus alba* and *P. euphratica*, on good soils, may be much more productive and comparable to exotics.

Plantations of exotics are much more productive than natural stands as these species are fast growers and as they are also, in general, planted in much better soils. Productions of up to 10 000 kg dry wood/ha/yr have been recorded with *Eucalyptus camaldulensis*, at Hashian in Libya, under particularly favourable conditions of deep sandy soils benefiting from runoff and under 250–300 mm annual rainfall; an average value would be closer to one third of that figure on deep sandy soils. We have seen that *Tamarix stricta* could produce about as much.

Table 10.4 Average fuelwood production over the first four years of the experiment, with an initial density of 1000 plants/ha (2 X 5 m).

	kg DM/ha/yr
Casuarina cunninghamiana	5400
Atriplex canescens subsp. *linearis*	4100
Atriplex halimus var. *schweinfurthii*	4100
Atriplex nummularia	2400
Parkinsonia aculeata	2200
Acacia saligna	1500
Acacia salicina	1500
Acacia ligulata	1500
Atriplex canescens subsp. *canescens*	1400
Colutea istria	1300
Acacia victoriae	950
Cassia sturtii	950
Acacia cyclops	900
Periploca laevigata	600
Casuarina equisetifolia	500
Acacia aneura	400

Acacia saligna produces an average of about 3500 kg dry wood/ha/yr in the arid zone of Tunisia and Libya. The results of a four year trial in the arid zone of Libya with a long term mean precipitation of 150 mm (170 mm during the four years of the experiment) on deep sandy loam alluvia receiving some run-off, with mild winters ($m \rangle 4°C$) are shown in Table 10.4, p. 137.

The following species have negligible production: *Acacia pendula, A. pycnantha, A. karoo, A. nilotica, A. tortilis, A. gerrardii, A. brachystachya, A. farnesiana, Prosopis cineraria, P. juliflora, P. chilensis, Colutea arborescens, Eucalyptus albens, Morus alba, Atriplex mollis, A. undulata, A. rhagodioides* and *Simmondsia chinensis.*

The productivity of the native steppe of *Artemisia herba-alba* in good condition on the same site is 1000–1200 kg DM/ha/yr, about 50 per cent of it woody.

Conclusions

The present paper has summarily reviewed some 160 species of potential economic value as fodder or fuel in North Africa, the Near and Middle East, about 80 per cent of them native. Seventy-five species are herbaceous fodder species, mainly perennial grasses and legumes, 45 are fodder shrubs and trees and some 40 may constitute a source of firewood. Exotics are sometimes more productive, faster growers; but it is not necessarily so; *Tamarix stricta* or local populations of *Festuca arundinacaea,* for instance, have so far outgrown all exotics in their natural environment, and one could cite many other examples. The present review does not claim to be exhaustive, however, as there may be local species of particular value the present writer may not be aware of.

The species reported herein were selected with the view to their possible use in revegetation programmes, so that species of theoretical value which are difficult to multiply have been discarded, e.g. *Stipa, Aristida, Hyparrhenia,* etc. The species reported make it possible to meet virtually all possible ecological conditions within the region, in terms of climates and soils. There is a large spectrum of genetic diversity within most species both in terms of ecological adaptation and in traits of economic significance. Many species have become rare and some are threatened or endangered, such as: *Calligonum arich, Hedysarum argentatum, Tamarix stricta, T. aphylla* cv. erecta, *Digitaria commutata* subsp. *nodosa, Cupressus dupreziana, Olea laperrinei, Ceratonia oreothauma.* Others like *Hedysarum carnosum* could add new perspectives to the use of brackish water or saline land. Many of the species of potential economic value come from semi-arid mountain forests and shrublands, particularly in the Near and Middle East, which are being cleared for cultivation at the rate of about two per cent per annum.

A far sighted research programme on Mediterranean legumes has made possible a 'green revolution' in southern and western Australia; but, far more remains to be done; and time is running short for the salvage of this rich Mediterranean patrimony. It is suggested that the International Board for Plant Genetic Resources be entrusted the duty and given the means to carry out a closer evaluation of the endangered resources and take the most urgent conservatory measures.

References

Baumer, M. 1983. *EMASAR phase II. Notes on trees and shrubs in arid and semi-arid regions.* Rome: FAO.

Bhimaya, C.P. 1971. *Sand dune stabilization. Report to the Government of Iran.* Rep. TA 2959. Rome: FAO.

Bhimaya, C.P. 1974. *Sand dune fixation. Report to the Government of Iran.* Rep. TA 3252. Rome: FAO.

Borja-Carbonnel, J. 1962. *Las mielgas y "carretones" espanoles.* Madrid: Inst. Nac. Inv. Agron.

Corriols, F. 1965. Essai d'adaptation de plantes fourragères en Tunisie. *Ann. Inst. Nat. Rech. Agron. (Tunis)* 38: 1–318.

Daget, P. 1971. Le quotient pluviothermique d'Emberger et l'evapotranspiration global. *Bull. Rech. Agron. Gembloux,* no. H.S.: 87–94.

Daget, P. 1977. Le bioclimat Méditerranéen. *Vegetatio* 34, 1: 1–20; 2: 87–103.

Delhaye, R., H.N. Le Houérou and M. Sarson 1974. *Amélioration des pâturages et de l'élevage dans le bassin du Hodna (Algerie).* AGS: DP/Alg 66/509. Rome: FAO.

Delisle, D.G. 1964. Chromosome numbers in *Cenchrus. Am. J. Bot.* 51(10): 1133–1134.

Diguet, L. 1928. *Les Cactées utiles du Mexique.* Paris: Archives Hist. Nat. Soc. Nat. d'Acclimatation.

Dumancic, D. 1975. *Progress in agricultural and pastoral development in the project pilot zone. Caspian Forests Development Project, Iran.* FO: SF/IRA/66/519. Rome: FAO.

Dumancic, D. and H.N. Le Houérou 1981. *Acacia cyanophylla* Lindl. as supplementary feed for small stock in Libya. *J. Arid Environ.* 4: 161–167.

Fallon, L.E. 1972. *Rapport preliminaire sue l'amélioration de l'élevage et des terrains de parcours au Maroc.* Rabat: USAID. (mimeo).

Forti, M. 1971. *Introduction of fodder shrubs and their evaluation of use in semi-areas of the north western Negev.* Beer Sheva: Negev Arid Zone Res. Inst.

Franclet, A. and H.N. Le Houérou 1971. *Les Atriplex en Afrique du Nord et en Tunisie.* UNDP/SF/TUNIS, Rapp. Techn. No. 3. Rome: FAO.

Froment, D. 1970. *Aménagement des parcours et leurs relations avec les cultures fourragères en Tunisie centrale.* AGS: SF/TUN 17. Rome: FAO.

Gallacher, R. 1972. *Fodder crops and pasture lands in Algeria.* AGP: SF/ALG 16. Rome: FAO.

Gupta R.K. and S.K. Saxena 1970. Some ecological aspects of improvement and management of Sewan (*Lasiurus scindicus*) rangelands. *Ann. Arid Zone* 19(3): 193–208.

Heyn, C.C. 1963. *The annual species of Medicago.* Scripta Hierosolymitana vol. 12. Jerusalem: Hebrew University.

Hillcoat, H., G. Lewis and B. Verdcourt 1980. A new species of *Ceratonia* (Leguminosae-Caesalpinioideae) from Arabia and the Somali Republic. *Kew Bull.* 35: 261–271.

Katznelson, J. and F.H.W. Morley 1965. Speciation processes in *Trifolium subterraneum* L. *Isr. J. Bot.* 14: 15–35.

Kaul, R.N. 1970. *Afforestation in arid zones.* The Hague: Junk.

Kernick, M.D. 1978. *EMASAR phase II, vol. IV. Indigenous arid and semi-arid forage plants of North Africa, the Near and Middle East.* Rome: FAO.

Lapeyronie, A. 1965. Le Fétuque élevée en Tunisie. *Bull. Ec. Nat. Sup. Agric. (Tunis)* 8/9: 165–193.

Lapeyronie, A. 1968. Présence d'un cycle endogène de germination d'*Oryzopsis miliacea. C.R. Hebd. Seances Acad. Sci. Paris* (D) 267 (21): 1724–1726.

Lapeyronie, A. and A. Semadeni 1968. Caractères biologiques culture et valeur fourragère d' *Oryzopsis miliacea. Bull. Ec. Nat. Sup. Agric. (Tunis)* 21: 77–90.

Le Houérou, H.N. 1958. Note sur un arbre nouveau de Grand Erg Oriental: *Calligonum arich. Bull. Soc. Hist. Nat. Afr. Nord* 49: 297–301.

Le Houérou, H.N. 1959. *Recherches écologiques et floristiques sur la végétation de la Tunisie méridionale.* 2 vols. + annexes. Inst. Rech. Sahar. Univ. Alger Mém. H.S. No. 6.

Le Houérou, H.N. 1965. *Les cultures fourragères en Tunisie.* Doc. Techn. No. 13. Tunis: Inst. Nat. Rech. Agron.

Le Houérou, H.N. 1969a. Quatrième contribution à la flore Libyenne. *Bull. Soc. Bot. Fr.* 116: 279–284.

Le Houérou, H.N. 1969b. La végétation de la Tunisie steppique. *Ann. Inst. Nat. Rech. Agron. (Tunis)* 42 (5): 1–643.

Le Houérou, H.N. 1971. *Bioclimatologie des cultures fourragères: les bases écologiques d'amélioration de la production fourragère et pastorale en Algérie.* Rome: FAO. (mimeo).

Le Houérou, H.N. 1973. Fire and vegetation in the Mediterranean Basin. *Proc. Ann. Tall Timbers Fire Ecology Conf.* 13: 237–277.

Le Houérou, H.N. 1974a. *Principles methods and techniques of range managements and fodder production in the Mediterranean,* 2nd edn. Rome: FAO.

Le Houérou, H.N. 1974b. Etude préliminaire sur la compatibilite des flores Nord-Africaine et Palestinienne. Coll. Intern. CNRS No. 235. *La flore du Bassin Mediterranéen essai de systématique synthétique:* 345–350. Paris: CNRS.

Le Houérou, H.N. 1975. *Report on a consultation mission to the Range Organization of Iran.* Rome: FAO.

Le Houérou, H.N. 1977. Plant sociology and ecology applied to grazing lands research, survey and management in the Mediterranean Basin. In *Handbook of vegetation science,* vol. 13, W. Krause (ed.): 213–274. The Hague: Junk.

Le Houérou, H.N. 1979. Resources and potential of the native flora for fodder and sown pasture production in the arid and semi-arid zones of North Africa. In *Arid land plant resources,* J.R. Goodin and D.K. Northington (eds): 384–401. Lubbock: Texas Tech Univ.

Le Houérou, H.N. 1980. Browse in northern Africa. In *Browse in Africa,* H.N. Le Houérou (ed.): 55–82. Addis-Ababa: Int. Livest. Centre for Africa.

Le Houérou, H.N. 1981. *Report on a study tour to central and southern Tunisia.* Tripoli: FAO.

Le Houérou, H.N. 1983. *A list of native forage species of potential interest for pasture and fodder crop research and development programmes.* Tripoli: FAO. (mimeo).

Le Houérou, H.N. 1984a. Salt tolerant plants of economic value in the Mediterranean Basin. In *Reclamation and revegetation research,* C.V. Malcolm (ed.). Amsterdam: Elsevier (in press).

Le Houérou, H.N. 1984b. The desert and arid zones of northern Africa. In *Hot deserts and arid shrublands,* M. Evenari and D.W. Goodall (eds). Amsterdam: Elsevier (in press).

Le Houérou, H.N. and M.S. Barghati 1982. *An evaluation of fodder shrubs in the Benghazi plains.* Tripoli: FAO.

Le Houérou, H.N., D. Dumancic, M. Eskileh, D. Schweisquth and T. Telahique 1982. *Anatomy and physiology of a browsing trial: a methodological approach to fodder shrubs evaluation.* Tripoli: FAO.

Le Houérou, H.N., D. Dumancic and M. Eskileh 1983. *Feeding shrubs to sheep in Libya: intake, performance and feed-value.* Tripoli: FAO.

Leone, G. 1924. Consolidamento ed imboschimento delle zone dunoze della Tripolitania. *L'Agricolt. Col.* 9: 299–308.

Mahdizadeh, P. and C.P. Bhimaya 1974. Sand dune afforestation in Khuzistan, Iran; and the economics of *Tamarix stricta* plantations raised under this programme. In *Sand dune fixation. Report to the Government of Iran,* C.P. Bhimaya (ed.): 35—44. Rep. TA 3252. Rome: FAO.

Maignan, F. 1970. *Les essais fourragères entrepris dans la zone Nord du project.* FAO/FS/ALG/ 16 (1967—70) a' SI. Lakhdar. Rome: FAO.

Maignan, F. 1971. *Developpement des pâturages et des cultures fourragères: Zones de SI. Lakhdar et de Bou Saada-Djelfa.* AGP: SF/ALG/16. Rapp. Techn. No. 1. Rome: FAO.

Malcolm, C.V. 1969. *Plant collection for pasture improvement in saline and arid environments.* Perth: W. Austr. Dept Agric. Techn. Bull. No. 6.

Malcolm, C.V. and A.J. Clarke 1973. *Plant collection for salt land revegetation and soil conservation.* Perth: W. Austr. Dept Agric. Techn. Bull. No. 21.

Malcolm, C.V. and S.T. Smith 1965. *Puccinellia* – outstanding saltland grass. *J. Dept Agric. W. Aust.* 6(3): 153—156.

Mann, H.S. and S.K. Saxena 1980. *Khejri (Prosopis cineraria) in the Indian Desert.* Jodhpur: Centr. Arid Zone Res. Inst.

Monjauze, A. and H.N. Le Houérou 1965. Le role des *Opuntia* dans l'economie agricole Nord-Africaine. *Bull. Ec. Nat. Sup. Agric. (Tunis)* 8/9: 85—164.

Nemati, N. 1976. Range rehabilitation problems in the steppic zone of Iran. *J. Range Manag.* 30(5): 339—342.

Nichaeva, N.T. and S.Y. Prikhod'ko 1968. *Sown winter ranges in the foothill deserts of Soviet Central Asia.* Jerusalem: Israel Prog. for Scient. Translations.

Niknam, F. and B. Ahranjani 1975. *Dunes and development in Iran.* Tehran: Min. of Agric. Nat. Res.

Pabot, H. 1967. *Pasture development and range improvement through botanical and ecological studies. Report to the Government of Iran.* UNDP/FAO, TA Rep. No. 2311. Rome: FAO.

Parr-Smith, G.A. 1982. Biogeography and evolution in the shrubby Australian species of *Atriplex.* In *Evolution of the flora and fauna of arid Australia,* W.P. Barker and P.J. Greensdale (eds): 291—299. Frewvile, S. Austr.: Peacock Publications.

Robson, M.J. 1967. A comparison of British and North African varieties of tall fescue *(Festuca arundinacea). J. Appl. Ecol.* 4: 475—484.

Robson, M.J. and O.R. Jewiss 1968. A comparison of British and North African varieties of tall fescue *(Festuca arundinacea). J. Appl. Ecol.* 5: 179—190; 191—204.

Rodin, L.E. 1979. Productivity of desert communities of Central Asia. In *Arid land ecosystems, structure, functioning and management,* D.W. Goodall and R.A. Perry (eds): 273—298. London: Cambridge University Press.

Semadeni, A. 1965. Etude de deux écotypes de fétuque élevée. *Bull. Ec. Nat. Sup. Agric. (Tunis)* 8/9: 195—230.

Shoenenberger, A. 1971. *Premiers ensignements des Arboretum forestiers.* FAO: SF/TUN/11. Rapp. Techn. No. 5. Rome: FAO.

Stutz, H.C. and S.C. Sanderson 1979. The role of polyploids in the evolution of *Atriplex canescens.* In *Arid land plant resources,* J.R. Goodin and D.K. Northington (eds): 615—621. Lubbock: ICASALS, Texas Tech Univ.

Villax, E.J. 1963. *La culture des plantes fourragères dans la région mediterranéenne occidentale.* Cahier Rech. Agron. No. 17. Rabat: Inst. Nat. Rech. Agron.

11 Forage and browse – the northern Australian experience

J. J. Mott[1] and R. Reid[2]

[1]CSIRO Division of Tropical Crops and Pastures, Cunningham Laboratory, 306 Carmody Road, St. Lucia, Queensland 4067, Australia

[2]CSIRO Division of Tropical Crops and Pastures, Davies Laboratory, Private Mail Bag, Post Office, Aitkenvale, Queensland 4814, Australia

Introduction

In this review of the implications of the introduction of exotic forage and browse species into the Australian pastoral industry, the area under consideration is that defined by Henzell (1974) as northern Australia. This includes the whole of Queensland and the Northern Territory and that part of Western Australia north of latitude 26°S. This is a region of over 450 million ha and comprises 40 per cent of the continent. The vegetation (Fig. 11.1, p. 145) varies from the savanna-like tall grasslands of the semi-arid tropics and sub-tropics to the arid *Acacia* shrublands and tussock grasslands of the interior (Mott *et al.* 1981). The dividing line between these two broad systems can be taken as the arid/humid line proposed by Perry (1968).

In this region the general pattern of pastoral utilisation is determined by property ownership or tenancy of land, with grazing being on enclosed fixed areas or a ranch type system. The individual properties may be large, ranging from 5000 ha in more productive regions in the southeast to more than 250 000 ha in the poorer arid or more northern areas. The use of fixed area management with defined and fenced areas enables a reduction of the high manpower input needed in the herding type of animal management practised in many countries. However, the extremely high labour costs in Australia, together with a dependence on the export market for sales make it essential that all management systems are planned to maximise output per man.

This requirement for more cost efficient pastoral production has led to the use of much higher energy inputs for animal production in Australia compared to Africa (Bremen & De Wit 1983). As well as making more efficient use of manpower by the use of technology, the possibility of increasing the productivity and reliability of the forage base has led to the evaluation and use of introduced plant species in the pastoral industry. Thus although the native vegetation has been the basis of the pastoral industry and still accounts for more than 85 per cent of the animal production in the region (Mott *et al.* 1981), there is continuing research and development input into techniques aimed at increasing and stabilising animal production by the use of exotic plant material. The greater productivity and more reliable rainfall of the semi-arid tropics and sub-tropics, together with specific nutrient deficiencies in the native pastures of this zone has meant that the majority of research has taken place in this region, rather than the more stressed arid pastures.

In recent years there have been a number of reviews dealing specifically with both the range of material tested, its nutritional value and the detailed techniques used for both introduction and management (e.g. 't Mannetje *et al.* 1980, Williams 1983, Jones 1984, Mott & Tothill 1984, Tothill *et al.* 1985, Winter *et al.* 1985). It is not proposed to repeat this coverage. Instead, this paper examines the basic philosophies behind the use of exotic plant material in the varied environments of northern Australia, and compares them with those in other countries.

Semi-arid tropics and subtropics

NATURAL FORAGE RESOURCE

The main vegetation communities are the tallgrass savanna woodlands defined as black speargrass, tropical tallgrass and *Aristida–Bothriochloa* lands by Mott *et al.* (1981). These communities (Fig. 11.1) form an arc across the north of Australia extending down the east coast to the Queensland border. Broadly confined to a region with greater than 700 mm annual rainfall in the tropics or over 500 mm in the subtropics. The climate is distinctly seasonal, characterised by a single hot, wet growing season and a milder, dry season. There may be high seasonal variability in rainfall leading to marked variation in pasture yield from season to season (McCown 1982). Although lack of soil water limits pasture growth in the tropical systems, temperature also plays an important part in reducing pasture growth in the subtropics (Mott & Tothill 1984).

All tallgrass savanna systems have a dense graminoid understorey with a predominantly eucalypt upper storey. In the subtropics, speargrass (*Heteropogon contortus*) forms a dense sward over large areas of central Queensland, while *Themeda triandra* (= *T. australis*) and annual *Sorghum* spp. are dominant over large areas of the tropical tallgrass pastures. On the more fertile clay soils the native vegetation is either a tall shrubland dominated by *Acacia harpophylla* (brigalow) or grasslands dominated by *Dichanthium* spp. (bluegrasslands). Throughout this region, man-induced or natural fires during the dry season are an almost annual event. They have the effect of removing excess dry roughage and promoting accessibility of new growth after the start of the wet season.

Throughout this region, the low fertility soils and seasonal distribution of rain result in pastures which have a low nutritional quality in terms of animal production over much of the year (Fig. 11.2a). Initial growth after the commencement of the wet season is of high quality, but it declines rapidly as the grasses become reproductive. With the onset of the dry season, crude protein content of

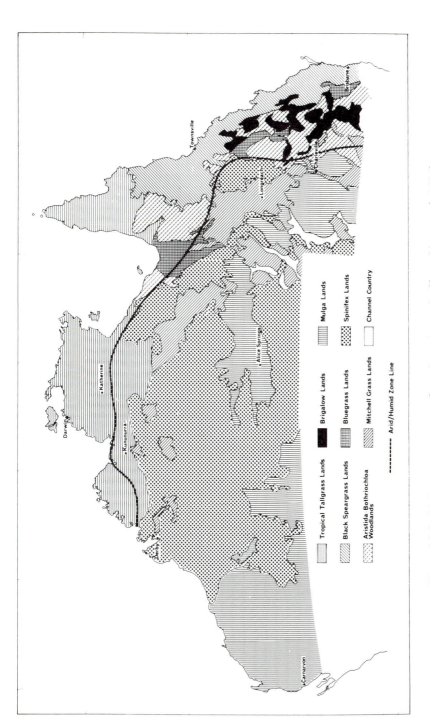

Figure 11.1 Major pasture types of northern Australia (from Mott *et al*. 1981).

the grasses can fall below 6% with a dry matter digestibility of less than 50% (McIvor 1981). Animal production on these poor quality pastures requires low stocking rates to enable selective grazing of the most nutritous herbage; stocking rates vary from one cattle beast to 3 ha in the better quality pastures of south-eastern Queensland to 1:35 ha in the northern tropics (Mott *et al*. 1981). Throughout the region, stocking rate depends on the quality (principally digestibility and protein content) rather than the amount of vegetation available during the poorest part of the year. Animal production reflects the seasonal quality differences, i.e. rapid growth following the start of the wet season continues until the pastures mature, senesce and dry off (Fig. 11.2b). Cattle gain weight in the wet season, but once the dry season begins they initially maintain condition, but finally lose weight towards the end of the season. The more severe nutritional stress occurring in the tropics compared with the subtropics is reflected in a greater loss of weight over the dry season. In the tropics steers may take up to 5–6 years to reach a slaughter weight of 500 kg compared to 2–3 years on the better quality *Heteropogon contortus* pastures of the subtropics.

Given the above constraints, the beef industry is based on the use of large fixed areas of lightly stocked land. In the poorer northern pastures, production systems are aimed at lean beef for transport away to areas of fattening.

IMPACT OF EXOTIC SPECIES

The recognition that the major deficiencies in the native tallgrass pastures of the region are both the low digestibility and nitrogen (protein) content of the forage for a large part of the year has led to the development of a series of options for increasing nutritional status of the feed. The use of dietary supplements to provide nitrogen can improve utilisation of the existing feed resource by increasing the ability of the rumen micro-organisms to digest the high fibre portion of the feed, and thus help reduce the decline in productivity occurring during the dry season (McLennon *et al*. 1981, Winks *et al*. 1982). Also other mineral deficiencies in the pasture (e.g. P, S, Na) can be overcome by year-round supplementation (Holm & Payne 1980). At a more intensive level of input and management, introduced exotic species can also be used in a number of ways to increase the quality of the forage available to the animals. Adapted tropical legumes have been found to provide feed which is both of higher protein level than the native grasses and which has high digestibility at maturity.

The poor nutrient status of many soils in northern Australia means that phosphatic fertiliser is usually required for adequate growth and persistence of almost all exotic species currently in use for pasture improvement (Russell 1978). Given sufficient of this soil nutrient requirement the introduction and establishment of both legumes and grass pastures is technically feasible over all of subhumid northern Australia (Shaw & Bryan 1976). In addition to the requirement of legume

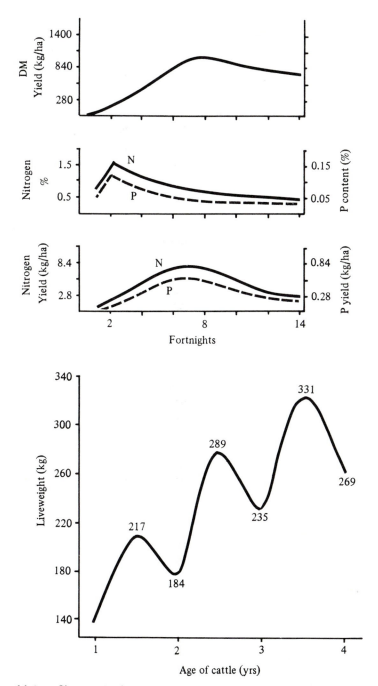

Figure 11.2a Changes in dry matter yield and protein content of native pasture at Katherine, N.T., Australia (after Norman 1963).
11.2b Seasonal liveweight changes in cattle one to four years old grazing native pasture at Katherine (after Norman 1966).

persistence, the need to use superphosphate in improved pastures may also have implications for animal production. In unfertilised grazed pastures of *Stylosanthes* spp. animal weight gains may be poor despite the fact that legumes are dominant (Gardener 1984). In some areas the cattle may avoid eating a high portion of the legume even in the dry season (McLean *et al.* 1981) with resultant poor animal production.

At the lowest input level, the oversowing of legumes into native pasture has been carried out at widespread sites, with *Stylosanthes* spp. (mainly *S. hamata* cv. Verano, *S. humilis* and *S. scabra* cv. Seca and *S. scabra* cv. Fitzroy) in the drier tropical region and *Macroptilium atropurpureum* cv. Siratro as well as *S. guianensis* cv. Schofield in the subtropics. Experimental paddocks established with these species have shown a considerable improvement in the animal production both per head and per unit area (Gillard & Winter 1984; Fig. 11.3).

At a high level of management input the complete replacement of the original pasture with either introduced grass or a grass/legume mixture has been carried out. High productivity has been achieved with mixtures of *Urochloa mosambicensis* and *S. hamata* at Katherine in the northwest and Townsville in the northeast. In the subtropics *M. atropurpureum* and *S. guianensis* again provide highly productive pastures in conjunction with *Cenchrus ciliaris, Panicum maximum* or *Chloris gayana.* Various strains of *C. ciliaris* show widespread adaptation over the region and have formed productive pastures on the areas of cracking clay soils once these are cleared of their original dense forest cover of brigalow.

Under more controlled systems of animal management small areas of dominantly legume pasture have been grazed in conjunction with unimproved native pasture. Early work by Norman (1970) with the annual *Stylosanthes humilis* at Katherine showed that increase in animal productivity was directly related to the amount of time cattle had access to the better quality pasture.

Initially all the above sown pasture systems have led to markedly improved animal production in both experimental and commercial pastures (e.g. Shaw & 't Mannetje 1970). These results prompted optimistic statements on the potential impact of sown pastures on animal productivity in the pastoral industry of the semi-arid tropics of northern Australia. Reports have ranged from that of Begg (1972), who estimated that up to 45 million ha in northern Australia could potentially be sown to Townsville stylo (*S. humilis*), to those of Winter (1978) who estimated that tallgrasslands have the potential to increase animal production six-fold by improved animal husbandry and the introduction of legumes into the native pastures. Most reviewers have followed the path used by 't Mannetje (1984), who used both the area reportedly sown to pastures together with exerimental animal production figures, to estimate that in the early 1980's sown pastures had the potential to contribute 24 per cent of the annual beef production in Queensland. But recent field results have shown these estimated to be too optimistic.

Some experimental pastures have proved stable and highly productive over considerable periods, with similar reports for many completely sown pastures in

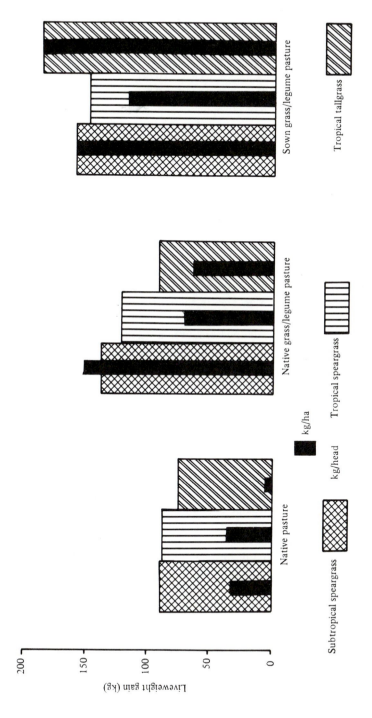

Figure 11.3 Summarised animal production data obtained from experimental grazing trials in the semi-arid tropics and subtropics of northern Australia (adapted from Tothill *et al.* 1985).

higher rainfall areas and isolated properties in drier regions (e.g. Edye & Gillard 1985). However, in recent years, there have been a series of biological constraints to production into sown grass and legume pastures on pastoral properties. The first problems were noted in the tropical tallgrass areas of the far north in the late 1960's. Although *S. humilis* pastures were found to be continuously productive in the subtropical black speargrass region (Shaw 1978), there was considerable instability in dominant *S. humilis* pastures in the tropical tallgrass areas. In this region there was a major invasion of nitrophilous annual grass weeds as soil fertility increased (Torssell 1973). These caused a large drop in the proportion of the legume in the pastures. With the release of perennating, shrubby legumes such as *S. hamata* cv. Verano and *S. scabra* cv. Seca in the early 1970's, high legume productivity and good animal production was again reported (Fig. 11.3). But in the mid 1970's another major biological problem arose to limit the productivity from many *Stylosanthes* spp. based pastures. This was the occurrence of the fungal pathogen anthracnose (*Colletotrichum gloeosporioides*) (Irwin *et al.* 1984). This pathogen spread rapidly through *Stylosanthes* pastures in both the tropics and sub-tropics, severely damaging or destroying almost all the large areas of naturalised *S. humilis* as well as sown areas of other *Stylosanthes* spp, apart from the tolerant *S. hamata* cv. Verano and resistant *S. scabra* cv. Seca. The impact of this pathogen is evident in data obtained from an area in the northern black speargrass region. Here after the first report of anthracnose in 1977 over 50 per cent of the pre-dominantly *S. humilis* pastures disappeared over a 12-month period (P. Lloyd, pers. comm.). A joint effort between CSIRO and the Queensland Dept. of Primary Industries involving both plant pathologists and breeders is underway to examine methods of producing more resistant lines and strategies for management of this important genus (I. Staples, pers. comm.).

As well as this serious and widespread pathogen, two other major problems have been found to constrain the continued high production of sown pastures. The first has been the difficulty that many pastoralists have experienced in maintaining reasonable legume content under their normal management regime for sown pastures. In a survey of farm pastures of humid southern areas of the black spear-grass lands, Anderson *et al.* (1983) showed that pastures with only 34 per cent of sown legume still maintained a legume content which could be classed as giving a substantial increase in animal production over unimproved native pasture. Similarly in sown grass pastures on clay soils in the area, only 30 per cent still maintained a highly productive pasture due to a decrease of fertility associated with a reduction in available nutrients (Catchpoole 1982) or to shrub invasion (Anderson *et al.* 1984).

The above managerial and biological constraints to high levels of animal production from sown pastures in northern Australia, together with a loss of sown pastures on clay soils to crop production (P. Lloyd, pers. comm.), mean that the 24 per cent cited by 't Mannetje (1984) as the proportion of beef production from sown pastures appears an overestimate. Although the majority of dairy production is certainly from sown pastures it is difficult to justify that more than 10 per cent of beef is produced from sown pastures.

In addition to the above problems with exotic species introduced into the native systems of northern Australia, most of the options for improved animal production result in increased grazing pressure on native pastures incorporated in the management system. The major impact of this increased pressure on the native grasses is during the summer growing period. Under this regime there have been significant regional differences in the response of the grasses to this increased defoliation. The dense disclimax *Heteropogon contortus* pastures of the subtropical black speargrass pastures show remarkable stability (Tothill & Jones 1977), but in the more arid western fringes of the region, and in the tropical north, they are proving unstable under heavy grazing (Mott & Tothill 1984). In these eastern regions the introduced species *Bothriochloa pertusa* and *Digitaria didactyla* have shown the ability to replace *Heteropogon contortus* under heavy grazing and recolonise degraded areas (Bissett 1980, Tothill & Hacker 1983). The *Themeda triandra* dominated pastures of the tropical tallgrass lands are also sensitive to large increases in grazing pressure and breakdown of the grass cover may lead to soil exposure, surface sealing and slow regeneration of the grass cover. This differential response to heavy grazing appears to be due to a greater sensitivity of the species to increased levels of defoliation in a more extreme environmental situation. The potential changes occurring in these systems are illustrated in Figure 11.4, p. 152 in which the subtropical *Heteropogon contortus* pasture of south-eastern Queensland and the tropical *Themeda triandra* pastures in the Northern Territory are compared following legume introduction.

These biological constraints on the management of pastoral systems based on sown pasture species show that there is still considerable research needed to develop long-term productive and stable systems. Recent studies have shown that shrub legumes may be more persistent and easier to manage than some of the herbaceous species [e.g. *Leucaena leucocephala* (Rickert & Winter 1980)]. The potential of this shrub has been further emphasised by the resolution of a recurrent toxicity caused by mimosine break-down products (Jones 1979 & 1981). As well as examining the disease anthracnose, research is being carried out on both the problem of fertility run-down in clay soils and the principles determining the response of legumes to the heavy grazing in commercial management systems.

However, in all potential options for improvement of animal production the economic base of the management system must be considered because the real price of animal products has shown a progressive decline over recent years (Daly 1983). This economic constraint has limited the total area of sown pastures to less than five per cent of the potential area available (Weston *et al.* 1981). In the continuing constrained economic climate, it appears certain that low cost management systems incorporating large areas of native pasture will remain over large areas. At this latter level a greater understanding of the native pasture systems is required, together with the subsequent application of specific management strategies, along with the potential introduction of appropriate genotypes to increase animal production.

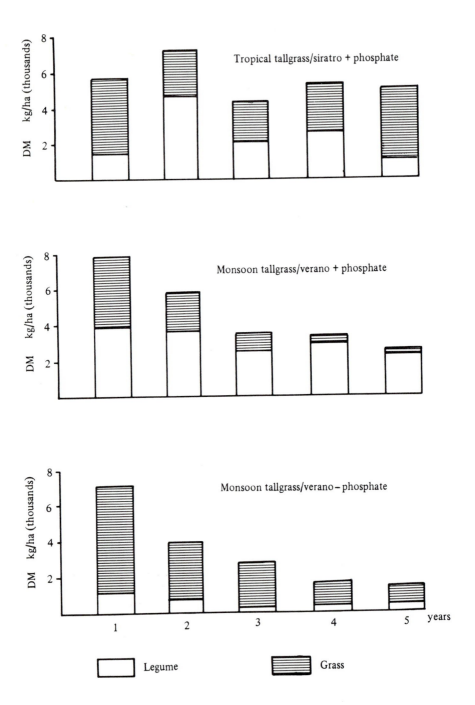

Figure 11.4 Differential response of pasture grazed at approximately one beast to two hectares in northern tallgrass lands (Katherine, N.T.) and in eastern tallgrass lands (Narayen, Qld) (from Mott & Tothill 1984).

Arid pastures

EXISTING FORAGE RESOURCE

In the arid regions there are three dominant vegetation types. Mitchell grasslands on fertile heavy clay soils with *Astrebla* being the dominant genus; the coarse spinifex grasslands occurring on the low fertility lighter soils and dominated by *Triodia* spp.; and the large expanses of mulga lands with shrubby *Acacia* spp. and an understorey of mixed perennial and annual grasses.

Rainfall has a dominant summer component, but the most characteristic factor of the rainfall pattern is both the lack of total volume and the variability of precipitation both within and between seasons. Under these conditions the year to year variation in forage supply can be greater than 5:1 (Christie & Hughes 1983). For economic and logistical reasons, it is not possible to vary stock numbers to this extent so the feed supply varies from being under-utilised in good rainfall years to over-utilised in dry years. Unlike the semi-arid region, plant protein is not the main deficiency. In most vegetation types the protein content is above 9 per cent (Wilson & Harrington 1983) and it is only in the Mitchell grasslands with their higher productivity that a response to supplementary nitrogen might be expected (Stephenson *et al.* 1981).

In these arid lands there are a variety of plant types available as forage ranging from ephemeral grasses and herbs through perennial grasses to a variety of browse shrubs (Wilson & Harrington 1984). The predominant grazing animals are sheep and they select available forage in the order ephemerals/perennials/browse. The main animal production and reproductive success comes from the high quality and digestible ephemerals (Lorimer 1978), with the coarser perennials *Astrebla* and *Triodia* species as well as browse shrubs acting as reserve feed during drought periods. Although this reserve feed may be of reasonable nutritional quality in the case of the browse species (approx. 9% protein), digestibility is low (less than 50%) and this reserve is only a maintenance feed with little value for increased production (Wilson 1977).

As with the poorer northern pastures of the subhumid zone, the maintenance of animal production over these areas of low plant production is only possible at extremely low stocking rates and these range from the equivalent of one cattle beast per 40 ha in the poorer quality *Acacia* shrublands to one beast per 15 ha in the higher productivity tussock grasslands.

The maintenance of a continuum between stable browse species to better quality ephemerals is important for both long term productivity and viability of the systems in such variable arid environments. Both facets of the vegetation can react to different rainfall regimes and they are both necessary for optimum animal production. Since damage to soils and plants may start before animals are seriously affected, the manager has the responsibility of assessing future problems (Woolcock 1979). In cases where excessive stock numbers are maintained through a drought period by the injudicious use of available top feed, permanent damage may be done to the ecosystem by the destruction of the ground flora, leading to degradation of the soil or invasion of woody species (Burrows 1980, Harrington *et al.* 1984). As well as native species, exotic browse shrubs may be potential weeds. In western Queensland, *Acacia nilotica* was widely planted in grasslands as a shade and drought fodder tree. Due to a combination of circumstances including a change from sheep to cattle raising and a sequence of good seasons — dense thickets of up to 1000 ha developed (Anon. 1983a). Also, as browse and herbaceous species may compete for the same moisture and nutrient store, increases in the former can lead to a reduction in higher quality herbaceous feed (Beale 1973) and a consequent long term reduction of animal production (Tunstall *et al.* 1981).

IMPACT OF EXOTIC SPECIES

The main aim of pasture management in this region is amelioration of the effects of long term and seasonal drought on the vegetation, concurrent with the stabilising of animal production. Thus, plant introduction has been predominantly aimed at making available species which increase the stability of the whole vegetation complex. In this context the main species considered have been those which serve a multiple purpose of increasing animal production, stabilising the degraded soil systems and also produce biomass to allow fire to be used as management tool in the control of woody weed invasion (Harrington *et al.* 1984). Throughout the better watered parts of the region three species, *Cenchrus ciliaris* (buffel grass), *C. setigerus* (birdwood grass) and *C. pennisetiformis* (cloncurry buffel) have been recognised as of great importance in fulfilling many of the above aims. The latter has, over a 50 year period, spread along parts of the frontages of all major rivers and watercourses throughout north-west Queensland, replacing lower yielding and poorer quality native grasses (Hall 1978). Since its inadvertent introduction into the northern areas of Australia early in this century (Humphreys 1967) buffel grass has become widespread on many fertile clay soils in areas with an average annual rainfall of greater than 300 mm. However, the low fertility of many other arid rangeland soils has limited the use of buffel over a large area. Buffel grass requires at least 25 ppm phosphorus for satisfactory growth and subsequent drought survival (Christie 1975) and this level is rarely approached in either the mulga lands or *Triodia* grasslands (Noble *et al.* 1984). Economics mitigate against widespread use of fertiliser, but techniques such as seed pelleting are currently under investigation (Silcock & Smith 1982). Continued screening of a wide variety of grass accessions has shown that two Indian species (*Lasiurus scindicus* and *Panicum antidotale*) and three African species (*Eragrostis curvula*, *Anthephora pubescens* and *Schmidtia pappophoroides*) may have potential, especially in the rehabilitation of degraded areas, but there are few data on long term persistence (Cameron 1977, Noble *et al.* 1984). In general the sowing of exotic species is not regarded as a major method for increased stability or productivity, but rather of exploiting favourable niches, either natural or man-made. In this context the introduced kapok bush (*Aerva javanica*) has been found successful in stabilising extensive areas of degraded land in the Ord River catchment of Western Australia (Fitzgerald 1976) by operating as the initial step in a succession, with native grasses replacing this species as the catchment was stabilised (Noble *et al.* 1984).

The lesser importance of specific nutrient needs such as seasonal protein deficiency in this environment reduce the potential impact of introduced high quality browse species on animal production. Experience in the drier areas of the subhumid region has also shown that those species likely to be of use, e.g. shrubby *Stylosanthes* spp. will either require fertilisation or phosphatic supplementation of animals to increase production.

Recent studies have shown that some exotic leguminous species may have the potential for both drought reserve and as a high quality forage supplement to increase reproductive success. Their use will ultimately place increased grazing pressure on the more ephemeral grasses. As well as higher grazing pressure the competition between the introduced browse and native grasses for scarce moisture and nutrients would also be expected to reduce the availability of the productive grasses.

In the most stable and resilient of the arid zone vegetation types, e.g. the *Astrebla* spp. dominant Mitchell grasslands, there remains some potential for the introduction of exotic legumes (Orr & Holmes 1984).

Transfer of Australian technology

In attempting to evaluate the significance of Australian technology for improving animal production in comparable regions in other continents it is essential to evaluate socio-economic as well as technical problems. Thus, although animal production, both per head as well as per unit area, is much lower in many countries than is possible in northern Australia, considerable research may still be needed before techniques of use in Australia is relevant in other areas.

Even in regions where both the management systems and the animal base are similar to Australia, problems have occurred. Early work in South America by Centro International de Agricultura Tropical (CIAT) was largely based on varieties developed in Australia. These were generally unsuited to the acid soils of the Llanos and the *Stylosanthes* spp. which did succeed were decimated by anthracnose. This early failure has had to be followed by a major programme to collect and evaluate new genetic material before promising species could be made available (Anon. 1982, Thomas 1984). As in Australia, experimental paddocks have shown high yields of *Stylosanthes* spp. and research programmes have been aimed at answering the specific nutritional and climatic problems of the region. But there is a major need to generate more long term data on animal production, especially the effects of stocking rate (Thomas 1984).

The above technical problems, although significant, may be less important than the great socio-economic differences existing between Australia and many developing countries. Within northern Australia, the main aim of the pastoral industry is the production of livestock or wool for export sale. In this region there is little emphasis on either secondary bovine products or sheep for carcass production respectively. With a background of set stocking, fixed land tenure, high labour costs and high technology, the primary objective of increased inputs for improvement of the forage base is to increase profits. In contrast in many developing countries social pressures may favour more animals rather than increased production per animal unit. Under such pressures the rangeland resource may be severely overgrazed. Secondary animal products such as milk may also be of great importance.

Throughout Africa, India and China, grazing livestock may be owned individually but grazed collectively on communal ground. Under these conditions it is difficult for an individual to apply techniques for pasture improvement or to attempt to improve the existing forage base by reducing his cattle numbers. In the first case his neighbours would share in the improved forage while in the second increased herd numbers by others in the community would rapidly nullify any improvement in the native forage. Some advances in this field have been made in the subhumid regions by the use of small (2 ha) fenced blocks of dense legume in the Kaduna region of Nigeria. These have been recognised as a 'crop' by the pastoral community at large and left for use by the owner. Improved animal production has been obtained by daily pastorage of herded stock in these areas (Anon. 1983b).

In arid lands the concept of increasing the availability of browse species has been advocated by several authors and a considerable amount of information is available on factors such as the nutritional quality of browse (e.g. Le Houérou 1983). However, in both these cases stock must also have access to areas of grass pasture for increased productivity. With the lack of livestock control, care must be taken so that long term stability of the landscape is not impaired. Few data are available on potential overgrazing problems following the introduction of browse species into the existing pasture systems, but the similarity of many of the native grasslands systems in Africa to those existing in Australia gives cause for extreme caution.

In view of the potential problems involved in the transfer of exotic material evaluated in Australia into the extensive rangelands systems of developing countries, alternative methods of utilisation should be considered. Systems relevant to the semi-arid tropics are the use of sown pastures as short term leys in a cropping sequence or the use of legumes as intercrop species (e.g. Anon. 1983b). In ley pasture systems, legumes may provide valuable high quality fodder, increase the yield of subsequent crops, and aid in the restoration of soil fertility and structure after long periods of cropping (Jones & McCown 1984). These systems would allow closer integration of crops and livestock and also allow the continued use of traditional methods as well as needing the introduction of new principles such as minimal tillage.

Conclusions

Considerable research effort has been expended on the evaluation and release of exotic pasture species over the last 50 years, with almost 100 cultivars released commercially from some 30 000 introductions (Williams 1983). However, of the numbers released only 26 varieties can be considered to be used widely in northern Australia with less than five per cent of the potential land area being planted to sown pastures (Cameron 1977). More recent collecting expeditions have tended to concentrate on increasing the pool of legume genotypes adapted to the subhumid and arid regions, with emphasis has been placed on collecting germplasm from the genera *Desmanthus, Leucaena, Macroptilium* and *Stylosanthes* (Reid 1983).

In the savanna woodlands of the semi-arid tropics and sub-tropics, experimental studies have shown the great potential of using exotic species to remove the specific constraint of poor feed quality in the dry season. These results have been employed by some pastoralists with a similar increase in animal productivity (e.g. Edye & Gillard 1985). However, the widespread extrapolation of this technology to the commercial pastoral scene has been affected both by economic and biological constraints. For future improvement in animal production in this zone research must be aimed at understanding the implications of these factors in both the sown and native pasture components of the animal production system. It is only with this information, as well as the provision of more adapted species for such specific constraints, such as tolerance of pathogens, that robust productive and stable management systems can be evolved.

The more severe management and environmental constraints of the arid lands have meant that apart from buffel grass on higher fertility soils, the main use for exotic species has been where economic considerations have allowed the relatively high inputs necessary to revegetate degraded land systems. Although the continued research input in the evaluation and production of exotic plant material is producing higher quality species adapted to arid conditions, the employment of such material in the vast expanses of arid lands must be approached with caution.

Although there may be scope for the utilisation of high nutritional quality ephemeral species to optimise animal production from minor rainfall events, the incorporation of new drought reserve species into any ecosystem must be made in the context of both the forage base and the stability of the landscape. Since the main forage lack in these lands is for grasses to stabilise landscapes as well as increase animal production, there are serious questions as to the advisability of introducing additional browse species.

The extrapolation of Australian technology on both forage and browse species to other countries may be of great benefit for improved animal production. However, as in Australia, it is important to recognise that the stability and productivity of the entire forage resource must be considered in any management system, not only that of the introduced species. Also the great socio-economic differences existing between Australia and many developing countries indicate that the types of management systems employing similar genetic material may have markedly different results and new concepts of agronomic and pastoral use may have to be developed.

Acknowledgements

The senior author would like to acknowledge his helpful discussions with many officers from Queensland Department of Primary Industries and CSIRO Division of Wildlife and Rangeland Research, as well as access to unpublished data which is cited in the text. He also acknowledges his debt to his colleague, Dr J. Tothill, with whom he has had many and spirited discussions on the philosophy of pastoral use in Australia's savanna lands.

References

Anderson, E.R., F.J. Russell, J.C. Scanlan and G.W. Fossett 1983. *Pastures under development in central Queensland. Part 1. Mackay Region.* Brisbane: Dept Primary Industries, Qld Govt.

Anderson, E.R., J.C. Scanlan, G.W. Fossett and F.J. Russell 1984. *Pastures under development in central Queensland. Part 2. Northern Brigalow Region.* Brisbane: Dept Primary Industries, Qld Govt.

Anon. 1982. *CIAT report 1982.* Cali, Colombia: Centro Internacional de Agricultura Tropical.

Anon. 1983a. *Annual Report Land Administration Commission 1978–79.* Queensland Department of Lands.

Anon. 1983b. *ILCA Annual Report 1982*: 6–11. Addis Ababa: International Livestock Centre for Africa.

Beale, I.F. 1973. Tree density effects on yields of herbage and tree components in south-west Queensland mulga (*Acacia aneura* F. Muell.) scrub. *Trop. Grassl.* 7: 135–142.

Begg, J.E. 1972. Probable distribution of Townsville stylo as a naturalised legume in tropical Australia. *J. Aust. Inst. Agric. Sci.* 38: 158–162.

Bisset, W.J. 1980. Indian bluegrass has special uses. *Qld Agric. J.* 106: 507–517.

Bremen, H. and C.T. de Wit 1983. Rangeland productivity and exploitation in the Sahel. *Science* 221: 1341–1347.

Burrows, W.H. 1980. Range management in the dry tropics with special reference to Queensland. *Trop. Grassl.* 14: 281–287.

Cameron, D.G. 1977. Pasture plant introduction in Queensland – a continuing need. *Trop. Grassl.* 11: 107–119.

Catchpoole, V.R. 1982. Regeneration of grass pasture. In *CSIRO Tropical Crops and Pastures, Annual Report 1980–81*: 60. Melbourne: CSIRO.

Christie, E.K. 1975. A study of phosphorus nutrition and water supply on the early growth and survival of buffel grass grown on a sandy red earth from south-west Queensland. *Aust. J. Exp. Agric. Anim. Husb.* 15: 239–249.

Christie, E.K. and P.G. Hughes 1983. Interrelationships between net primary production, ground storey condition and grazing capacity of the *Acacia aneura* rangelands of semi-arid Australia. *Agric. Syst.* 12: 191–211.

Daly, J.J. 1983. The Queensland beef industry from 1930 to 1980: lessons from the past. *Qld Agric. J.* 109: 61–67.

Edye, L.A. and P. Gillard 1985. Pasture development in northern Queensland semi-arid tropics with a practical example. In *Proceedings of the international savanna symposium, Brisbane 1984*, J.C. Tothill and J.J. Mott (eds). Canberra: Aust. Acad. Science (in press).

Fitzgerald, K. 1976. Management and regeneration of degraded catchments and eroded pastoral land with particular reference to range reseeding. In *Conservation in arid and semi-arid zones*: 41–60. Rome: FAO.

Gardener, C.J. 1984. The dynamics of *Stylosanthes* undergrazing. In *The biology and agronomy of Stylosanthes*, H. Stace and L.A. Edye (eds): 335–351. London: Academic Press.

Gillard, P. and W.H. Winter 1984. Animal production from *Stylosanthes* based pastures in Australia. In *The biology and agronomy of Stylosanthes*, H.M. Stace and L.A. Edye (eds): 405–430. London: Academic Press.

Hall, T.J. 1978. Cloncurry buffel grass. (*Cenchrus pennisetiformis*) in north-western Queensland. *Trop. Grassl.* 12: 10–19.

Harrington, W.G., D.M.D. Mills, A.J. Pressland, and K.C. Hodgkinson 1984. Semi-arid woodlands. In *Management of Australian rangelands*, G.N. Harrington, A.D. Wilson and M. Young (eds): 189–208. Australia: CSIRO.

Henzell, E.F. 1974. Research for the livestock industries of northern Australia. *Proc. Aust. Soc. Anim. Prod.* 10: 322–331.

Holm, A.McR. and A.L. Payne 1980. Effects of mineral supplementation on the performance of steers grazing birdwood grass (*Cenchrus setiger*) pastures in north-western Australia. *Aust. J. Exp. Agric. Anim. Husb.* 20: 398–405.

Humphreys, L.R. 1967. Buffel grass (*Cenchrus ciliaris*) in Australia. *Trop. Grassl.* 1: 123–134.

Irwin, J.A.G., D.F. Cameron, and J.M. Lenne 1984. Responses of *Stylosanthes* to anthracnose. In *The biology and agronomy of Stylosanthes*, H.M. Stace and L.A. Edye (eds): 295–309. London: Academic Press.

Jones, R.J. 1979. The value of *Leucaena leucocephala* as a feed for ruminants in the tropics. *World Anim. Rev.* 31: 13–22.

Jones, R.J. 1981. Does ruminal metabolism of mimosine and DHP explain the absence of *Leucaena* toxicity in Hawaii. *Aust. Vet. J.* 57: 55.

Jones, R.J. 1984. Improving the nutrition of grazing animals using legumes, fertiliser and mineral supplements. In *ACIAR (Australian Centre for International Agricultural Research) proceedings of the East-Africa-ACIAR consultation on agricultural research, Nairobi, Kenya*: 122–136. Nairobi: National Council for Science & Technology, and Canberra: ACIAR.

Jones, R.K. and McCown, R.L. 1984. Research on a no-till, tropical legume-ley farming strategy In *ACIAR proceedings of the East-Africa-ACIAR consultation on agricultural research, Nairobi, Kenya*: 108–121. Nairobi: National Council for Science & Technology, and Canberra: ACIAR.

Le Houérou, H.N. (ed.) 1983. *Browse in Africa – The current state of knowledge*. Addis Ababa: International Livestock Centre for Africa.

Lorimer, M.S. 1978. Forage selection studies. 1. The botanical composition of forage selected by sheep grazing *Astrebla* spp. pastures in N.W. Queensland. *Trop. Grassl.* 12: 97–108.

McCown, R.L. 1982. The climatic potential for beef cattle production in tropical Australia. Part IV – Variation in seasonal and annual productivity. *Agric. Systems* 8: 3–15.

McIvor, J.G. 1981. Seasonal changes in the growth, dry matter distribution and herbage quality of three native grasses in northern Queensland. *Aust. J. Exp. Agric. Anim. Husb.* 21: 600–609.

McLean, R.W., W.H. Winter, J.J. Mott and D.A. Little 1981. The influence of superphosphate on the legume content of the diet selected by cattle grazing. *Stylosanthes* – native grass pastures. *J. Agric. Sci., Camb.* 96: 247–249.

McLennon, S.R., P.E. Dunster, P.K. O'Rourke and G.M. Murphy 1981. Comparison of dry season urea supplements containing salt, sulfur or molasses for steers grazing native pasture in the dry tropics of northern Queensland. *Aus. J. Exp. Agric. Anim. Husb.* 21: 457–463.

't Mannetje, L. 1984. Pasture development and animal production in Queensland since 1960. *Trop. Grassl.* 18: 1–18.

't Mannetje, L., K.F. O'Connor and R.L. Burt 1980. The use and adaptation of pasture and fodder legumes. In *Advances in legume sciences,* R.J. Summerfield and A.H. Bunting (eds): 537–651. England: Kew: Royal Botanic Gardens.

Mott, J.J. and J.C. Tothill 1984. Tropical and subtropical woodlands. In *Management of Australian rangelands,* G.N. Harrington, A.D. Wilson and M. Young (eds): 255–270. Melbourne: CSIRO.

Mott, J.J., J.C. Tothill and E. Weston 1981. The native woodlands and grasslands of northern Australia as a grazing resource for low cost animal production. *J. Aust. Inst. Agric. Sci.* 47: 132–141.

Noble, J.C., G.M. Cunningham and W.E. Mulham 1984. Rehabilitation of degraded land. In *Management of Australian rangelands,* G.N. Harrington, A.D. Wilson and M. Young (eds): 171–188. Melbourne: CSIRO.

Norman, M.J.T. 1963. The pattern of dry matter and nutrient content changes of native pastures at Katherine, N.T. *Aust. J. Exp. Agric. Anim. Husb.* 3: 119–124.

Norman, M.J.T. 1965. Seasonal performance of beef cattle on native pasture at Katherine, N.T. *Aust. J. Exp. Agric. Anim. Husb.* 5: 227–231.

Norman, M.J.T. 1970. Relationships between liveweight gain of grazing beef steers and Townsville lucerne. *Proc. XI Int. Grassl. Cong.*: 829–832. Surfers Paradise: Queensland.

Orr, D.M. and W.E. Holmes 1984. Mitchell grasslands. In *Management of Australian rangelands,* G.N. Harrington, A.D. Wilson and M. Young (eds): 241–254. Melbourne: CSIRO.

Perry, R.A. 1968. The need for rangelands research in Australia. *Proc. Ecol. Soc. Aust.* 2: 1–14.

Reid, R. 1983. Pasture plant collecting in Mexico with emphasis on legumes for dry regions. *Aust. Plant Introd. Rev.* 15: 1–11.

Rickert, K.G. and W.H. Winter 1980. Integration of feed sources in property management extensive systems. *Trop. Grassl.* 14: 239–245.

Russell, J.S. 1978. Soil factors affecting the growth of legumes on low fertility soils in the tropics. In *Mineral nutrition of legumes in tropical and subtropical soils,* C.S. Andrew and E.J. Kamprath (eds): 75–92. Melbourne: CSIRO.

Shaw, N.H. 1978. Superphosphate and stocking rate effects on a native pasture oversown with *Stylosanthes humilis* in central coastal Queensland. *Aust. J. Exp. Agric. Anim. Husb.* 18: 800–807.

Shaw, N.H. and L. 't Mannetje 1970. Studies on a speargrass pasture in central coastal Queensland – effect of fertiliser, stocking rate and oversowing with *Stylosanthes humilis* on beef production and botanical compositions. *Trop. Grassl.* 4: 43–56.

Shaw, N.H. and W.W. Bryan (eds) 1976. *Tropical pastures research – principles and methods.* Comm. Bur. Past. Fld Crops Bull. No. 51. Farnham Royal: Comm. Agric. Bur.

Silcock, R.G. and F.T. Smith 1982. Seed coating and the localised application of phosphate for improving seedling growth of grasses on acid sandy red earths. *Aust. J. Agric. Res.* 33: 285–802.

Stephenson, R.G.A., J.C. Edwards and D.S. Hopkins 1981. The use of urea to improve milk yields and lamb survival of merinos in a dry tropical environment. *Aust. J. Agric. Res.* 32: 497–509.

Thomas, D. 1984. Global ventures in *Stylosanthes.* I. South America. In *The biology and agronomy of Stylosanthes,* H.M. Stace and L.A. Edye (eds): 451–464. London: Academic Press.

Torssell, B.W.R. 1973. Patterns and processes in the Townsville stylo – annual grass pasture ecosystem. *J. Appl. Ecol.* 10: 463–478.

Tothill. J.C. and R.M. Jones 1977. Stability in sown and oversown siratro pastures. *Trop. Grassl.* 11: 55–68.

Tothill, J.C. and J.B. Hacker 1983. *The grasses of Southern Queensland.* St Lucia: University of Queensland Press.

Tothill, J.C., H.A. Nix, P. Stanton and M.J. Stanton 1985. Land use and productive potential of Australian savanna lands. In *Ecology and management of tropical savannas,* J.C. Tothill and J.J. Mott (eds): Canberra: Aust. Acad. Sci. (in press).

Tunstall, B.R., B.W.R. Torssell, R.M. Moore, J.A. Robertson and W.F. Goodwin 1981. Vegetation change in a poplar base (*Eucalyptus populnea*) woodland. Effects of tree killing and domestic livestock. *Aust. Rangel. J.* 3: 123–132.

Weston, E.J., J. Harbison, J.K. Leslie, K.M. Rosenthal and R.J. Mayer 1981. *Assessment of the agricultural and pastoral potential of Queensland.* Brisbane: Dept Primary Industries. Qld Govt.

Williams, R.J. 1983. Tropical legumes. In *Genetic resources of forage plants,* J.G. McIvor and R.A. Bray (eds): 17–38. Melbourne: CSIRO.

Wilson, A.D. 1977. The digestibility and voluntary intake of leaves of trees and shrubs by sheep and goats. *Aust. J. Agric. Res.* 28: 501–508.

Wilson, A.D. and G.N. Harrington 1983. Nutritive value of Australian browse plants. In *Browse in Africa – The current state of knowledge,* H.N. Le Houérou (ed.): 291–298. Addis Ababa: International Livestock Centre for Africa.

Wilson, A.D. and G.N. Harrington 1984. Grazing ecology and animal production. In *Management of Australian rangelands,* G.N. Harrington, A.D. Wilson and M. Young (eds): 63–78. Melbourne: CSIRO.

Winks, L., P.K. O'Rourke and S.R. McLennon 1982. Liveweight of grazing steers supplemented with molasses, urea and sulfur in northern Queensland. *Aust. J. Exp. Agric. Anim. Husb.* 22: 252–257.

Winter, W.H. 1978. The potential for animal production in tropical Australia. *Proc. Aust. Soc. Anim. Prod.* 12: 86–93.

Winter, W.H., T.H. McCosker, D. Pratchett and J.D.A. Austin 1985. Intensification of beef production. In *Agro-research for the semi-arid tropics of north-west Australia.* Melbourne: CSIRO (in press).

Woolcock, B.A. 1979. Animal management decisions. In *Rangeland ecosystem evaluation and management,* K.M.W. Howes (ed.): 57–69. Perth: Aust. Rangel. Soc.

12 Bees and honey in the exploitation of arid land resources

Eva Crane

International Bee Research Association (IBRA), Hill House, Gerrards Cross, Bucks SL9 0NR, UK

Introduction

My own interest in the subject of this chapter is bee-orientated, and it is necessary to explain briefly the background, and the history of IBRA's World Survey of Honey Sources (WHOS), before presenting some of the first results to come from it.

The estimated world's annual production of honey, according to official statistics, is a little short of a million tonnes, and the actual production – the amount harvested – is doubtless larger than this. Over 200 000 tonnes are exported on to the world market. The three major exporters are China, Mexico and Argentina, all of which include subtropical arid regions. Such areas can be important for honey production: when skies are clear, energy from the sun is freely transmitted to plants, and some of this energy is converted into sugar which plants secrete as nectar. If sufficient bees are present, this nectar can be converted into honey and harvested.

Areas made agriculturally unproductive by deforestation or improper farming usually become drier. But some of the plants that colonize such areas can be good honey sources, so that bees do well there. This is no new observation. Plato bemoaned the deterioration of the land by excessive farming *c.* 400 BC, in a passage in *Critias*. He referred to 'mountains in Attica which can now support nothing but bees, but which were clothed, not so very long ago, with fine trees . . . The country [then] produced boundless pasturage for cattle'.

Plants supply all the bees' food resources, i.e. nectar, honeydew and pollen, – and also the propolis they use in building their nest. Beeswax, venom and bee milk (used for feeding the young) are metabolized by adult worker bees, beeswax largely from carbohydrates in nectar and honey, and venom and bee milk largely from proteins in pollen. Immature bees cannot be reared without pollen, or an adequate pollen substitute provided by the beekeeper.

Using bees to crop the land is usually not an alternative, but an additional way of getting a harvest. And if the honey sources are plants that yield seed or fruit, the bees may increase yields by their pollinating activities.

Traditionally, honey is the harvest from bees that beekeepers work for, although some of them also harvest the various other substances mentioned above. The bees' foraging behaviour in collecting the raw materials of honey is keenly attuned to obtaining the maximum energy reward for a minimum energy expenditure. In European countries they disregard dilute nectars such as pear or plum when dandelions, with their more concentrated nectar, are growing nearby.

The pattern of bees that produce the world's honey is much more complex than anything Plato knew. As used here, the word 'bees' refers mainly to honeybees, genus *Apis*, for which there are four species; some taxonomists subdivide them, but there is no need to do so in the present context. The European hive bee, temperate-zone *Apis mellifera*, is indigenous to the Old World and has been introduced to many parts of the New World. Except in some parts of Asia, and parts of Latin America, the world's large-scale honey production is based on beekeeping with this bee. There are tropical ecotypes in Africa south of the Sahara, one of which was introduced to South America in 1956 and has spread widely there and into Central America as far as Costa Rica. In Asia, *Apis cerana* is the counterpart of tropical *Apis mellifera*; it is smaller, and generally less productive. In many countries bee-keeping with *Apis cerana* remains traditional, with fixed-comb hives, but in India especially it has been developed with modern movable-frame hives.

The other two species of *Apis* cannot be kept in hives, as their nest consists of a single comb built in the open. Honey is obtained from them by traditional honey-hunting methods. *Apis dorsata* has the largest body size of the four species. In India, and probably also in some other countries of tropical Asia, more honey is still produced from honey-hunting with *Apis dorsata* than from beekeeping with *Apis cerana*. *Apis florea*, the smallest of the species, is 'managed' to a certain extent for honey production in Oman, and elsewhere honey is harvested from wild nests.

In much of the tropics and subtropics honey is also obtained from colonies of stingless bees (Meliponini). In some regions, especially parts of Central and South America, certain species are kept in traditional hives, and can yield more honey than *Apis florea* colonies.

Historical background, and WHOS Phase I

In the present context honey sources are plants that yield nectar and/or honeydew from which bees produce honey. For historical and economic reasons most studies of them have been made on a local, subnational or national level.

In the early 1970s, when planning a comprehensive survey on honey (Crane 1975), a chapter on the world's honey plants was considered essential. There was no specialist in this subject, so I wrote the chapter, and my first task was to try to identify the most important honey sources on a world scale. Brief characteristics of 211 plant species or genera and their honeys were finally included, the selection being on a rather *ad hoc* basis. This material was augmented by a 17-page list of the best available publications on honey plants of individual countries. In Crane (1980), the same plants, with 21 important additions, were listed in relation to their world distributions. From 1975 onwards unpublished material was collected from institutions and individuals in as many countries as possible, attention being paid especially to the developing countries of the tropics and subtropics.

At several International Beekeeping Congresses discussions were initiated on the possibility of a Survey of World Honey Sources (WHOS). But in spite of much interest, and a general appreciation of the importance of the concept, no funding was available until 1979 when the International Commission for Bee Botany obtained a small subvention from its parent body, the International Union of Biological Sciences. It enabled a start to be made, and Phase I of WHOS was carried out at IBRA in 1980–1981. From published and unpublished material, 2569 plant species were recorded on multi-copy paper slips, each plant being reported as a honey source from at least one country, and some from many.

Funds for Phase I were too small to allow any publication of the results, except for a list of the 160 plant families recorded, with the number of plants species in each (Crane 1983a). Three families, Leguminosae (327), Myrtaceae (261) and Compositae (226) contributed nearly one-third of all species.

WHOS Phase II: Directory of important world honey sources

The next step was clearly to make some of our information on honey sources available. It was considered essential to prepare and store the information for publication in such a way that programmed searches could be made for honey sources that have specific characteristics. In 1982 IBRA obtained funds from the International Development Research Centre, Ottawa, for the preparation and publication of an international directory of honey sources (Crane, Walker & Day 1984). The first programmed searches, after those run for this directory, were used in preparing this Chapter.

The directory is something new in concept, in content, and in method of preparation. It identifies 467 plants, out of the preliminary selection 2569, that are reported to be major sources of the honey in a particular country or area. Some of these honey sources are geographically widespread, such as lucerne (*Medicago sativa*) and many of *Eucalyptus* species; others are confined to a single area, such as plectranthus (*Rabdosia rugosa*) which grows only on certain slopes of the western Himalayas.

The major part of the directory contains the main entries for the honey sources selected; 452 nectar-producing plants and 15 honeydew-producing plants. The information in these entries is supported by a bibliography of 820 references; almost all of them are held by IBRA, and each is coded with its library shelf mark. Table 12.1, p. 166, identifies the characteristics that were sought for each plant.

Seven lists of important honey sources with special characteristics are printed in

Table 12.1 Summary of information in main entries of the Directory.

For each plant, information (as available) is printed in the order below. Data on similar characteristics are grouped together in paragraphs or 'blocks', as shown. The 51 search fields are indicated by *.

---

entry number   Botanical name of plant, authority; family
any synonyms

common names
vegetative form of plant* (tree/shrub/herb); floral description
Distribution (tropical*, subtropical*, temperate*); and where native
Habitat
Soil (salt-tolerant*). **Temperature** (damaged by frost*). **Rainfall** (drought-tolerant*)

Economic* and other uses
Food*. Fodder*. Fuel*. Timber*. Land use (hedges*, windbreaks*, shade*, afforestation*, amenity*). **Soil benefit*** (erosion control*, enrichment*). **Other uses**

Warning about the plant* e.g. toxicity of plant, including nectar, to man or animals (see also under pollen, honey), invasiveness. **Alert to beekeepers***: difficulties in bee management, honey handling, etc.

Nectar rating + honeybee species
Nectar rating of a plant in a country is: **N1** = major honey source; **N2** = medium honey source; **N3** = minor honey source; **N** = honey source, importance unrated.

Apis species: ac = *A. cerana,* ad = *A. dorsata,* af = *A. florea,* am = *A. mellifera,* tm = tropical *A. mellifera*

Blooms (dates). **Nectar flow** (dates or duration). **Nectar secretion. Sugar concentration*** (low ⟨ 21%/medium 21–60%/high 61+%). **Sugar value***, mg/flower/day (low ⟨ 0.1/medium 0.1–2/ high 3+). **Sugar analysis.** Other characteristics.

Honey flow
Specimen **honey yield***, kg/colony/season (high 30+/moderate ⟨ 30).
Honey potential*, kg/ha/season (high 500+/moderate ⟨ 500).

Pollen
value to bees*: high: Pl rating reported (e.g. high nutritive value to bees; produced in large quantities); pollen recorded as collected by bees; warnings – toxic pollen, sticky pollen, pollen inadequate for brood rearing

representation of grains in honey*: under-represented (⟨ 20 000 per 10 g)/over-represented (⟩100 000 per 10 g)

Honeydew
honeydew produced on plant*; honey yield from honeydew: surplus recorded either numerical or not

Table 12.1 — continued

Recommendation for propagation as a bee plant
Recommendation for planting has been published*

Honey composition
water*: low $<$ 16/medium 16−20/high 21 (FAO max) and over; **glucose**: low $<$ 31/medium 31−39/high 40+; **fructose***: low $<$ 35/medium 35−42/high 43+; **sucrose***: low $<$ 1/medium 1−4/high 5 (FAO max) and over; **reducing sugars***: low $<$ 65 (FAO min)/high 65 and over; **ash***: low $<$ 0.1/medium 0.1−1.0/high 1.0 (FAO max) and over; **free acids*** (meq/kg): low $<$ 15.0/medium 15−39/high 40 (FAO max) and over; **amylase*** (Gothe): low $<$ 3 (FAO min)/ high 3 and over; **HMF*** (ppm): high 40 (EEC max) and over/low below 40 (HMF not included in FAO/WHO Codex); **fermentation** on storage*: likely/unlikely/never; **vitamins*** present; **toxicity*** any information on adverse reactions, e.g. when fed to bees; **colour**: any unusual tinges*; **Pfund*** (mm): white 0−34 (includes grades water white to white), amber 35−114 (includes grades extra light amber, to amber), dark 115+ (grade dark amber); **granulation*** complete within: $<$ 2 wks = rapid/2−52 wks = medium/$<$ 1 yr = slow; **flavour** descriptions: bland or strong*, very sweet*, objectionable*, unusual flavour*.

the directory. The lists were made by running programmes to search the data base, using some of the 51 coded fields, e.g. Table 7b in Crane, Walker and Day (1984), identifies the 37 honey sources showing some degree of salt tolerance.

The directory also includes a separate list of 196 candidate plants, which might have been eligible for inclusion if more information had been available.

Honey sources for arid lands

Data presented below are derived from searches such as that mentioned above; all the data refer to plants in the directory, that is, plants reported to be a major honey source in one or more parts of the world.

Of the 467 plants, 77 are listed in Table 7a of the directory as drought-tolerant, including 11 that are very drought-tolerant. Of the 77, 73 are found in the directory from at least one place in the tropics and subtropics. Table 12.2, p. 168, lists 47 relevant to the arid or semi-arid tropics, and Table 12.3, p. 169, contains 58 relevant to the arid or semi-arid subtropics, plus four recorded only in temperate zones. Eleven of the 77 were new additions to Kew's SEPASAT data bank.

Tables 12.2 and 12.3 identify plants that could form a useful basis for bee-keeping in arid or semi-arid lands: to add substance and savour to the food, or to add to the income by the sale of honey to the poorer indigenous peoples or, at a higher economic level, to stimulate a modest beekeeping industry. Among the important world honey sources, we rated as high any yield of 30 kg/colony/year or more (from a single source). Table 12.4, p. 171, gives examples of honey yields from some of the plants in Tables 12.2 and 12.3. It shows that a fair proportion of the plants listed are among the best of the world's honey sources.

Table 12.2 Drought-tolerant plants recorded in the tropics that are important world honey sources. *very drought-tolerant, T=tree, S=shrub, H=herb.

002*	*Acacia caffra*; Leguminosae	S
005	*Acacia mellifera*; Leguminosae	S
008*	*Acacia senegal*; Leguminosae	T
009	*Acacia seyal*; Leguminosae	T
010	*Acacia tortilis*; Leguminosae	T
022	*Agave americana*; Agavaceae	H
024	*Aloe dichotoma*; Liliaceae	T
030	*Anacardium occidentale*; Anacardiaceae	T
039	*Azadirachta indica*; Meliaceae	T
066	*Caesalpinia coriaria*; Leguminosae	T
067	*Cajanus cajan*; Leguminosae	S
070	*Calliandra calothyrsus*; Leguminosae	S
078	*Cassia siamea*; Leguminosae	T
093	*Citrus limon*; Rutaceae	T
106	*Combretum celastroides*; Combretaceae	S
122	*Dalbergia sissoo*; Leguminosae	T
126	*Dialium engleranum*; Leguminosae	T
156	*Eucalyptus camaldulensis*; Myrtaceae	T
172	*Eucalyptus leucoxylon*; Myrtaceae	T
176*	*Eucalyptus melliodora*; Myrtaceae	T
180	*Eucalyptus paniculata*; Myrtaceae	T
187	*Eucalyptus sideroxylon*; Myrtaceae	T
205	*Gleditsia triacanthos*; Leguminosae	T
207	*Glycine max*; Leguminosae	H
208	*Gmelina arborea*; Verbenaceae	T
217	*Gymnopodium antigonoides*; Polygonaceae	S
221	*Helianthus annuus*; Compositae	H
237	*Ipomoea batatas*; Convolvulaceae	H
244	*Jacquemontia nodiflora*; Convolvulaceae	H
246	*Julbernardia paniculata*; Leguminosae	T
290	*Medicago sativa*; Leguminosae	H
296	*Melilotus alba*; Leguminosae	H
313	*Olea europaea subsp. africana*; Oleaceae	S
315	*Opuntia engelmanii*; Cactaceae	H
319	*Parkinsonia aculeata*; Leguminosae	T
330	*Pithecellobium dulce*; Leguminosae	T
335	*Pongamia pinnata*; Leguminosae	T
336*	*Prosopis cineraria*; Leguminosae	T
338*	*Prosopis glandulosa*; Leguminosae	T
339*	*Prosopis juliflora*; Leguminosae	T
340*	*Prosopis pallida*; Leguminosae	T
349	*Rhigozum trichotomum*; Bignoniaceae	S
397	*Tamarindus indica*; Leguminosae	T
441	*Viguiera helianthoides*; Compositae	H
448*	*Ziziphus mauritiana*; Rhamnaceae	T
452*	*Ziziphus spina-christi*; Rhamnaceae	T

Table 12.3 Drought-tolerant plants recorded in the subtropics that are important world honey sources. (4 grow exclusively in temperate zones – see end of Table, while others extend into temperate zones). *very drought-tolerant; T=tree, S=shrub, H=herb.

001	*Acacia berlandieri*; Leguminosae	S
002	*Acacia caffra*; Leguminosae	S
004	*Acacia greggii*; Leguminosae	T
008	*Acacia senegal*; Leguminosae	T
009	*Acacia seyal*; Leguminosae	T
022	*Agave americana*; Agavaceae	H
024	*Aloe dichotoma*; Liliaceae	T
026	*Aloysia gratissima*; Verbenaceae	S
030	*Anacardium occidentale*; Anacardiaceae	T
039	*Azadirachta indica*; Meliaceae	T
066	*Caesalpinia coriaria*; Leguminosae	S
067	*Cajanus cajan*; Leguminosae	S
076*	*Carnegiea gigantea*; Cactaceae	S
078	*Cassia siamea*; Leguminosae	T
084	*Centaurea solstitialis*; Compositae	H
085	*Parkinsonia florida*; Leguminosae	T
093	*Citrus limon*; Rutaceae	T
122	*Dalbergia sissoo*; Leguminosae	T
139	*Echium lycopsis*; Boraginaceae	H
149	*Eriobotrya japonica*; Rosaceae	T
153	*Eucalyptus anceps*; Myrtaceae	T
154	*Eucalyptus caleyi*; Myrtaceae	T
156	*Eucalyptus camaldulensis*; Myrtaceae	T
158	*Eucalyptus cladocalyx*; Myrtaceae	T
160	*Eucalyptus crebra*; Myrtaceae	T
167	*Eucalyptus gomphocephala*; Myrtaceae	T
168	*Eucalyptus gracilis*; Myrtaceae	S
170	*Eucalyptus incrassata*; Myrtaceae	T
172	*Eucalyptus leucoxylon*; Myrtaceae	T
176*	*Eucalyptus melliodora*; Myrtaceae	T
178*	*Eucalyptus oleosa*; Myrtaceae	T
180	*Eucalyptus paniculata*; Myrtaceae	T
181	*Eucalyptus platypus*; Myrtaceae	T
182	*Eucalyptus polyanthemos*; Myrtaceae	T
185	*Eucalyptus rubida*; Myrtaceae	T
187	*Eucalyptus sideroxylon*; Myrtaceae	T
191	*Eucalyptus wandoo*; Myrtaceae	T
205	*Gleditsia triacanthos*; Leguminosae	T
207	*Glycine max*; Leguminosae	H
221	*Helianthus annuus*; Compositae	H
237	*Ipomoea batatas*; Convolvulaceae	H
272	*Lotus corniculatus*; Leguminosae	H
280	*Mahonia trifoliata*; Berberidaceae	S
290	*Medicago sativa*; Leguminosae	H
296	*Melilotus alba*; Leguminosae	H

Table 12.3 – continued

314	*Onobrychis viciifolia*; Leguminosae	H
319	*Parkinsonia aculeata*; Leguminosae	T
330	*Pithecellobium dulce*; Leguminosae	T
335	*Pongamia pinnata*; Leguminosae	T
336*	*Prosopis cineraria*; Leguminosae	T
338*	*Prosopis glandulosa*; Leguminosae	T
340*	*Prosopis pallida*; Leguminosae	T
354	*Robinia pseudoacacia*; Leguminosae	T
397	*Tamarindus indica*; Leguminosae	T
426	*Trifolium alexandrinum*; Leguminosae	H
450	*Ziziphus nummularia*; Rhamnaceae	S
452*	*Ziziphus spina-christi*; Rhamnaceae	T

Drought-tolerant plants recorded in the directory only in temperate zones
that are important world honey sources.

220	*Hedysarum coronarium*; Leguminosae	H
297	*Melilotus officinalis*; Leguminosae	H
317	*Paliurus spina-christi*; Rhamnaceae	S
405	*Thymus capitatus*; Labiatae	S

The first SEPASAT Newsletter (SEPASAT 1983) stresses the problems caused by salinity, and states that an estimated 381 million hectares of arid and semi-arid lands of the world are saline. A programmed search showed that 15 of the 73 drought-tolerant important world honey sources recorded in the tropics and sub-tropics are also to some degree salt-tolerant (Table 12.5, p. 172). The possibility of using the dry saline areas for honey production should certainly be investigated.

Beekeepers get their honey mainly from the plants that are growing within the bees' flight range of a home apiary; those who make their living from honey production usually take up one or both of two further options. Either to base their operations in an area with a good honey potential, and/or to migrate their bees from one honey flow to another. A third option is to grow plants as honey sources for the bees. This is not usually cost-effective in the moderate or good growing areas where most professional beekeepers live – the land is more valuable for other purposes. Whereas in arid and semi-arid areas, the yield per hectare in terms of food produced is often very low and the return from using bees to crop the land may be comparable with that obtained in other ways.

A further category of drought-tolerant honey sources warrants consideration;

Table 12.4 Some honey yields reported from drought-tolerant plants recorded in the tropics and subtropics.

Yields are quoted in kg/colony/year unless otherwise stated. 'Honey potential' is a term in Eastern Europe for the estimated weight (kg) of honey that could be obtained in the course of a season from 1 hectare of land covered with the plant, assuming optimal conditions (Crane 1975).

---

001   *Acacia berlandieri*, 27 in USA

004   *Acacia greggii*, 72 (also 10) in USA

010   *Acacia tortilis*, 2–3 in Oman, using *Apis florea*

022   *Agave americana*, 41 in Mexico

066   *Caesalpinia coriaria*, 'much honey in the hives in October' in the Caribbean area

070   *Calliandra calothyrsus*, 1–3 kg/colony/month in Java, Indonesia, with *Apis carana*

086   *Cicer arietinum*, 20–2.5 in Uttar Pradesh, India, with *Apis carana*

093   *Citrus limon*, 30–60 (mixed with other *Citrus* spp.) in Israel

122   *Dalbergia sissoo*, 4–9 in India, using *Apis cerana*?; 27 (mixed with honey from clovers) in Pakistan

149   *Eriobotrya japonica*, 20 in Lebanon; in Pakistan 3.6 using *Apis mellifera*, also 1 using *A. cerana*

Eucalyptus spp.:

156   *E. camaldulensis*, 55, 60 in Australia; 100–120 (mixed with honey from *E. cladocalyx*) in Morocco

158   *E. cladocalyx*, 12–25, maximum 90, in Cape, S. Africa; 100–120 (often with honey from *E. camaldulensis*) in Morocco

160   *E. crebra*, 82 in Queensland, Australia, but good yield only every 3 years

168   *E. gracilis*, mean 27–36, maximum 113, in Victoria, Australia

176   *E. melliodora*, 25, maximum 75, in Australia

178   *E. oleosa*, 54 in Victoria, Australia

180   *E. paniculata*, 100 in Australia; 50 in S. Africa

191   *E. wandoo*, 90 in Australia

205   *Gleditsia triacanthos*, honey potential 250 kg/ha in Romania

208   *Gmelina arborea*, mean 20, up to 100, in Gambia

217   *Gymnopodium antigonoides*, 136 (with *Viguiera helianthoides*), in Mexico

315   *Opuntia engelmanii*, 30 in south-west USA, best during partial drought in Texas

338   *Prosopis glandulosa*, main source in Punjab and Sind, Pakistan; mean 27, maximum 90, in Texas, USA; most flowers are produced when soil moisture is low

340   *Prosopis pallida*, 120–150, 227–363, in 2 Hawaiian islands

354   *Robinia pseudoacacia*, 8–10 kg/colony/day in Romania; honey potential up to 1600 kg/ha in eastern Europe

426   *Trifolium alexandrinum*, using *A. cerana* in Pakistan 9, also (with *T. resupinatum*) 27; 'important in Baluchistan, Punjab, Sind, NWFP'; honey potential 165 kg/ha in Bulgaria

441   *Viguiera helianthoides*, 30% of honey crop in Yucatan, Mexico.

Table 12.5 Plants recorded in the tropics and subtropics that are drought- and salt-tolerant, and important world honey sources.
This list includes plants reported to show any degree of salt tolerance.

022 *Agave americana*; Agavaceae	272 *Lotus corniculatus*; Leguminosae
067 *Cajanus cajan*; Leguminosae	319 *Parkinsonia aculeata*; Leguminosae
122 *Dalbergia sissoo*; Leguminosae	330 *Pithecellobium dulce*; Leguminosae
156 *Eucalyptus camaldulensis*; Myrtaceae	335 *Pongamia pinnata*; Leguminosae
167 *Eucalyptus gomphocephala*; Myrtaceae	336 *Prosopis cineraria*; Leguminosae
180 *Eucalyptus paniculata*; Myrtaceae	340 *Prosopis pallida*; Leguminosae
205 *Gleditsia triacanthos*; Leguminosae	426 *Trifolium alexandrinum*; Leguminosae
221 *Helianthus annuus*; Compositae	

Table 12.6 Drought-tolerant plants recorded in the tropics and subtropics that are specifically recommended for planting to increase honey production. All have other economic uses.

039 *Azadirachta indica*; Meliaceae	181 *Eucalyptus platypus*; Myrtaceae
070 *Calliandra calothyrsus*; Leguminosae	182 *Eucalyptus polyanthemos*; Myrtaceae
122 *Dalbergia sissoo*; Leguminosae	185 *Eucalyptus rubida*; Myrtaceae
156 *Eucalyptus camaldulensis*; Myrtaceae	187 *Eucalyptus sideroxylon*; Myrtaceae
158 *Eucalyptus cladocalyx*; Myrtaceae	221 *Helianthus annuus*; Compositae
160 *Eucalyptus crebra*; Myrtaceae	290 *Medicago sativa*; Leguminosae
167 *Eucalyptus gomphocephala*; Myrtaceae	296 *Melilotus alba*; Leguminosae
172 *Eucalyptus leucoxylon*; Myrtaceae	330 *Pithecellobium dulce*; Leguminosae
176 *Eucalyptus melliodora*; Myrtaceae	354 *Robinia pseudoacacia*; Leguminosae
180 *Eucalyptus paniculata*; Myrtaceae	

those that have other economic uses than honey production. Of the 73 drought tolerant honey sources under discussion, 67 are known to have economic uses; Table 12.6 lists 19 of them that are specifically recommended for planting.

It is important when making recommendations for planting, to know if any of the plants being assessed have potentially undesirable characteristics of which users should be aware. In the directory two types of such characteristics are indicated. A warning is given, e.g., entries 039, 070 and 354 can be invasive under some conditions, and for 158 and 354 whose leaves may be toxic to animals (see Tables 12.2 & 12.3). An alert to beekeepers is also given where appropriate, e.g. if a plant yields no pollen, or pollen that is inadequate for brood rearing, colonies working it may dwindle and die. The beekeeper must therefore check whether other pollen sources are available and, if not, feed pollen substitutes. A number of *Eucalyptus* species have this characteristic, including 6 of the 17 listed in Tables 12.2 and 12.3, entries 168, 172, 176, 180, 182 and 187.

Some nectar flows, that from *Helianthus annuus* (221) for example, seem to stimulate swarming in colonies. Other plants, that are very good honey sources, flower so early in the reproductive cycle of the honeybee colony that the bee population is too small to take full advantage of the nectar available, unless the beekeeper follows special management procedures. This situation is usually a result of the introduction of a crop plant whose growth cycle and flowering are out of phase with the local growth cycle. Autumn-sown oilseed rape (*Brassica napus* var. *oleifera* (060)) is a well known example, and *Robinia pseudoacacia* (354) in eastern Europe is another.

Beekeepers face a different type of problem if the honey from a particular plant granulates (crystallizes) very rapidly. If granulation occurs within a few days, while the honey is still in the hive, the honey cannot be extracted from the combs. Honey from *Eucalyptus paniculata* (180) in Table 12.6 is reported to granulate in the hive in cold weather, and that from *E. polyanthemos* (182) is reported to be 'difficult to extract'.

Information on honey composition is known for only 108 out of the 467 plants listed in the directory, and 77 (71%) of these are temperate-zone plants. We hope that honey chemists will take note of this imbalance and extend their studies to more of the tropical and subtropical honeys.

The future of the WHOS programme: satellite directories and Phase III (data bank)

IBRA intends to use further search programmes to prepare satellite directories of selective entries on honey sources that have certain important characteristics, e.g. drought-tolerant plants that give especially high honey yields are of great interest for semi-arid parts of the world. Details to be published about each of the plants would include, for example, botanical and ecological information, economic uses, honey-producing capability, and experience in propagation specifically for honey production. We should welcome suggestions as to satellite directories likely to be in greatest demand and although commissioned searches for specific information can be made, no funds are available at present for a free service.

We can also use search programmes to show which data are lacking, and thus where new research should be focussed. Programmes that involve the growing of good honey plants can be monitored, and areas of the world where a specific plant might be especially profitable can be identified.

IBRA regards the survey of the world honey sources as a continuing project. Storage of the information on discs allows the future addition of further data for plants included in the directory. It will also allow the upgrading of some of the candidate plants to the status of a full entry, as information becomes available.

The 40 000 plant species visited by bees are too numerous to be absorbed into the present honey-oriented project; they must be left for environmentalists to document. The aim of IBRA is to use the information bank primarily in ways that will advance apiculture in different parts of the world, especially in developing countries.

Pollination of arid land crops

Another project, the preparation of a pollination directory for world crops (Crane & Walker 1984), has been funded by the New Zealand Ministry of Foreign Affairs, and the book will be published by IBRA. It is on a much simpler level than the directory of honey sources (Crane, Walker & Day 1984), and is intended especially for use by crop growers in developing countries. It will contain a single sequence of 400–500 entries under common name of crop, with an index of botanical names. Each entry gives brief details, as available, of the uses and distribution of the crop, and of the plant's floral structure and pollination mechanism. Pollinating agents are identified where possible; where there are difficulties in ensuring adequate pollination these are set out and recommendations made for overcoming them. For many insect-pollinated crops the most practical action is to take hives of bees to the crop, the number of hives per hectare depending on the crop. The isolation distance required for production of 'certified' seed is also quoted where applicable.

Many crops are among the important world honey sources, and about 100 of those in the pollination directory (Crane & Walker 1984) are cross-indexed to entries in the honey sources directory (Crane, Walker & Day 1984). For these, the use of bees may give a dual benefit; a honey harvest and increased crop yields.

Conclusions

The use of bees to crop food resources can provide a harvest additional to that for which plant crops are primarily grown. Moreover the primary harvest from many insect-pollinated crops is increased in quantity and quality by the presence of foraging bees. Beekeeping at an appropriate technological level, low, intermediate or high, can add usefully to food and income in developing countries of the tropics and subtropics. Furthermore beekeeping can be undertaken by men, women and children, on a part-time or full-time basis.

The above statements are generally accepted but constraints to the most effective use of beekeeping still need to be studied quantitatively. One constraint has certainly been a lack of knowledge about plants that are important honey sources in different habitats in the tropics and subtropics. The 'Directory of important world honey sources' now helps to remove this particular constraint.

In the context of the Conference, 73 important honey sources in the arid and semi-arid lands of the tropics and subtropics are identified, 67 of which are recorded as having other economic uses. They are examined from various points of view: degree of drought-tolerance; habit (tree, shrub, herb); distribution (tropics, subtropics); salt-tolerance coupled with drought-tolerance (15 plants); recommendation to plant for honey production (19 plants). These 19 plants are of special interest because other uses of arid land may give such poor yields of food, fodder or timber, etc. that the yield from beekeeping may be more profitable.

It is considered most important that, in future work, information related to bee and honey production should be linked with broader plant data bases for developing countries. IBRA is certainly willing to take responsibility for this specialized aspect of such work, provided funds can be found to support it.

Acknowledgements

I much appreciate assistance from the Royal Botanic Gardens, Kew with plant taxonomy and other matters, and especially Dr G E Wickens's interest in WHOS as a potential component of SEPASAT. I take this opportunity also to pay special tribute to my colleagues Penelope Walker and Rosemary Day for their valuable contributions to the project.

References

Crane, E. 1975. *Honey: a comprehensive survey.* London: Heinemann in co-operation with International Bee Research Association.

Crane, E. 1980. *A book of honey.* London: Oxford University Press.

Crane, E. 1983. *Surveying the world's honey plants.* Ilford, Essex: Central Association of Bee-Keepers.

Crane, E. and P. Walker 1984. *Pollination directory for world crops.* London: International Bee Research Association.

Crane, E., P. Walker and R. Day 1984. *Directory of important world honey sources.* London: International Bee Research Association.

SEPASAT 1983. *SEPASAT Newsletter* No. 1, Sept. 1983. Survey of Economic Plants for Arid and Semi-Arid Tropics, irregular publication, Royal Botanic Gardens, Kew.

13   Economic halophytes – a global review

James Aronson

Boyko Institute for Agriculture and Applied Biology, Ben-Gurion University of the Negev, Beer-Sheva 84110, Israel

Introduction

Not less than about 400 million ha and perhaps as much as 950 million ha of land in arid and semi-arid regions may be salt-affected from natural and anthropogenic causes (Massoud 1974, Ponnamperuma 1977). Definitive data on the annual world-wide loss of farmland due to salinization and related causes is lacking. However salinity is unquestionably the most important problem of irrigated agriculture (Dregne 1977), and one-fifth of the irrigated lands of the world, approximately 47 million ha, is salt-affected today (Maas & Hoffman 1977).

Reliable estimates suggest that in the Mediterranean region alone, some 15 million ha have become seriously salt-affected due to man's activities (Le Houérou 1984). In India over 6 million ha, of a national total of only some 40 million ha of irrigated farmland, have been made useless by salinity and waterlogging, while about 40 thousand ha are being abandoned each year (Chapman 1975a). Well over 500 thousand ha of Californian soils (USA), particularly in the Coachella, Imperial and Sacramento – San Joaquin Valleys had been strongly affected and made useless to agriculture by 1963 (Weir 1963, cited in Kelley *et al.* 1979); and the United States as a whole reportedly loses 80–120 thousand ha of cropland each year due to salt build-up (Yensen *et al.* 1981).

The problem is hardly new. Early civilizations in the Indus and Nile valleys and in the Peruvian coastal region were all apparently drastically affected as the build-up of salts in their fields gradually led to yield losses and eventually to crop failure. A similar phenomenon is thought to have contributed to the collapse of the early Mesopotamian civilization in the Tigris–Euphrates valleys (Jacobsen & Adams 1958). Today, in the attempt to feed the world's burgeoning population of people, increasingly huge areas are being brought under irrigation in arid and semi-arid lands. As a result, the problems related to salinity have once again begun to have overwhelming, detrimental effects on both traditional and modern agriculture and animal husbandry.

Since most irrigated lands lie in arid and semi-arid areas, salinization is not only one of the leading causes of desertification, but is also one of the most frustrating results of man's attempts to expand cultivation into these regions. The commonly sought solution of providing more water for arid lands is fraught with danger. For every effort to bring more land under cultivation with irrigation provided by the harnessing of rivers and the building of canals, an equivalent or greater amount of land is going out of production as a result of salinization of over-irrigated lands (White 1973).

In the fight against salinization, three basic approaches are available. First, improve drainage so that salt build-up is slowed or halted. This typically requires huge investments in sub-surface tiling, drainage wells, and tube wells, combined with frequent leaching of the upper soil layers. This engineering approach has been undertaken on a large scale in areas like western Pakistan and the Imperial Valley of California. Nevertheless salinization is still increasing in such areas. Second, select or breed for greater tolerance in conventional crops showing some genetic variability for salt tolerance, e.g. wheat and barley (Epstein & Norlyn 1977, Kurian & Iyengar 1971); tomatoes (Tal 1971); or cotton and sugar beets (Ahmad & Abdullah 1981). A third and fundamentally different approach: start looking for potentially useful, wild halophytes. The last approach, which only began to attract attention in the past ten years, also raises the possibility of utilizing vast areas previously considered unsuitable for any sort of agriculture, such as natural salt flats, desert dunes, and coastal deserts. Over 32 000 km of desert coasts and equally huge areas of sand dunes in interior deserts could theoretically be brought under cultivation with the right halophytes and appropriate new methods of agromanagement. Problems with these new systems will be numerous, yet in the long run halophytes may be the most widely applicable and significant of the three approaches to cultivating saline lands.

A comprehensive survey of the halophytic vascular plants of the world reveals that more than 1250 species of true halophytes in at least 100 families occur in saline coastal environments and desert interior (Aronson in preparation). Increasing attention has been paid to research and development on them since Hugo Boyko (1966, 1968) first proposed utilizing undiluted seawater for irrigation. Of course the fodder value of various halophytic chenopods, notably *Atriplex,* has been observed and exploited for quite some time. Yet it is only within the last ten years that the broader agronomic and economic potential of halophytes as a group has been recognized (Mudie 1974, Chapman 1975b, Somers 1975, O'Leary 1979, Felger 1979).

Most significantly, actual field screening trials of a wide variety of halophytes have been initiated in the past few years in several countries. These trials are not only helping establish criteria for selecting halophytes for specific purposes and regions, but are also increasing our understanding of how to manage them agriculturally (see O'Leary *et al.* 1984a, Pasternak *et al.* 1984a, Gallagher 1984). A great deal of new information is also being accumulated on the actual levels of salinity tolerated by different species and populations within species under controlled conditions, as well as the range of productivity and quality of plant products that can be obtained from cultivated or managed halophytes. We are also learning which halophytes adapt well when introduced and cultivated in environments such as coastal sand dunes irrigated with sea-water. Although a universally accepted definition of what is a halophyte has not yet been achieved, these large-scale screening trials suggest as a practical definition: a halophyte is any plant that can complete its life cycle and reproduce itself under conditions of soil–water salinity of 8–10 dS/mECe (approximately 20% seawater) or more. However, it should be noted that germination and early seedling survival under such saline conditions may be a much more restricted occurrence.

It remains to be seen which avenues of development and which halophytes will prove to be the most useful. The following, then, is intended to suggest the range of uses for which halophytes have been and are being used. Finally, I will attempt to indicate the most promising schemes for halophyte development and utilization in the future.

Past uses of halophytes

MANGROVES: MULTI-PURPOSE TROPICAL HALOPHYTES

Mangroves as a group have probably been put to the largest number of uses of any halophyte. An abundant source of timber, firewood and charcoal, as well as of tannin and many minor products, they have been exploited quite extensively in virtually all the tropical coastal regions and estuaries where they occur. Various authors define mangroves differently, but at least 50 tree and shrub species in 10 families and 14 genera are commonly accepted as mangroves, with a biogeographic centre in the Indo-Pacific region (Teas 1982) and one in northeast Australia (Mepham 1983a; 1983b). The numerically important families are the Rhizophoraceae, Avicenniaceae and Combretaceae; and the major economic genera of world-wide importance are *Avicennia, Brugieria, Conocarpus, Heritiera, Laguncularia* and *Rhizophora*.

As a result of overexploitation, formerly dense mangrove forests have been eliminated or greatly reduced in many areas such as West Africa, northwest India, and the Red Sea. In Indonesia, Malaysia and other areas, mangrove forests are still largely intact and in use. Chapman (1976) has given a detailed account of the utilization of various mangrove species.

The multi-purpose potential of mangroves under managed conditions should also be emphasized: besides the above products, many species have palatable foliage that can be a major source of animal feed. Indeed in many arid coastal areas where mangroves occur sporadically, but little additional vegetation is available, the foliage of such species as *Avicennia marina, Sonneratia alba* and *Rhizophora mangle* has served as camel and cattle feed (Chapman 1976); the leaves of the latter species have even been suggested as human food (Morton 1965). However, the high tannin content of mangrove leaves does present some problems. In Bhavnagar, India, Dr T. Kurian was investigating the integrated fuel and fodder potential of local *Avicennia marina* stands at the time of his unfortunate death in January 1984. Clearly, more work is needed in this area.

SALTBUSHES AND OTHER CHENOPODS

Another group of halophytes important worldwide for forage and fodder in continental and coastal desert regions is the large group of Chenopodiaceae commonly called saltbushes. The most widespread and best known of these is *Atriplex*, with 245 species distributed in virtually all the world's arid and semi-arid regions (Osmond *et al.* 1980). Other important genera in the family are *Chenopodium*, *Kochia, Maireana, Salicornia, Salsola* and *Suaeda.* The potential for increasing and maintaining year-round fodder production in arid and saline regions, as well as in other drastically disturbed lands such as mine tailings, has long been recognized. Various attempts have been made for at least a century to systematically manage wild stands and increase their potential as well as to introduce new species, reseed depleted regions, and otherwise increase the fodder potential of traditional grazing lands. Most of this work has been carried out in Australia, the western USA, and Israel (Nelson 1904, Trumble 1932, Koller *et al.* 1958, Malcolm 1969). More recently, the rate of developmental research on *Atriplex* and related shrubs has accelerated and expanded to other countries. A major reason is the ability of many *Atriplex* species to not only survive but also produce exceptionally large quantities of biomass under extremely saline conditions. This and their high nutritional value to livestock make *Atriplex* and some related taxa among the most promising of all economic halophytes. More details of the current work on the group will be given in the last section.

HALOPHYTES TRADITIONALLY USED FOR FOOD

The only major food crop derived from a halophyte gene pool is the sugar beet, *Beta vulgaris,* a cultigen derived from and probably synonymous with the Euro-Siberian and Mediterranean *B. vulgaris* subsp. *maritima* (= *B. maritima*). Various races are grown as root or leaf crops in temperate zones of the Old and New Worlds, and major commercial breeding programs have been carried out for over a century to increase yields, improve disease resistance, etc. However, the natural salt tolerance of the wild progenitor has mostly been bred out of the commercial varieties.

Eelgrass and Palmer's saltgrass from the Gulf of California in North America are seed plants that grow in ocean water and have been used as major food staples (Felger & McRoy 1975). The seed of eelgrass, *Zostera marina,* was harvested by the Seri Indians each spring from the Sonoran coast of the Gulf, toasted, ground and prepared as a gruel. Usually the ground seed was mixed with sea turtle oil or fat or with the seed of the columnar cactus *Pachycereus pringlei.* The resulting mixture was very rich in protein, fats and carbohydrates (Felger *et al.* 1980). Similarly, Palmer's saltgrass (*Distichlis palmeri*) was regularly collected by the Cocopa and other Yuman Indians who lived along the delta and lower Colorado River at the head of the Gulf of California. The tidal flats of the Colorado Delta once supported thousands of acres of this coarse perennial grass which were frequently inundated by tidal seawater. Its grain is similar in size to wheat (6.5 mm long), and was used by the Cocopa to make bread (Felger 1979). Dr N. Yensen of the Environmental Research Laboratory, University of Arizona has recently selected large-seeded lines of *Distichlis palmeri* which may be similar to the 'improved' strains that were apparently once managed by the Cocopa but are now thought to be extinct (Yensen *et al.* 1981, Felger & Mota-Urbina 1982).

Although the cultivated date palm, *Phoenix dactylifera,* is not a halophyte by the definition given above, there seems good evidence that it was domesticated from wild halophytic populations somewhere in the Middle East. Indeed the cultivated date can be irrigated with moderately brackish water without serious loss of yield.

BARILLA

From the ashes of certain chenopodiaceous halophytes, particularly from *Arthrocnemum, Haloxylon, Salicornia, Salsola* and *Suaeda,* a carbonate of soda called barilla was at one time obtained in large quantities and exported for the soap and glass industries. In the 18th and 19th centuries, the major exporters of barilla were India and North Africa (Watt 1889), as well as the Camargue and adjacent regions of the Mediterranean littoral of France and Spain (Martinez 1983). To this day, *Suaeda fruticosa* and other chenopods are collected from the saline ranns of Rajasthan to make soap and baking soda for local use (Aronson 1982). Of course, sodium carbonate and bicarbonate are now manufactured commercially with the Solvay process.

RUSHES

The stems of a number of Old and New World rushes have been used since ancient times for the manufacture of mats, baskets and cordage. Many species are significantly salt-tolerant, and much infraspecific variability in salt tolerance is known. In the early 1960's Hugo and Elizabeth Boyko initiated the development of a *Juncus*-based industry utilizing highly saline water in the Dead Sea Valley. Despite backing from the Israeli government, however, this project never came to fruition. In recent years, interest in *Juncus* has revived in Egypt and some other countries (Zahran *et al.* 1979).

MINOR FOODS

The sea fennel, *Crithmum maritimum* is collected from wild populations in southern Italy and Greece and used in salads (Franke 1982). This perennial umbellifer is widely distributed along the shores of the Mediterranean and the Black Seas, and along the western shoreline of Europe and Britain. The young leaves and shoots of *Batis maritima, Portulaca oleracea, Tetragonia tetragonoides, Salicornia* spp., and *Suaeda torreyana* have also all been used for salads and pickles in various parts of the world.

Salvadora oleoides and *S. persica* yield edible fruits rich in oil and fat. Both are multi-purpose trees in NW India, providing wood, shade, fodder as well as the fruits. The potential use of *Salvadora* fruits as the source of an industrial non-edible oil has also been reviewed (Gupta & Saxena 1968).

At least one species of *Euterpe* grows in saline marshes in western South America and has been exploited for its edible palm heart (Strudwick pers. comm.).

Major current uses of halophytes

Major current uses of halophytes fall into three categories which are discussed below. Several other uses of selected species are, e.g. for minor food products, medicines, and so-called biological desalinization.

FODDER

As mentioned in the previous section, the fodder value of many halophytic and xerophytic chenopods has been exploited in arid and semi-arid regions for millenia. Within the past hundred years, however, major advances have been made through introduction, selection, proper range management, and new applications. Most of these species combine high biomass and high protein and mineral levels with outstanding adaptive ability to a wide range of environmental stresses. Considerable success has been achieved in Australia, North Africa, the Middle East, the Americas and elsewhere in cultivating chenopods, especially *Atriplex* species in areas subject to total summer drought or on badly salt-affected lands. The management and selection of *Atriplex* spp. is well documented and work to produce improved forms suiting local conditions in many countries is proceeding. Much must still be learned about the proper agromanagement and conditions necessary to maximize yields, and perhaps more important, palatability to livestock (Goodin 1979). Relatively soon, it may be economically feasible to cultivate *Atriplex* alone or in combination with other species with direct seawater irrigation and obtain fodder yields comparable to those of alfalfa grown on sweet water (Pasternak 1982).

Another taxon which has attracted much attention as a fodder plant on soil affected by salinity and water logging is Kalar grass, *Leptochloa fusca* (SEPASAT

1983). In India, Pakistan and elsewhere, this salt and alkali-tolerant grass has been utilized for fodder production on previously unproductive lands.

LAND RECLAMATION

Halophytes of agricultural or horticultural value are also being increasingly used for land reclamation. In the hypersaline Lago Texcoco northeast of Mexico City the saltgrass *Distichlis spicata* has been successfully introduced and irrigated with Mexico City sewage water to stabilize some 6000 ha of the lake bed. Forage for approximately four head of cattle per hectare for nine months annually is thereby produced, and more importantly, stabilization of the seasonally dry lake bed also helps reduce the dust storms that plague Mexico City each dry season (Urbina 1979 and unpublished).

Considerable interest also lies in various salt-tolerant species for revegetation and rehabilitation of surface mines or other drastically disturbed lands in arid and semi-arid regions. The USDA Soil Conservation Service has selected several species for use in the western United States (Thornburg 1982). The Australians have also made great progress.

In Kenya, various halophyte trees and shrubs, e.g. *Conocarpus lancifolius,* have been planted directly in seaside quarries to produce fuel and fodder from otherwise useless land (René Haller, pers, comm.).

One rather intriguing way of using halophytes in land reclamation is the production, from various species of *Plantago,* of a gum or glue used as a mulch binder to hold fibre and seed in place on steep slopes and sandy soils. Thus erosion from wind and rain can be eliminated or at least reduced until new vegetation is re-established (Hubbs Bros. 1982).

LANDSCAPING

Many halophytes have been selected for amenity planting and landscaping in arid and semi-arid regions where saline or brackish water is available more readily or cheaply than fresh water. Several commercial enterprises in the USA and elsewhere specialize in offering such 'plants and seeds for adverse conditions' (Pecoff Bros. 1980, Kimberley Seeds 1980). In the southern part of Israel, salt-tolerant ornamentals have been used for gardening and landscaping for over two decades (Pasternak *et al.* 1984b).

Some of the outstanding species selected and disseminated by The Institutes for Applied Research, Israel are: trees: *Melaleuca halmaturorum, Eucalyptus sargentii, Tamarix* spp., *Conocarpus erectus;* shrubs: *Borrichea frutescens, Clerodendrum inerme, Maireana pyramidata, M. sedifolia, Scaevola* spp.; ground covers: *Crithmum maritimum, Galenia secunda, Drosanthemum floribundum, Halimione portulacoides, Limonium* spp., *Phyla nodiflora, Sesuvium* spp. (Pasternak *et al.* 1977).

One exceptionally attractive halophyte widely used as an ornamental in southern Florida is the silver buttonwood, *Conocarpus erectus* var. *sericeus,* which is a naturally occurring silvery-grey form of the widespread tree or shrub of the mangrove transition zone of south Florida, the West Indies and West Africa (Workman 1980).

The future of economic halophytes

Over the past 30 years research on halophytes has proceeded rapidly from the initial stage of observation that wild halophytes survive and reproduce under conditions of high salinity, through the stage of proposing ways of using halophytes economically, and on to the final stage of testing the economic practicality of various innovative applications of halophytes. Several promising avenues of research and development on halophytes are now continuing in various institutions, and these will no doubt lead to other, as yet unsuspected, possibilities. Four of the more exciting current developmental research efforts are mentioned below.

FODDER

The choice, management, and breeding of vigorous, high-yielding palatable fodder shrubs is becoming more sophisticated (Stutz *et al.* 1975, Stutz 1982, Pasternak 1982, Aronson & Forti 1983). The leading candidate species at this time are *Atriplex barclayana, A. canescens, A. halimus* and *A. nummularia.* New efforts are being made to assemble large germplasm banks of these highly variable taxa to allow for selection and breeding for the desired traits. Since all four species produce plentiful nutritious fodder under adverse conditions, major new applications and large-scale plantings should be undertaken on one or more of them soon.

FUEL

Despite a huge and growing literature on biomass production for fuel in arid lands, the prevailing opinion seems to be that it is still uneconomical to grow any crop anywhere for biomass alone. I contend that irrigation of selected halophytes with cheaply available seawater or other highly saline water unsuitable for any other use may be the exception. Some woody species of *Atriplex,* e.g. *A. lentiformis,* may be suitable for rapid production of kindling and medium-sized firewood with seawater irrigation.

Certain salt-tolerant trees such as *Casuarina, Tamarix,* some *Prosopis* species and even one species of *Eucalyptus* could theoretically be grown for fuel with saline water. In many cases, the multi-purpose use of such biomass species could improve their economic viability. *Atriplex lentiformis,* for example, also yields large quantities of foliage for fodder; *Prosopis* and other salt-tolerant legumes yield copious amounts of edible pods in addition to their high quality wood (Felker *et al.* 1981). Mangroves of course represent a wealth of valuable wood and multi-purpose crop candidates whose potential for development remains to be fully recognized. In addition the mangrove associate nipa palm, *Nypa fruticans,* appears to be a very promising source of fuel alcohol in tropical countries such as Malaysia, New Guinea and the Philippines (Mercer & Hamilton 1984).

BREEDING FOR INCREASED SALT TOLERANCE

The possibility of incorporating significantly greater salt tolerance into certain conventional cereal, vegetable and fodder crops by the use of wild halophytic relatives should gain greater attention in the future. With wheat, barley, tomatoes, cotton, melons, beans, aubergines and several others, this approach has already begun to be recognized and applied (Epstein *et al.* 1980).

OILSEEDS

The main cause for the low palatability of halophytes used for fodder is the accumulation of salts in the vegetative tissues. Significantly, the seeds of these same plants usually have salt levels below 10 per cent by dry weight (O'Leary *et al.* 1984). Often the nutritional value of halophyte seed oils is as high or higher than that of conventional seed oils, and they could perhaps be used by various food industries. Some leading candidates for further investigation are the salt grasses *Distichlis* and the closely related *Jouvea,* and the chenopods *Allenrolfea occidentalis, Salicornia europea, Suaeda fruticosa* and *S. torreyana.* In all these cases, the multi-purpose potential is great.

Conclusion

Increased utilization of halophytes in arid and semi-arid regions seems very likely in the near future. Both the rapidly escalating problems of desertification and salinization and the failure of traditional efforts to obtain water of the quality required by conventional crops in these areas favor this conclusion. However, without adequate funding, both regional and international research and development efforts on economic halophytes and novel systems incorporating them will be unnecessarily delayed.

References

Ahmad, R. and Z. Abdullah 1981. Biomass production of food and fibre crops using highly saline water under desert conditions. In *Biosaline research*, A. San Pietro (ed.): 149–164. New York: Plenum Press.

Aronson, J.A. 1982. *Preliminary report on halophyte and desert plant collections in northwest India, March–April, 1982* (unpublished manuscript).

Aronson, J. in prep. *A global master list of halophytes with emphasis on economic uses past, present and future.*

Aronson, J.A. and M. Forti 1983. The development of new crops for arid and semi-arid lands. *Genetika*, Supplementum III: 265–277.

Boyko, H. (ed.) 1966. *Salinity and aridity. New approaches to old problems.* The Hague: Junk.

Boyko, H. (ed.) 1968. *Saline irrigation for agriculture and forestry.* The Hague: Junk.

Chapman, V.J. 1975a. The salinity problem in general, its importance and distribution with special reference to natural halophytes. In *Plants in saline environments*, A. Poljakoff-Mayber and J. Gale (eds): 7–24. New York: Springer.

Chapman, V.J. 1975b. Terrestrial halophytes as potential food plants. In *Seed-bearing halophytes as food plants*, G.F. Somers (ed.): 75–90. Newark: University of Delaware.

Chapman, V.J. 1976. *Mangrove Vegetation.* Vaduz: J. Cramer.

Dregne, H.E. 1977. Desertification in the United States. *Nature and Resources* XIII: 10–13.

Epstein, E. and J.D. Norlyn 1977. Seawater based crop production: a feasibility study. *Science* 197: 249–251.

Epstein, E., J.D. Norlyn, D.W. Rush, W. Kingsbury, D.B. Kelley, G.A. Cunningham and A.F. Wrona 1980. Saline culture of crops: a genetic approach. *Science* 210: 399–404.

Felger, R.S. 1979. Ancient crops for the 21st century. In *New agricultural crops*, G.A. Ritchie (ed.): 5–20. Boulder, Colorado: Westview.

Felger, R.S. and C.P. McRoy 1975. Seagrasses as potential food plants. In *Seedbearing halophytes as food plants*, G.F. Somers (ed.): 62–68. Newark: University of Delaware.

Felger, R.S. and J.C. Mota-Urbina 1982. Halophytes: new sources of nutrition. In *Biosaline research*, A. San Pietro (ed.): 473–477 New York: Plenum Press.

Felger, R.S., M.B. Moser and E.W. Moser 1980. Seagrasses in Seri Indian Culture. In *Handbook of seagrass biology, an ecosystem perspective*, R.C. Phillips and C.P. McRoy (eds): 260–276. New York: Garland STPM.

Felker, P., G.H. Cannell, P.R. Lark, J.F. Osborn, and P. Nash 1981. *Screening Prosopis (mesquite) for biofuel production on semi-arid lands. Final Report to U.S. AID.* Kingsville: Texas A & I University.

Franke, W. 1982. Vitamin C in sea fennel (*Crithmum maritimum*), an edible wild plant. *Econ. Bot.* 36: 163–165.

Gallagher, J.L. 1985. Improving quality and productivity of halophytic crops grown at seawater salinity. In *Biosaline research – A step into the future*, D. Pasternak and A. San Pietro (eds). The Hague: Junk (in press).

Goodin, J.R. 1979. *Atriplex* as a forage crop for arid lands. In *New agricultural crops*, G.A. Ritchie (ed.): 133–148. Boulder, Colorado: Westview.

Gupta, R.K. and S.K. Saxena 1968. Resource survey of *Salvadora oleoides* Decne. and *S. persica* Linn. for non-edible oil in western Rajasthan. *Trop. Ecol.* 9: 140–152.

Hubbs Bros. 1981. *Muciloid tac, a natural soil binder.* Phoenix, Arizona: Hubbs Bros. Seed Co.

Jacobson, T. and R.M. Adams 1958. Salt and silt in ancient Mesopotamian agriculture. *Science* 128: 1251–1258.

Kelley, D.B., J.D. Norlyn and E. Epstein 1979. Salt-tolerant crops and saline water: resources for arid lands. In *Arid land plant resources*, J.R. Goodin and D.K. Northington (eds): 326–334. Lubbock: Texas Tech University.

Kimberley Seeds Pty. Ltd. *Catalogue* 1980. Osborne Park, Australia.

Koller, D., N.H. Tadmore and D. Hillel 1958. Experiments in the propagation of *Atriplex halimus* L. for desert pasture and soil conservation. *Ktavim* 9: 83–106.

Kurian, T. and E.R.R. Iyengar 1971. Evaluation of seawater tolerance of crop plants. *Ind. J. Agric. Res.* 5: 145–150.

Le Houérou, H.N. 1984. *Salt-tolerant plants of economic value in the Mediterranean basin.* Paper presented at a 'Research for development' Seminar: Forage and fuel production from salt-affected wasteland. Aust. Dev. Assist. Bur. and W. Aust. Dept. Agric. May, 1984. Canberra.

Maas E.V. and G.J. Hoffman 1977. Crop salt tolerance: evaluation of existing data. In *Managing saline water for irrigation*, H.E. Dregne (ed.): 187–198. Lubbock: Texas Tech University.

Malcolm, C.V. 1969. Use of halophytes for forage production on saline wastelands. *J. Aust. Inst. Agric. Sci.* 35: 32–49.

Martinez, M. 1983. Contribution à l'histoire de la fabrication de la soude vegetal à partir des 'salicors'. In *Les zones palustres et le littoral mediterranéen de Marseilles aux Pyrénées*, :143–153. Montpellier: Fédération historique du Languedoc Mediterranéen et du Roussillon.

Massoud, F.I. 1974. *Salinity and alkalinity as soil degradation hazards*, FAO/UNEP Expert Consultation on Soil Degradation. Rome: FAO.

Mepham, R.H. 1983a. Mangrove floras of the southern continents. Part 1. The geographical origin of Indo-Pacific mangrove genera and the development and present status of the Australian mangroves. *S. Af. J. Bot.* 2(1): 1–8.

Mepham, R.H. 1983b. *Mangroves.* Cape Town: Balkema.

Mercer, D.E. and L.S. Hamilton 1984. Mangrove ecosystems: some economic and natural benefits. *Nature and Resources* 20, 2: 14–19.

Morton. J.F. 1965. Can the red mangrove provide food, feed and fertilizer? *Econ. Bot.* 19(2): 113–123.

Mudie, P.J. 1974. The potential economic use of halophytes. In *Ecology of halophytes*, R.J. Reimold and W.H. Queen (eds): 565–596. New York: Academic Press.

Nelson, E. 1904. Native and introduced saltbushes: three season's trials. *Wyoming Expt. Sta. Bull.* 63.

O'Leary, J.W. 1979. Yield potential of halophytes and xerophytes. In *Arid land plant resources*, J.R. Goodin and D.K. Northington (eds): 574–581. Lubbock: Texas Tech University.

O'Leary, J.W. 1984. The role of halophytes in irrigated agriculture. In *Salinity tolerance in plants*, R.C. Staples and G.A. Toenniessen (eds): 397–414. New York: Wiley.

O'Leary, J.W., E.P. Glenn and M.C. Watson 1984. Agricultural production of halophytes irrigated with seawater. In *Biosaline research – A step into the future*, D. Pasternak and A. San Pietro (eds). The Hague: Junk (in press).

Osmond, C.B., O. Bjorkman and D. Anderson 1980. *Physiological processes in plant ecology. Towards a synthesis with Atriplex.* Ecological Studies, Vol. 36, New York: Springer.

Pasternak, D. 1982. Biosaline research in Israel: alternative solutions to a limited fresh water supply. In *Biosaline research*, A. San Pietro (ed.): 39–58. New York: Plenum Press.

Pasternak, D., Y. Ben-Dov and M. Forti 1977. *Recommended list of drought and saline tolerant ornamentals.* Beer-Sheva, Israel: Applied Research Institute.

Pasternak, D., A. Danon, J. Aronson and R. Benjamin 1984a. Seawater agriculture in Israel: research, development and prospects. In *Biosaline research – A step into the future*, D. Pasternak and A. San Pietro (eds). The Hague: Junk (in press).

Pasternak, D., J. Aronson, Y. Ben-Dov, A. Danon, M. Forti, S. Mendlinger and D. Siton 1984b. Development of arid zone crops for the Negev Desert of Israel. (Submitted to *J. Arid Environ.*).

Pecoff Bros. Nursery & Seed Inc. catalogue 1981. *Plants and Seeds for Adverse Environments.* Escondido, California.

Ponnamperuma, F.N. 1977. Varietal tolerance for salt in rice. In *Plant response to salinity and water stress*, Abstracts. Mildura, Australia.

SEPASAT 1983. Draft Dossier: *Leptochloa fusca*. Kew: Royal Botanic Gardens, (mimeo).

Somers, G.F. 1975. *Seed-bearing halophytes as food plants*, DEL–SG–3–75, College of Marine Studies, Univ. of Delaware, Newark, DE.

Stutz, H.C. 1982. *Breeding superior plants for disturbed lands*. Paper presented at Western mined-land rehabilitation research workshop (U.S. Forest service, Fort Collins, Colorado, June 10–11, 1982).

Stutz, H.C., J.M. Melby and G.K. Livingston 1975. Evolutionary studies of *Atriplex*: a relic gigas diploid population of *Atriplex canescens*. *Am. J. Bot.* 62: 236–245.

Tal, M. 1971. Salt tolerance in the wild relatives of the cultivated tomato: response of *Lycopersicon esculentum, L. peruvianum* and *L. esculentum minor* to sodium chloride solution. *Aust. J. Agric. Res.* 22: 631–638.

Teas, H.J. 1982. Saline silviculture. In *Biosaline research,* A. San Pietro (ed.): 369–381. New York: Plenum Press.

Thornburg, A.A. 1982. *Plant materials for use on surface-mined lands in arid and semi-arid regions*. Washington D.C.: USDA Soil Conservation Service.

Trumble, H.C. 1932. Preliminary investigations on cultivation of indigenous saltbushes (*Atriplex* spp.) in an area of winter rainfall and summer drought. *J. CSIR* 5: 152–161.

Urbina, J.C.M. 1979. Determinacion del rango de tolerancia al ensalitramiento por el pasto salado *Distichlis spicata* (L.) Greene, en suelos del lago de Taxcoco. *Ciencia Forestal* 22: 21–44.

White, G.F. 1973. The changing role of water in arid lands. In *Coastal deserts,* D.K. Amiran and A.W. Wilson (eds): 37–44. Tucson: University of Arizona.

Watt, G. 1889. *A dictionary of the economic products of India* (2nd edn 1972) Vol. I: 394–399. Delhi: Cosmo.

Workman, R.W. 1980. *Growing native: native plants for landscape use in coastal south Florida*. Sanibel: Sanibel-Captiva Conservation Foundation, Inc.

Yensen, N.P., M.R. Fontes, E.P. Glenn and R.S. Felger 1981. New salt tolerant crops for the Sonoran Desert. *Desert plants* 3: 111–118.

Zahran, M.A., A.A. Wahid and M.A. El-Demerdash 1979. Economic potentialities of *Juncus* plants. In *Arid land plant resources,* J.R. Goodin and D.K. Northington (eds): 244–260. Lubbock: Texas Tech University.

14   Present and potential economic usages of palms in arid and semi-arid areas

Dennis V. Johnson

3311 Stanford Street, Hyattsville, Maryland 20783, USA

Introduction

Although palms are very conspicuous when they occur in arid and semi-arid areas, the scientific literature concerned with better utilization and development of plant resources of such areas pays relatively limited attention to the Palmae. Even among the numerous articles and books which focus on irrigated agriculture, palms are seldom among the crops discussed. This apparent neglect may be explainable, in part, by two related factors. First, palms are not true xerophytes and could not survive in most arid or semi-arid areas were it not for the presence of underground water sources. Secondly, because the palms exhibit a scattered and highly variable distribution, and form atypical vegetation associations when they do occur, they are unrelated to the characteristic climatic climax vegetation formations. It is anticipated that the forthcoming volume on hot desert and arid shrubland in the 'Ecosystems of the World' series will contribute to a better understanding of the overall role of palms. They are already dealt with quite effectively in the recently-published volume on tropical savannas (Bourlière 1983).

Palm distribution

Among scientists who have studied the origin of palms in arid and semi-arid areas, there is a consensus that their occurrence can be explained either as relict distributions or as a result of human agency, or a combination of both. Substantial fossil records have been found to support the thesis that ancestral palms once formed a basic component of the natural vegetation of geographic areas which today are deserts or semi-deserts. For example, Corner (1966) postulates that the present distribution of *Phoenix*, which extends from West Africa across to East Asia and includes the world's major deserts, is a relict pattern. The *Washingtonia* palms of the desert areas of southwestern North America also survive as relics of an arid subtropical forest of the Eocene period (Axelrod 1950). Therefore current patterns may be representative of taxa which have survived through evolution and become adapted to favorable niches found around springs, along geologic faults where water seepage occurs and on the floodplains of intermittent streams where ground water is present.

In semi-arid areas which grade into savannas, where rainfall is higher but still strongly seasonal, palms grow most commonly along watercourses and may represent the dominant species in gallery forests. There, too, palms may well be an example of relict distributions. Fire is a major environmental factor in savannas and the palms which grow there are as adapted to periodic burning, and to seasonal flooding, as are the tropical grasses which characterize the formation. Palms in arid and semi-arid areas possess the same general adaptations.

In order to determine which palms should be considered in this discussion, a listing of candidate species was first compiled from the literature and their reported distributions then compared to the maps which accompany the study of arid and semi-arid homoclimates by Meigs (1953). Although it would have been desirable to differentiate the candidate species according to the subtypes used by Meigs — extremely arid, arid and semi-arid — the natural distribution of most palms is not known to the degree necessary to permit such a detailed examination. Rather than attempt to be exhaustive, a representative number of species were selected for study which provide examples from North and South America, Africa, Asia and Australia. The eight species chosen are listed in Table 14.1, in approximately descending order of economic importance. Excluded from consideration here, but meriting attention in a broader context, are certain palms which occur on xeric sites outside the areas delimited by Meigs. The guriri palm (*Allagoptera arenaria*), for example, grows on sandy strands along the coast of southeastern Brazil, an area which enjoys a tropical rainy climate (Hodge 1964).

It is interesting to note that the palms in Table 14.1 consist of four Coryphoid genera: *Livistona, Copernicia, Washingtonia* and *Sabal*; one Phoenicoid: *Phoenix*; and three Borassoid: *Borassus, Hyphaene* and *Medemia*. According to Moore (1973), these three groups represent the three least specialized of the 15 groups of palms he recognized, and comprise a major evolutionary line. However, it should be pointed out that *Livistona, Phoenix, Borassus* and *Hyphaena* each has species occurring in more humid areas.

Palm utilization

This section will discuss in sequence the eight palms selected for detailed study. In addition to examining the historical, contemporary and potential uses of each palm, reference will be made to any current development programmes and research.

Table 14.1 Selected palms of arid and semi-arid areas.

Botanical name	Common name	Origin and native habitat
Phoenix dactylifera	date palm	Middle East: subtropical desert and semi-desert, especially oases; unknown wild
Copernicia prunifera	carnauba wax palm	South America: tropical forest into savanna, especially floodplains
Borassus flabellifer	palmyra palm	South Asia: tropical dry forest into savanna; naturalized in Southeast Asia and widely cultivated
Hyphaene thebaica	doum (dum) or gingerbread palm	Africa: desert and semi-desert, especially oases
Sabal mexicana	Mexican palmetto	North America: subtropical and tropical semi-desert
Washingtonia filifera	California fan palm	North America: Southern California desert, especially canyons and oases
Medemia argun	argoun (argun) palm	Africa: desert and semi-desert, especially oases
Livistona mariae	Central Australian cabbage palm	Australia: desert canyons

These eight palms together embrace a full range of conditions as far as both the degree of plant domestication and value of economic products are concerned. It is instructive to use these two criteria to segregate the palms into three groups before proceeding with the discussion.

The first group consists of palms which have undergone genetic improvement through selection, possess recognized named varieties and are of major commercial importance. Only the date palm fulfills these requirements. The second group is made up of palms which are unimproved genetically and either cultivated, or natural stands managed to some degree. Their economic products are of minor significance. The carnauba, palmyra and doum palms fall into this group. The third group consists of palms which are classified as either semi-wild or wild and have very limited, if any, current economic value in terms of their products. This group includes the Mexican palmetto, California fan palm, argoun palm and the Central Australian cabbage palm. The first two of these have some standing as ornamental plants, but that is beyond the criteria used. In the aggregate, the palms considered in this study exemplify what has been achieved in terms of a developed palm such as the date, and the potential which exists among the less-developed species.

Phoenix dactylifera

What is probably the most familiar and most valuable tree of the desert is the date palm. It ranks among the three most important of the world's economic palms, in company with the coconut and oil palm. Thus as far as arid and semi-arid land palms are concerned, the date is a developed tree crop and may serve as an example of what potentially could be achieved with other species. The large volume of published material on the date makes it difficult to deal with in a summary fashion.

A large dioecious palm reaching 30 m in height, the date has a relatively thick trunk covered with persistent leaf bases. Suckers occur at the base of the tree. The leaves are pinnate, average 3–6 m in length and form a crown of more than 100 leaves. A new leaf is produced approximately each month. Upon reaching sexual maturity at about five years, the female palm blossoms once each year. Fruits require some six months to ripen. An average of a dozen inflorescences are borne on a single tree. The date fruit is cylindrical in shape, measures about 5 cm in length and 2 cm in width and contains a single hard seed. Annual fruit yield per adult tree varies from as little as 5 kg to more than 100 kg.

The date palm thrives in hot, dry climates of the subtropical and tropical regions, providing there is an abundant supply of water. One decided advantage possessed by the date is that its needs can be met by brackish water without adverse effects. It can also withstand cold temperatures of -7°C. Essential to high fruit production is the absence of rainfall during the period of pollination. Date growing is covered in detail by Nixon (1951), Munier (1973) and FAO (1982).

According to Carpenter and Ream (1976), 19 species of *Phoenix* are recognized. All are native to tropical or subtropical Africa or South Asia. The taxonomy of the genus is greatly complicated by the fact that the species are all interfertile and produce natural hybrids when they grow or are grown together. The precise origin of *P. dactylifera* cannot be determined with certainty because clear-cut examples of wild palms have never been found, although it is subspontaneous in many areas. On the basis of fossil remains of cultivated dates, Zohary and Spiegel-Roy (1975) place the origin of the date in Mesopotamia at least 6000 years ago. Thus it is one of the oldest of cultivated plants.

The only palm mentioned in the Bible, the date is frequently referred to in the cuneiform sources of the third millenium BC (Landsberger 1967). Of ancient ritual significance, the date palm, its leaves, fruits and inflorescences are represented widely in early monuments and in decorative art. Danthine (1937), in a massive study of the iconography of West Asia, discovered more than 1200 depictions. Fascinating historical accounts of the date palm are to be found in Täckholm and Drar (1950), Goor (1967) and Popenoe (1973). From such accounts it is known that over 4000 years ago date growing had advanced to the point of having sophisticated practices of propagation by separation and transplantation of basal suckers as well as hand pollination of female trees to improve fruit yield. Over the long history of its cultivation and selection, more than 1000 date varieties have been named.

Very probably the total is excessive for it is common to assign local names to varieties which in fact are grown in another location. At present there are considered to
be less than 100 major varieties.

The cultivation of dates evolved in oases and elsewhere in combination with
other crops. Thus there is a long tradition of inter-cropping with citrus, figs and
other perennial crops; within sparsely planted palm groves, annual crops such as
wheat, beans, etc. are grown as well as forage grasses for livestock.

Date varieties are classified as being soft, semi-dry or dry on the basis of moisture content; Medjool, Deglet Noor and Thoory are examples, respectively. A dry,
fresh ripe date contains 75–80% sugar and is a good source of iron and potassium.
Dates eaten with dairy products make an acceptable diet and some desert groups
subsist on that combination for months at a time. Dried dates can be stored almost
indefinitely. Whether fresh or dried, dates are consumed raw, chopped and fried in
butter, boiled and then fried, preserved whole, and made into preserves, syrup or
date butter. Macerated pulp can be made into a beverage with water or milk and
drunk or allowed to ferment. Date wine can be made into vinegar or distilled to
produce spirits. Cull dates commonly are utilized for the latter.

Livestock can be fed date pulp and softened or ground seeds. According to El-
Shurafa *et al.* (1982), on a dry weight basis, date seeds contain 20.64% starch,
4.38% sugars, 6.43% protein and 9.2% oil. Date seed oil is of good quality, but does
not occur in sufficient quantity to justify commercial extraction.

Date growing today is still heavily concentrated in the North African and Near
East region. In 1982 (Table 14.2), world date production totalled 2640 thousand
tons. Currently Iraq and Saudi Arabia are leading producers, each with 15.2 per
cent of the total. They, together with Egypt and Iran, account for 56.7 per cent
of the world's dates. Comparison of the average production for the period 1974–76
with 1982 totals reveals that there has been an increase of 9.8 per cent.

Not reflected in statistics are the numerous other products furnished by this
'tree of life'. Literally hundreds of domestic uses are reported, with the greatest
number occurring in remote parts of the date-growing areas of the Old World. In
brief, the tree can be tapped for sap which can be drunk fresh, reduced to palm
sugar or fermented into palm wine. When a date palm is felled, the palm heart is
extracted and consumed. The leaves are used for thatch and to construct fences.
To both Jew and Christian, the date palm leaf has ritual significance. Leaflets are
woven into mats, baskets, fans, etc. and fiber extracted from the leaves makes a
strong rope. The midribs serve to make crates, simple furniture, chicken coops and
are burned as fuel. Leaf bases also are used as fuel and as floats for fishing nets.
The trunk can be cut into boards and employed in construction as rafters, for
walls and to fashion shutters and doors. Young green dates are strung as necklaces
by children; fresh flower heads are distilled into tarah water for flavouring sherbet;
and date seeds used in traditional medicine. Finally, the date palm is an excellent
provider of shade in the hot deserts and is appreciated for its landscape value.

Table 14.2   Date production, leading countries (1000 tons).

Country	1974–76 (average)	1980	1981	1982
Iraq	409	395	400	400
	(17.1%)	(15.4%)	(15.8%)	(15.2%)
Saudi Arabia	318	342	350	400
	(13.3%)	(13.3%)	(13.8%)	(15.2%)
Egypt	409	446	391	393
	(17.1%)	(17.3%)	(15.5%)	(14.9%)
Iran	300	300	301	301
	(12.5%)	(11.7%)	(11.9%)	(11.4%)
Algeria	162	201	195	207
	(6.8%)	(7.8%)	(7.7%)	(7.9%)
Pakistan	176	202	205	205
	(7.3%)	(7.8%)	(8.1%)	(7.8%)
Sudan	102	113	114	115
	(4.2%)	(4.4%)	(4.5%)	(4.4%)
All others	520	574	574	609
	(21.7%)	(22.3%)	(22.7%)	(23.2%)
World	2396	2573	2530	2630

Source: FAO production year-book, Vol. 36, 1982. Rome.

Despite its long history of use for subsistence purposes, it was not until the 1920s that large-scale orchards began to be planted. A considerable amount of research was done on the date in succeeding years, much of it linked to the establishment of commercial date growing in the United States in the desert of Southern California. There, superior varieties were planted and cultivation techniques developed which have included a degree of mechanization. Tissue culture of date palms also has shown very encouraging recent results (Tisserat 1983). These achievements have tended to focus on commercial growing almost exclusively, with minimal carry-over to subsistence cultivation in the Old World. At present, however, there is increasing concern for improving traditional date growing (see Carpenter 1981). Food technology research also is being directed toward broader utilization of dates. Khatchadourian et al. (1983) investigated date butter and dates in syrup; Shubbar (1981) and Samarawira (1983) date syrup, sugar, alcohol and vinegar; and Al Obaidi and Berry (1981) the production of citric acid from dates.

Internationally, further development of the date palm is benefiting from establishment of the Palm and Dates Research Centre in Baghdad. Begun in 1978 under FAO auspices, the objective is to create a permanent regional center to solve major problems of date production, processing and marketing. Selection of

Baghdad as a site is ideal since that region of Iraq contains the largest known number of good date varieties. Increased interest in the date also is evidenced by the First Symposium on Date Palm held in Saudi Arabia in 1982 (Al-Hassa King Faisal University 1983).

As already mentioned, the United States has a small but profitable date growing region in California. In 1982 there were just over 1900 ha under cultivation, which yielded 20 000 tons of dates valued at 14.5 million dollars. About 75 per cent of the dates grown are of the Deglet Noor variety and 12 per cent of the Medjool. The former are the typical dates sold in American supermarkets and also are exported. The relatively small production of Medjool dates goes to speciality shops or is sold by mail order by the producers within the United States. To promote dates, Indio, California has an annual Date Festival complete with a date queen, camel races and an Arabian Nights pageant.

Copernicia prunifera

The carnauba wax palm is native to the semi-arid portions of northeast Brazil. It occurs in river valleys and in poorly drained coastal zones and tolerates periodic flooding. The palm also is pyrophytic, aided in that regard by adhering leaf bases which help insulate the trunk. The carnauba is a solitary, monoecious tree which grows to a height of 10–15 m. It has large fan-shaped leaves with a heavy coat of cuticular wax and bears a black ovoid fruit containing a small amount of edible mesocarp pulp and a single seed. Fruiting occurs throughout the year. Estimates have placed the number of carnauba palms in the hundreds of millions. Although nearly all are wild, large numbers have been brought under a degree of management to permit the planting of annual crops within the groves, and also livestock grazing. A total of 25 species of *Copernicia* are known (Dahlgren & Glassman 1961, 1963).

In Brazil, the carnauba has been called the 'tree of life' because of the many useful products it provides. Today, wax is the chief economic product. Leaves are cut from the trees, spread in the sun to dry and the wax particles dislodged by hand or the dried leaves are cut up in a portable machine. The collected wax particles are subsequently melted and filtered to remove impurities. Each leaf yields about 5 g of wax. Carnauba wax is amorphous, hard, tough, lustrous and edible. Its chemical and physical properties are: melting point 83–86°C; acid number 3–8; saponification number 72–85; iodine number 8–12; acetyl number 55; specific gravity at 25°C is 0.990–0.999; and hardness ASTM D–5, 1 (McLoud 1970). Brazil is the sole source of carnauba wax and annual production averages about 19 000 tonnes.

The leaves are a secondary economic product and support local cottage industries. They furnish the raw material for weaving hats, baskets, mats, etc., and yield a fiber made into such things as mesh bags and casting fish nets. When the wax is dislodged manually, the leaves afterwards can be used for some of these products. Petioles are used for fencing and to reinforce mud houses. Carnauba trunks are popular for building small bridges, as pilings and as roof beams. Canes and other small objects can be fashioned from the wood. The fruit is edible but the amount of pulp in each fruit is small; livestock eat the fruit, and chopped leaves when forage is scarce. The seed contains about 14 per cent oil (Eckey 1954). When a tree is felled, the heart is taken for food or fed to animals. Starch can be extracted from the pith of the upper trunk and the same part can be tapped for sap. The carnauba seed can be roasted and ground to make ersatz coffee. The roots of this palm have medicinal uses.

Earlier in this century, carnauba palms were formally cultivated on a few plantations. An American company carried out a 25-year research project in Brazil to develop a better waxy palm using carnauba and related species, but abandoned it in the early 1960s because of increasing industrial competition from synthetic waxes. Nevertheless, there are certain uses, such as in medicines and food products, where a substitute has not been found and therefore the carnauba continues as an important source of natural wax. As a managed tree crop it has some potential, especially in association with cattle grazing. Johnson (1972) made a study of the economic botany of this palm.

Borassus flabellifer

Surpassing the date palm, and possibly any other palm, in terms of reported utility is the palmyra palm. It is believed to be native to India, but today has a range which includes all of the drier parts of tropical Asia. The palmyra is the most important of the *Borassus* palms.

B. flabellifer is a solitary, dioecious fan palm which grows to a height of 25–30 m. It has a dense crown of leaves each measuring 1–1.5 m in diameter and bears spherical brown fruits 15–20 cm in diameter. The fruit contains a fleshy fibrous mesocarp and 1–3 fibrous seeds which have a hollow endosperm. About 20 years is required before the palm reaches sexual maturity.

The Tamils of South India are intimately associated with the palmyra, and it is widely cultivated. A Tamil poem dating back to the pre-history of India extols 801 uses of the palm. Blatter (1926) furnishes a long account of the great utility of the palmyra, and points out that first and foremost it is a food source. The sap, or toddy, obtained by tapping the unopened inflorescence, is of primary importance. Tapping does not normally begin until the palm is about 25 years of age but may be continued for up to 30 years. A single inflorescence yields about 2 liters of sap per day, which can be drunk fresh or set aside for a few hours to ferment into palm wine. By carrying the process further, vinegar is produced. The wine may be distilled into spirits. Fresh sap may also be evaporated into palm sugar. Toddy tapping is of such magnitude in South India that a particular caste has become associated with the activity (see PWDS 1978). Not surprisingly, the sap has a number of medicinal uses.

The palmyra fruit is edible and is eaten fresh or made into a preserve. The endosperm of immature seeds may be consumed, as well as the tuberous portion of the first juvenile leaves which are rich in starch. To obtain the latter, seeds are buried at the top of a mound of earth and allowed to germinate. The tubers are boiled and eaten with salt and spices in India (Padmanabhan *et al.* 1978). The leaves and leaf fiber are put to varied uses, one of the most fascinating being the use of leaflets as writing material. In addition to providing excellent thatch, leaves are used to make fans, hats, baskets, mats, etc. From the base of the leaf stalk a fiber is extracted which India exports. These exports amounted to over 9000 tonnes annually in the early 1970s. The major commercial use of palmyra fiber is in manufacturing scrubbing and carpet brushes. Wood from very old palmyra has wide use a rafters, pillars and posts, and can be fashioned into canes, fancy boxes, etc. Given the multitude of uses, the palmyra represents perhaps the best example of what some have called a "palm civilization".

Two other species of *Borassus* share the utility of the palmyra: the African fan palm, *B. aethiopum*, a common savanna palm of the continent, and the lontar palm, *B. sundaicus*, which grows under more humid conditions in Indonesia. Fox (1977) has done a detailed study of the human used of the latter. There is strong evidence to suggest that *B. sundaicus* is a synonym of *B. flabellifer* (J. Dransfield, pers. comm.).

The development potential of the palmyra palm is under investigation by FAO, which has recently published a monograph on the palm (Kovoor 1983).

Hyphaene thebaica

The doum palm shares the desert oasis habitat of subtropical and tropical Africa with the date palm. But unlike the date, it is known in the wild and its natural range includes Africa, the Middle East and West India. It has a relict distribution which is somewhat uncertain because there are about 40 species of *Hyphaene*. Furtado (1970) has made a preliminary study of the taxonomy of this genus. *Hyphaene* is unique in the Palmae in having a dichotomously branching stem, present in most species. Taxonomic studies in process on *Hyphaene* indicates there are no more than ten valid species, and that *H. thebaica* is confined to lands border-ing the Sahara; reports of *Hyphaene* in East Africa appear to refer to *H. compressa* (J. Dransfield, pers. comm.).

Because its presence is most often linked to poorly drained soils with a high water table, the doum palm is an indicator species in Africa. The palm reaches a height of 10 m under favourable conditions and branches to form four to sixteen crowns of fan-shaped leaves. It is dioecious and occurs as a solitary tree or may form pure stands along streams (Fanshawe 1966), and bears large, smooth brown fruits composed of a juicy pulp and a very hard endosperm. Each tree produces about 50 kg of fruit per year. Baboons and elephants eat the fruit.

In Egypt, the doum palm has been cultivated since ancient times and long has been considered to be a sacred tree symbolizing masculine strength. Täckholm and Drar (1950) furnish a lengthy account of its utilization and the depiction of the tree in Egyptian art. The most important product is the fruit, a common wild fruit of the Middle East. The pulp is edible and is described as having a taste suggestive of gingerbread or carob pods. Doum fruits have been found in many Egyptian tombs. In early times the fruits were made into cakes. At present in Egypt they are sold dried and reconstituted and eaten as a paste. Irvine (1961) lists various ways in which different African peoples consume the fruit. The palm heart is also edible.

A doum fruit weighs an average of 20 g. Fruits of the sweet type are composed of, by weight, 22–30% exocarp, 39–42% mesocarp and 34–44% endosperm; those of the bitter type of 30–41% exocarp, 23–27% mesocarp and 34–44% endocarp (Fanshawe 1966). When mature the endosperm has the hardness of vegetable ivory. For that reason the seed has been used for centuries to carve trinkets and other small objects, and in Egypt early in the present century was exploited commercially to make buttons. The endosperm contains about 10% oil (Eckey 1954).

The fan-shaped leaves are used widely for thatch and to weave mats, baskets, bags, etc. Fiber is extracted from the leaves and made into rope. Young leaves of the palm are eaten by camels. In the drier parts of East Africa, *H. compressa* and *H. coriacea* are widely tapped for palm wine (J. Dransfield, pers. comm.). Doum wood is strong and durable and has utility for posts, beams and can be hollowed out for water pipes. It has a chocolate-brown color streaked with black and makes attractive furniture. The fruit pulp and roots are employed in folk medicine (Boulos 1983). In West India, where *H. thebaica* and *H. indica* occur, Rolla (1964) reports that utilization of both palms is the same as described above.

Reportedly, there is currently a program is northern Kenya to develop the doum palm and cultivate it on river banks and in other areas of assured water availability.

Sabal mexicana

The Mexican palmetto, or Texas palmetto in the United States, is an attractive fan palm which may reach 15 m in height. It has large leaves and bears considerable quantities of moderate-sized fruit. The natural range of this semi-desert palm is from Texas to Guatemala. A total of 14 species of *Sabal* are known, all from the New World.

Throughout its range, this palm has a reputation of furnishing a number of useful products. The leaves serve for thatching and can be woven into hats, baskets, etc., and made into brushes. The trunk makes an excellent post and the fruit is sweet and edible and harvested for local markets. The heart is also eaten (Bailey 1944, Bomhard 1950, Latorre & Latorre 1977). It is reported by Girón (1966) that this palm is occasionally cultivated in Guatemala where the leaves support a hat-making industry. *S. mexicana* is frost-hardy and planted as an ornamental in Texas.

Washingtonia filifera

The only palm native to the western United States, the California fan palm is a solitary desert palm with a thick trunk which reaches 25 m in height. It grows in canyons adjacent to and along seepages within the Colorado Desert and extends southward into Mexico. A second species, *W. robusta*, which is taller and thinner, occurs exclusively in Mexico. The natural history of *W. filifera* has been studied in greatest detail by Bailey (1936) and by Vogl and McHargue (1966). Both species are widely cultivated as ornamentals, and occasionally as windbreaks, are frost-hardy and are easily naturalized when introduced to new locations. In Israel, for example, Oppenheimer (1960) states that *W. filifera* has escaped from cultivation.

As far as utility is concerned, these two species are equal. The large leaves can be used for thatching, for making sandals, and fiber extracted for basketry and other uses. Large numbers of small, black edible fruits are borne by this palm; in the wild they are consumed by coyotes and birds. The fruits can be eaten fresh or dried and subsequently ground whole into a meal. The petioles were used by California Indians to make several types of implements. Moran (1978) provides a summary of the utilization.

In its native habitat, the palm stands provide a habitat for birds and other wild-life. Potentially, the fruits could be gathered and fed to livestock when the trees are planted for other purposes.

Medemia argun

The argoun is a very rare palm which grows in association with the date and doum palms in some parts of Africa. It is a solitary, monotypic, dioecious fan palm which grows to a height of 10 m. A palm of great antiquity in Egypt, it was commonly planted in gardens; its fruits have been uncovered in tombs dating to about 2400 BC.

Information about this palm is derived chiefly from Täckholm and Drar (1950) and Boulos (1968). Leaves of the argoun are used for making mats, the fiber extracted to braid rope, and trunk wood employed in construction. The fruits are eaten although they possess a bitter flavour. Feared to have become extinct early in the present century through overexploitation, a small number of palms were found in oases in the Nubian Desert in the 1960s. The argoun deserves attention to make better use of its products and because of its status as an endangered plant (Moore 1977).

Livistona mariae

A relatively rare palm restricted in range to a canyon of the Finke River in the Simpson Desert of Australia, the Central Australian cabbage palm was named after the Australian cabbage palm *L. australis.* Latz (1975) and Lothian (1958 & 1959) have studied *L. mariae* and it is clearly another example of a relict distribution. A slender palm which reaches 25 m in height, it has leaves measuring 2–2.5 m in diameter and produces abundant fruit. Except for the use of leaves for thatch, and possibly the heart as food, the absence of a vernacular name suggests that the aboriginal people of the area made minimal use of this palm.

Other economic uses to which this palm could be put can be inferred by considering the other species of *Livistona* found in Australia, Southeast Asia and Oceania. According to accounts gleaned from Blatter (1926), Burkill (1966) and Whitmore (1973), *Livistona* leaves are used to weave hats, fans, umbrellas, mats and widely employed in thatching houses. The leaf fiber is of good quality and the trunk wood serves for construction purposes. The fruits of several species are edible. The Chinese fan palm, *L. chinensis,* and *L. australis* are both popular ornamentals in more humid areas.

Like the argoun palm, the Central Australian cabbage palm merits some attention so that its products can be utilized and the estimated 3000 trees remaining in the wild protected from extinction.

Conclusion

The eight species discussed in this paper clearly demonstrate the great utility of palms in arid and semi-arid areas. In their native habitats, these palms frequently represent the dominant tree of an agroforestry system, including, as is often the case, annual cropping and livestock grazing on the same plot of land. Such systems should be respected for their success and multiple productivity; this diversity should be maintained and overall productivity increased without any major shift toward monoculture.

Especially in locations where only brackish water is available, palms can be used successfully for revegetation purposes. Palm oases can be conservational by providing natural habitats for birds and other small desert animals. Also, they can function as germ-plasm banks for the species being threatened with extinction in the wild.

Given the scarce resources of arid lands, and the increasing pressure being placed on existing plant resources, palms have over a long history proven to be of considerable material value to humankind, and therefore they deserve to be actively considered in any development project or program involving arid or semi-arid areas.

References

Al-Hassa King Faisal University 1983. *Proceedings of the first symposium on the date palm in Saudi Arabia,* March 23–25 1982. Al-Hassa, Saudi Arabia: King Faisal University, College of Agricultural Sciences & Food.

Al Obaidi, Z.S. and D.R. Berry 1981. Production of citric acid from date syrup. *Date Palm J.* 1: 79–98.

Axelrod, D.I. 1950. *Evolution of desert vegetation in western North America.* Publ. 590. Carnegie Institution of Washington.

Bailey, L.H. 1936. *Washingtonia. Gent. Herb.* 4: 52–82.

Bailey, L.H. 1944. Revision of the American palmettoes. *Gent. Herb.* 6: 366–459.

Blatter, E. 1926. *The palms of British India and Ceylon.* Oxford: Oxford University Press.

Bomhard, M.L. 1950. *Palm trees in the United States.* Agr. Inf. Bull. 22. US Department of Agriculture.

Boulos, L. 1968. The discovery of *Medemia* Palm in the Nubian Desert of Egypt. *Bot. Not.* 121: 117–120.

Boulos, L. 1983. *Medicinal plants of North Africa.* Algonac, Michigan: Reference Publications.

Bourlière, F. (ed.) 1983. *Tropical savannas: ecosystems of the world,* vol. 13. Amsterdam: Elsevier.

Burkill, I.H. 1966. *A dictionary of the economic products of the Malay Peninsula.* 2 vols. Kuala Lumpur: Ministry of Agriculture and Co-operatives.

Carpenter, J.B. 1981. Improvement of traditional date culture. *Date Palm J.* 1: 1–16.

Carpenter, J.B. and C.L. Ream 1976. Date palm breeding, a review. *Ann. Date Grow. Inst.* 53: 25–33.

Corner, E.J.H. 1966. *The natural history of palms.* Berkeley: University of California Press.

Dahlgren, B.E. and S.F. Glassman 1961. A revision of the genus *Copernicia.* 1. South American species. *Gent. Herb.* 9: 1–40.

Dahlgren, B.E. and S.F. Glassman 1963. A revision of the genus *Copernicia.* 2. West Indian species. *Gent. Herb.* 9: 41–232.

Danthine, H. 1937. *Le palmier-dattier et les arbres sacrés dans l'iconographie de l'Asie occidentale ancienne.* 2 vols. Paris: Geuthner.

Eckey, E.W. 1954. *Vegetable fats and oils.* New York: Reinhold.

El-Shurafa, M.Y., H.S. Ahmed and S.E. Abou-Naji 1982. Organic and inorganic constituents of date palm pit (seed). *Date Palm J.* 1: 275–284.

Fanshawe, D.B. 1966. The doum palm – *Hyphaene thebaica* (Del.) Mart. *E. Afr. Agr. For. J.* 32: 108–116.

FAO. 1982. *Date production and protection.* FAO Plant Production and Protection Paper 35. Rome: FAO.

Fox, J.J. 1977. *Harvest of the palm.* Cambridge, Mass.: Harvard University Press.

Furtado, C.X. 1970. Some notes on *Hyphaene. Garcia de Orta* 15: 427–460.

Girón, J.I.A. 1966. *Relación de unos aspectos de la flora util de Guatemala.* 2nd edn. Guatemala City: Tipografia Nacional.

Goor, A. 1967. The history of the date through the ages in the Holy Land. *Econ. Bot.* 21: 320–340.

Hodge, W.H. 1964. A strand palm of southeastern Brazil. *Principes* 8: 55–57.

Irvine, F.R. 1961. *Woody plants of Ghana.* Oxford: Oxford University Press.

Johnson, D. 1972. The carnauba wax palm (*Copernicia prunifera*). *Principes* 16: 16–19; 42–48; 111–114; 128–131.

Khatchadourian, H.A., J.K. Khalil, A.S. Mashadi and W.N. Sawaya 1983. Processing of five major Saudi Arabian date varieties into 'date butter' and 'dates in syrup'. *Date Palm J.* 2: 103–119.

Kovoor, A. 1983. *The palmyrah palm: potential and perspectives.* FAO Plant Production and Protection Paper 52. Rome: FAO.

Landsberger, B. 1967. *The date palm and its by-products according to the cuneiform sources.* Graz: Weidner.

Latorre, D.L. and F.A. Latorre 1977. Plants used by the Mexican Kickapoo Indians. *Econ. Bot.* 31: 340–357.

Latz, P.K. 1975. Notes on the relict palm *Livistona mariae* F. Muell. in Central Australia. *Trans. R. Soc. S. Austr.* 99: 189–195.

Lothian, T.R.N. 1958. The livistonas of Australia, with particular reference to the Central Australian cabbage palm. *Principes* 2: 92–94.

Lothian, T.R.N. 1959. Further notes concerning the Central Australian cabbage palm – *Livistona mariae*. *Principes* 3: 53–63.

McLoud, E.S. 1970. Waxes. *Encyclopedia of chemical technology.* vol. 22, 2nd edn.: 156–173. New York: Wiley.

Meigs, P. 1953. World distribution of arid and semi-arid homoclimates. In *Arid zone hydrology, Unesco Arid Zone Res.* 1: 203–209.

Moore, H.E. Jr. 1973. The major groups of palms and their distribution. *Gent. Herb.* 11: 27–141.

Moore, H.E. Jr. 1977. Endangerment at the specific and generic levels in palms. In *Extinction is forever: the status of threatened and endangered plants of the Americas.* G.T. Prance and T.S. Elias (eds): 267–282. Bronx: New York Botanical Garden.

Moran, R. 1978. Palms in Baja California. *Principes* 22: 47–55.

Munier, P. 1973. *Le palmier-dattier.* Paris: Maisonneuve and Larose.

Nixon, R.W. 1951. The date Palm – 'tree of life' in the subtropical deserts. *Econ. Bot.* 5: 274–301.

Oppenheimer, H.R. 1960. Adaptation to drought: xerophytism. In *Plant-water relationships in arid and semi-arid conditions, Unesco Arid Zone Res.* 15: 105–38.

Padmanabhan, D., M. Gunamani, S. Pushpa Veni and D. Regupathy 1978. Tuberous seedlings of *Borassus flabellifer*. *Principes* 22: 119–126.

Popenoe, P. 1973. *The date palm.* Miami: Field Research Projects.

PWDS. 1978. *Report on the socio-economic conditions of palmyrah workers.* Martandam, Tamil Nadu: Palmyrah Workers' Development Society.

Rolla, S.R. 1964. The doum palms in India. *Principes* 8: 49–54.

Samarawira, I. 1983. Date palm, potential source for refined sugar. *Econ. Bot.* 37: 181–186.

Shubbar, B.H. 1981. Sugar extraction from dates. *Date Palm J.* 1: 61–78.

Täckholm, V. and M. Drar 1950. Palmae. *Flora of Egypt.* vol. 2: 163–355. Cairo: Fouad I University Press.

Tisserat, B. 1983. Tissue culture of date palms. *Principes* 27: 105–117.

Vogl, R.J. and L.T. McHargue 1966. Vegetation of California fan palm oases on the San Andreas fault. *Ecol.* 532–540.

Whitmore, T.C. 1973. *Palms of Malaya.* Oxford: Oxford University Press.

Zohary, D. and P. Spiegel-Roy 1975. Beginnings of fruit growing in the Old World. *Science* 187: 319–327.

15 Plants for conservation of soil and water in arid ecosystems

H. L. Morton

*Arid Land Ecosystems Improvement, Agricultural Research Service,
U.S. Department of Agriculture, 2000 E. Allen Rd, Tucson, AZ 85719, USA*

Introduction

Soil and water are the critical resources of the world, and they are especially critical in arid and semi-arid regions. Unfortunately they have been squandered through unwise and foolish use. Accelerated soil erosion has been a problem associated with dry land farming and grazing management in semi-arid areas of the world for as long as agriculture has been practiced. Successful ancient cropping practices were able to cope with the problem of soil erosion and lowered fertility mainly through a system of shifting cultivation (Leonard 1973). However increasing population pressure has resulted in grazing and cultivation, for all practical purposes, of all suitable land. Severe droughts, overgrazing, removal of grasses, trees and shrubs have caused severe losses of both soil and water in many regions of the world.

This paper will focus on some of the physical processes involved in soil erosion and water loss and on plants known to be useful for their conservation in arid ecosystems. Additional reviews of soil eriosion by wind and water with descriptions of control measures to which the reader can refer for greater detail are by Bennett (1931), FAO (1960 & 1965), and by Cannell and Weeks (1979).

Most technical problems concerning soil and water conservation have been solved or are solvable with continued research effort. The problems which are not so easily solved are those associated with man. As Perry (1978) points out in his discussion of rangeland resources 'Man is an integral part of all world ecosystems, rangelands are no exception. The future of rangeland resources and their use lies in adaptation and innovation in human aspects of rangeland ecosystems in association with their physical and biological limitations and potentials.'

Plants for conservation of soil

The importance of plants and other biotic factors in the formation of soils is well documented and accepted (Buol *et al.* 1980, Jenny 1980). Soil is formed through processes which require thousands of years, consequently soil is not a renewable resource when viewed from a human perspective. Water plays an important role in soil genesis. It is one of the primary ingredients in the soil forming processes; and, of course, it plays an essential role in the establishment, growth and development of plants and other organisms. Water in the form of rainfall is also responsible for much of the soil erosion in arid regions. Natural erosion is a widespread, centuries-old phenomenon. The deposition of alluvial material has been a key factor in the formation of many productive agricultural and forest soils. Unfortunately, activities of man coupled with natural phenomena such as drought, fire and intense rainfall have caused accelerated soil erosion and degradation of many ecosystems (Cloudsley-Thompson 1977, Cox *et al.* 1983). The importance of plants for the conservation of soil is recognized, has been shown experimentally, but is all too frequently not accepted in practice.

WATER EROSION

The more critical point for watershed managers is how to quantify the loss of soil from water erosion and how important plants are in preventing these losses. A universal soil loss equation has been developed by Wischmeier and Smith (1960) to help in quantifying the problem. The equation is $A=RKCLSP$, where A is the average annual soil loss in tons/ha; R is a rainfall-runoff factor; K is a soil-erodibility factor, C is a crop-management or plant cover factor; L and S are factors for slope length and slope steepness, respectively; and P is a conservation practice factor. Base value for the equation is an average annual loss of soil from tilled continuous fallow, expressed for specific field conditions by the product of the term R, K, L and S. This equation is useful for cultivated fields in more humid areas; however, as Renard (1980) has pointed out, sediment yield from water erosion on rangelands of the western United States (and this is also true of many other arid regions) is larger than might be expected. These high yields are due to: (a) the low density of vegetation which inadequately protects the soil against the erosive forces of raindrops and runoff; (b) steep land slopes and low infiltration, which results in high sheer from the water moving over the land surface; (c) high intensity thunderstorms and their associated high kinetic energy, which leads to excessive splash erosion and overland runoff; and (d) steep channel slopes which contain large amounts of alluvium for transport in the runoff. In the Sahel zone of Africa and most arid areas, the C factor is the most important term in the equation because it varies as a function of cover (Fauck 1977). Most mechanical practices are not suitable to the Sahel, thus, the P factor becomes a constant.

Complete plant cover will effectively control soil erosion as is shown by dense forests and grasslands in more mesic environments. Of the many classes of plants, grasses have been shown to be the most effective vegetative cover for retarding runoff (Connor *et al.* 1930). The ability of a crop to decrease runoff is partially due to its coverage of the soil and partially to its removal of water from the soil and subsequent influence on water infiltration rate.

In Sri Lanka Sandanam and Rasasingham (1982) used a mulch of grass residues at 37.65 tons/ha in a new tea planting with soil slopes of 25 to 39 per cent. They also planted smooth crotalaria, (*Crotalaria pallida (= C. mucronata)*), or weeping lovegrass (*Eragrostis curvula*). Soil losses were lowest on the mulched areas and greatest on those planted with smooth crotalaria. Tea yields were lowest on areas planted to weeping lovegrass and highest on areas with bare soil. During dry periods soil moisture was highest on the mulched areas.

Hoffman *et al.* (1983) using simulated rainfall studied the relationship of soil loss to vegetative ground cover in improved pastures dominated by tall-growing smooth bromegrass (*Bromus inermis*), crested wheatgrass (*Agropyron desertorum*), intermediate wheatgrass (*Elymus hispidus* subsp. *hispidus* (= *A. intermedium*)) and alfalfa (*Medicago sativa*), and on native pastures dominated by low-growing blue grama (*Bouteloua gracilis*) and sedges (*Carex* spp.) and taller growing species, e.g. green needlegrass (*Stipa viridula*), needle-and-thread (*S. comata*) and western wheatgrass (*Agropyron smithii*). In both vegetation types they found that live surface cover estimates were poorly related to runoff and soil loss and adequacy of soil protection can be best estimated by measuring per centage of bare soil.

WIND EROSION

Many arid areas, especially those with sandy, unaggregated soils and sparse vegetation, have a high potential for wind erosion. The wind erosion equation was developed to estimate the annual loss of soil from a given area by wind. The equation may also be used to estimate the physical properties of a soil surface, and vegetation or physical barriers needed to reduce soil loss to acceptable levels under different climatic conditions (Woodruff & Siddoway 1965). The relationship between average annual soil loss and five factors influencing wind erosion is given by: $E=f(I',K',C',L'V)$. E is amount of annual soil loss in tons/ha; I' is a soil erodibility index which depends upon the proportion of soil aggregates greater than 0.84 mm diam. and the slope of the surface; K' is soil surface roughness; C' is the climatic factor which depends on wind velocity and surface soil moisture; L' is unsheltered field length along the direction of the prevailing wind; and V is vegetative cover. The loss of soil for specific conditions can be estimated using the wind erosion equation and values published in tables (Skidmore 1965, Skidmore & Woodruff 1968, Woodruff *et al.* 1972).

SOIL LOSSES FROM WIND

Soil erosion from wind has been estimated by Skidmore and Siddoway (1978) as being a serious problem in many parts of the world and is a problem on 30 million ha in the United States where at least 2 million ha are moderately to severely damaged each year. Chepil (1962) studied the erosive force of wind in wind tunnel experiments and concluded that the most erodible soil particles were about 0.1 mm in equivalent diameter, while particles greater than 0.84 mm in equivalent diameter were nonerodible by most winds. Bagnold (1973) in studies conducted in the Sahara and Chepil (1945) working in the United States showed that most of the eroding material carried by wind remained within 30 cm of the surface, thus, barriers formed by low-growing plants are an effective means of preventing wind erosion.

Scientists on the Jornada Experimental Range, New Mexico marked the soil level on stakes in adjacent honey mesquite (*Prosopis glandulosa*) duneland and grassland areas in 1933 and 1935. Soil levels in these areas were remeasured in 1950 and 1955 and again in 1980 (Gibbens *et al.* 1983). These measurements showed extensive soil movement had occurred on the mesquite duneland site, where, maximum removal and deposition was 64.6 and 86.9 cm, respectively in 1980. On the site partially occupied by mesquite dunes in 1933 there was a net loss of 4.6 cm in soil depth and mesquite dunes had completely occupied the site in 1980. On a transect established across the mesquite-duneland ecotone in 1935, there was a net loss in soil depth of 3.4 cm and mesquite dunes had completely occupied the former grassland in 1980. Gross erosion rates on wind deflated areas were equivalent to 69 tonnes/ha/year on the area of large mesquite dunes and 52 tonnes/ha/year on the area partially occupied by mesquite. At the ecotone transect gross erosion rates were 45, 101, and 40 tonnes/ha/year for the 1935–1950, 1950–1955 and 1955–1980 periods, respectively. The yearly losses probably have little real meaning because in the mesquite dunelands soil loss rates will vary widely between years and in the mesquite dune areas, soils which moved by creep and saltation were trapped by the mesquite plants and remain on site. Therefore, gross soil loss can be best approximated by considering losses on areas of deflation only. The loss of silt and clay through suspension by wind erosion from dunelands could have a major influence on soil properties such as water holding and cation exchange capacities. Loss of silt and clay would also reduce the aggregate stability and soil binding properties imparted by these two size fractions.

PREVENTING WATER AND WIND SOIL EROSION

Numerous studies have shown that establishment of vegetation on degraded lands will improve the structure of the soil, increase infiltration rates and reduce soil erosion D'Egidio *et al.* 1981, Chheda *et al.* 1983, Bridge *et al.* 1983). Thus, revegetation is the only practical, permanent method of conserving soils and restoring productivity on degraded rangeland. At this time even revegetation is not considered to be cost effective for vast areas of the arid lands of the world because of the low yields and the high risks associated with revegetation practices. However, if stabilization of soils in rangelands is not accomplished the degradation will continue. For this reason it is essential that communities which base their economies on livestock reduce their herds to levels which can be supported by the plant resources.

Revegetation of arid lands has not always been successful and the knowledge needed to revegetate is still not adequate for many areas. Cox *et al.* (1982) reviewed the literature concerning the restoration of depleted rangelands in the Chihuahuan and Sonoran deserts of the southwestern United States and northern Mexico. During a 92-year period from 1890 to 1982, more than 300 forb, grass and shrub species and accessions were planted at 400 sites. From these studies 83 species have been recommended, but planting has been successful with these species in only one of ten years. Fourteen species showed greatest adaptation and planting has been successful in one of two or three years with these species. Eleven of the adapted species were introduced perennial grasses: kleingrass (*Panicum coloratum*), blue panicgrass (*P. antidotale*), Lehmann lovegrass (*Eragrostis lehmanniana*), boer lovegrass (*E. curvula* var. *conferta*), Wilman lovegrass (*E. superba*) cochise lovegrass (*E. lehmanniana* × *E. trichophora*), johnsongrass (*Sorghum halepense*), sorghum almum (*Sorghum* × *almum*), yellow bluestem (*Bothriochloa ischaemum* var. *ischaemum*), and buffelgrass (*Cenchris ciliaris*). Two of the adapted species were native perennial grasses: green sprangletop (*Leptochloa dubia*) and sideoats grama (*Bouteloua curtipendula*). One adapted species was a native shrub: fourwing saltbush (*Atriplex canescens*). The most widely adapted species were Lehmann lovegrass and boer lovegrass; however, no species was found to be adapted to all sites within either the Sonoran or Chihuahuan Deserts. However, when the area was divided into reference zones based on elevation, precipitation, number of frost-free and annual growing days, it was possible to identify species which were adapted to a reference zone. Based on these criteria the probability of successful establishment should improve markedly.

Cox and Jordan (1983) planted five Lehmann lovegrass accessions, cochise lovegrass, two boer lovegrass accessions, Wilman lovegrass and blue panicgrass in southeastern Arizona. Their data suggest that germination and emergence of these species follow single or groups of closely spaced rainfall events which deposit 20 mm or more. The probability of this type of summer precipitation total and distribution occurring in southeastern Arizona is one in ten (Smith 1956, Sellers & Hill 1974). This probability is typical of other low rainfall areas in the Chihuahuan Desert (Herbel *et al.* 1973). Thus, a successful rangeland seeding could be expected in only one of ten years. This makes seeding an exceedingly costly practice in the low precipitation areas of the Chihuahuan Desert. If the revegetation effort is to be based only on increased livestock numbers, then revegetation efforts should be discontinued (Jordan 1981). Revegetation based on other environmental considerations, such as soil erosion control, wildlife habitat, water quality, and stability of the watershed may justify the high monetary cost.

Frasier *et al.* (1984) showed that sideoats grama seed emerge during an initial wetting period of from one to five days but will suffer high seedling mortality during a 5-day period following the wet, resulting in less than 35 per cent survival rate. Cochise lovegrass was slower to germinate and less susceptible to the effect of the five-day dry period thus from 40 to 60 per cent of the seedlings survived. This information makes it possible to incorporate probabilistic aspects of precipitation and soil water relations into a description of seedling environment.

One of the most important factors in protecting soil from erosion is plant cover (Rauzi 1960). Gifford and Hawkins (1978), found that heavy livestock grazing reduced infiltration rates by about one-half that found on ungrazed areas, while light grazing reduced infiltration rates to about three-fourths that found on ungrazed areas. They conclude that grazing reduces infiltration and sometimes increases erosion. However, they further point out that grazing influences infiltration and erosion only when the total plant cover has been reduced to some critical level, the plant community is altered to the point where it can no longer produce and maintain mulch or litter cover, and soil structural characteristics have been significantly altered, particularly soil porosity (Orr 1975, Rauzi 1960 & 1963). In northern Australia, Bridge *et al.* (1983) found that heavily grazed pastures containing the legumes stylo (*Stylosanthes hamata*) and alyce clover (*Alysicarpus vaginalis*) had more macropore space and higher infiltration rates in the surface soil than a lightly grazed kangaroo grass (*Themeda triandra* (= *T. australis*)) native grassland.

Grazing lands which have been overgrazed and have reduced infiltration rates can be reclaimed (Orr 1975, Busby & Gifford 1981, Wood & Blackburn 1981) if given protection from grazing. Work in pinyon-juniper (*Pinus edulis–Juniperus osteosperma*) sites in southern Utah by Busby and Gifford (1981) showed that infiltration rates on grazed areas, and infiltration increased as the period of rest from grazing increased. Grazing systems and the intensity of grazing apparently had no effect on infiltration rate in areas which were covered with shrub canopy or in shortgrass interspaces between the shrubs (Wood & Blackburn 1981). However, in the midgrass interspaces grazing influenced infiltration rates, with ungrazed areas having the highest rate (16.5 cm/hr); rested, deferred rotation grazing area next (13.9 cm/hr); grazed, deferred rotation grazing area next (13.1 cm/hr); and heavily stocked continuous grazing having the lowest rate (8.1 cm/hr). In northern Australia, Bridge *et al.* (1983) attributed low hydraulic conductivities to high stocking rates and trampling during the wet season. Wood and Blackburn (1981) suggest that if the plant community and soils have not deteriorated too far, deferred rotation grazing may increase infiltration rate.

Apparently soils will recover from compaction and reduced infiltration rates through exclusion of livestock trampling associated with high grazing intensity (Wood & Blackburn 1981, Busby & Gifford 1981). While measurable improvement in soil macro volume can be measured in one or two years, Orr (1975) found that four years was needed for full recovery of a bluegrass (*Poa pratensis*) pasture in South Dakota and Bridge *et al.* (1983) found that macropore space in stylo pasture increased between the third and fourth wet seasons after establishment.

Black and Siddoway (1971) evaluated tall wheatgrass (*Elymus elongatus* (= *Agropyron elongatum*)) for reduction of wind speed, trapping of snow and protection of soils from wind erosion. Two rows of tall wheatgrass seeded in 90 cm rows

9 or 18 metres apart trapped from 86 to 116 per cent as much water from snow as crop-fallow without barriers. Wind speed at height of 30 cm from leeward of one barrier to windward of the next barrier increased from 17 to 70 per cent of open field in the 9 metre barrier spacing and from 19 to 84 per cent in the 18 metre barrier spacing. They found that short barrier spacing offers more protection on most soils than long barrier spacing when wind deviates from the perpendicular.

Bilbro and Fryrear (1983) found that cotton gin residues applied to the blank rows of two times two sowing pattern increased cotton yields and reduced soil losses. Likewise, a 70 cm wide band of Texas panicum (*Panicum texanum*) also reduced soil losses. A single row of weeping lovegrass or kleingrass formed efficient wind barriers without affecting cotton yields, but sunflower (*Helianthus maximiliani*) reduced cotton yields by 38 per cent and was a less effective wind barrier.

Reclaiming dunelands invaded by woody plants is possible by replacing the invading plants with grasses. In separate studies Gould (1982) and Gibbens (1983) treated honey mesquite in sand-dune areas of New Mexico with 2,4,5-T (2-[2,4,5-trichlorophenoxy]-acetic acid), killing from 48 to 84 per cent of the plants. Herbaceous plants became established on the interdunal areas of the treated sand-dune area, primarily mesa dropseed (*Sporobolus flexuosus*). Gibbens (1983) found that perennial grass production was 7-, 8-, and 4-fold greater on sprayed than on unsprayed areas in the first, second and third years following treatment, respectively. In the fourth year, the unsprayed area received 49 mm more precipitation than the sprayed and production of perennial grass was nearly equal on the two treatments. Gould (1982) found the amount of wind-blown particles during the windy season (February to May) was more than 15-fold greater in the unsprayed area than in the sprayed. Their studies illustrate the critical role played by plants and vegetative cover in preventing soil erosion.

Plants for conservation of water

Growing plants use large quantities of water, especially in arid areas where plants are adept at using water rapidly when it becomes available. With the notable and important exceptions of water harvesting and ground water recharge, to me conservation of water means retention and use of the precipitation where it falls on the land. I would like to briefly discuss why plants can survive in arid areas on naturally occurring precipitation by either escaping or enduring prolonged periods of drought.

DROUGHT ESCAPING PLANTS

Plants which escape drought are those which make adjustments in their life cycles so that they complete vegetative and reproductive development during periods of when moisture is available and become dormant during droughts. Ephemeral annuals such as needle grama (*Bouteloua aristidoides*) survive during drought as seeds. In crop plants the date of planting annual species is adjusted to take advantage of the available moisture and varieties have been developed which mature during the moist period. Perennial species such as buffalo gourd (*Cucurbita foetidissima*) and wild rhubarb (*Rumex hymenosepalus*) have a deciduous growth habit and survive during drought as fleshy roots. While these annual and perennial species are able to conserve water they provide little protection to soil.

DROUGHT ENDURING PLANTS

Plants have several mechanisms which enable them to survive moisture stress. One of the most important is the ability to maintain photosynthesis and growth at low cell-water potential. Mexican palo verde (*Parkinsonia aculeata*) has very small leaflets which are shed when moisture stress is encountered. The leaf rachises and stems contain chlorophyll which continue to produce photosynthate even under moisture stress. Another important mechanism is enhanced water uptake from the soil by an extensive root system. Mesquite, alfalfa and long-lived perennial grasses frequently have extensive root systems. Generally plants growing in arid habitats have higher root to above-ground ratios than plants growing in moist habitats. Another mechanism by which plants endure drought is control of water loss from above-ground parts. This is accomplished by reducing the transpirational area. Creosotebush (*Larrea tridentata*) sheds mature leaves when first exposed to moisture stress because the guard cells are unable to produce movement (Warskow 1965, Hull *et al.* 1971). Immature leaves dry out but are capable of resuming growth when moisture conditions improve. When stress continues, creosotebush sheds newly formed stems and if drought is prolonged branches are also abscissed. As a result of this shedding the mature plant has stems which are typically unbranched, smooth and leafless.

Cacti and other succulents endure drought by living on moisture stored in their tissue. The green stems produce photosynthate and the thick, tough, waxy epidermis prevents moisture loss. Stomata close during periods of moisture stress and the plants are essentially closed. Other plants such as honey mesquite (*Prosopis glandulosa* var. *glandulosa*) do not store large quantities of moisture in their tissue but are able to survive periods of water stress by closing stomata and re-opening stomata during periods of low water stress in order to continue photosynthesis (Haas & Dodd 1972).

The efficient use of precipitation by plants is one of the critical keys to their survival in arid ecosystems. Plants must be able to use water when it is available. If the plant is not able to use water efficiently it will not survive or it will not be as vigorous and productive as other plants. The introduction, establishment, spread and dominance of Lehmann lovegrass on pastures in southern Arizona illustrates this point. Precipitation at a study site on the Santa Rita Experimental Range in southern Arizona occurs in a bimodel pattern with approximately 67 per cent occurring in the summer, and 33 per cent occurring in the winter with dry periods in spring and fall (Green & Martin 1967). The native species occurring in this desert grassland are both warm-season and cool-season grasses; however, the cool-season species were eliminated or drastically reduced in density by heavy grazing pressure from domestic livestock. Consequently, the plants which remain are primarily warm-season species which grow when moisture is available in the summer but are dormant when precipitation falls in the winter. Lehmann lovegrass was introduced into southern Arizona in the 1930s where it has become well established on some pastures. It is a warm-season grass which is relatively more cold tolerant than the native grass species and will begin growing earlier in the spring and continue to grow later in the fall than the native, warm-season grasses (Elmi 1981). Thus, it is capable of exploiting winter precipitation which generally is not used by warm-season species. Table 15.1 shows forage production by Lehmann lovegrass and forage production by all species during a ten year period in which Lehmann lovegrass was invading a site in the Santa Rita Experimental Range, Arizona. Forage production increased more than threefold as the per centage of Lehmann lovegrass in the forage increased from 11 to 93 per cent. This increase in total forage production due to Lehmann lovegrass has been documented at other sites in southern Arizona (Martin & Morton 1980) and it is evident that this species is filling a void in the desert grassland left by the removal of cool-season grasses.

Table 15.1 Forage production by Lehmann lovegrass and all grasses during a ten year period on the Santa Rita Experimental Range, Arizona.

| Year | Forage production (kg/ha) | |
	Lehmann lovegrass	Total
1972	40	383
1973	38	195
1974	85	242
1975*	—	—
1976	446	664
1977	526	681
1978*	—	—
1979	863	969
1980	798	819
1981	1334	1432

*Forage production not measured because of livestock grazing in pasture.

References

Bennett, H.H. 1931. Problems of soil erosion in the United States. *Ann. Assoc. Am. Geogr.* 21: 147–170.

Bagnold, R.A. 1973. *The physics of blown sand and desert dunes.* London: Butler and Tanner.

Bilbro, J.D. and D.W. Fryrear 1983. Residue management and cultural practices for a semi-arid region. *J. Soil Water Conserv.* 38(3): 312–314.

Black, A.L. and F.H. Siddoway 1971. Tall wheat grass barriers for soil erosion control and water conservation. *J. Soil Water Conserv.* 26: 107–110.

Bridge, B.J., J.J. Mott, W.H. Winter and R.J. Hartigan 1983. Improvement in soil structure resulting from sown pastures on degraded areas in the dry savanna woodlands of northern Australia. *Aust. J. Soil Res.* 21(1): 83–90.

Buol, S.W., F.D. Hole and R.J. McCracken 1980. *Soil genesis and classification.* 2nd edn. Ames: Iowa State University Press.

Busby, R.E. and G.E. Gifford 1981. Effects of livestock grazing on infiltration and erosion rates measured on chained and unchained Pinyon–Juniper sites in southeastern Utah. *J. Range Manag.* 34: 400–405.

Cannell, G.H. and L.V. Weeks 1979. Erosion and its control in semi-arid regions. In *Agriculture in semi-arid environments,* A.E. Hall, G.H. Cannell and H.W. Lawton (eds): 238–256. Berlin: Springer.

Chepil, W.A. 1945. Dynamics of wind erosion: I. Nature of movement of soil by wind. *Soil Sci.* 60: 305–320.

Chepil, W.A. 1962. A compact rotary sieve and the importance of dry sieving in physical soil analysis. *Soil Sci. Soc. Am. Proc.* 26: 4–6.

Chheda, H.R., O. Babalola, M.A.M. Saleem and M.E. Aken'Ova 1983. Sown grassland as an alternative to shifting cultivation in lowland humid tropics. *Proc. XIV Int. Grassland Congr.* Lexington: Kentucky, June 15–24 1981, J.A. Smith and V.W. Hays (eds): 774–777. Boulder, Colorado; Westview Press.

Cloudsley-Thompson, J.L. 1979. *Man and the biology of arid zones.* Madison: University Park Press.

Connor, A.B., R.E. Dickson and D. Scoates 1930. *Factors influencing runoff and soil erosion.* Texas Exp. Sta. Bull. 411.

Cox, J.R. and G.L. Jordan 1983. Density and production of seeded range grasses in southeastern Arizona (1970–1982). *J. Range Manag.* 36(5): 649–652.

Cox, J.R., H.L. Morton, T.N. Johnsen, Jr., G.L. Jordan, S.C. Martin and L.C. Fierro 1982. *Vegetation restoration in the Chihuahuan and Sonoran deserts of North America.* ARM–W–28, Oakland, California: US Dept. Agric., Agric. Res. Serv.

Cox, J.R., H.L. Morton, J.T. LaBaume and K.G. Renard 1983. Reviving Arizona's rangelands. *J. Soil Water Conser.* 38(4): 342–345.

D'Egidio, B.P., L. Nistri and C. Zanchi 1981. Study of runoff, infiltration and soil losses on pasture under different management in the Appenine mountains using a field rainfall simulator. *Annali Ist. Sper. Studio Difesa Suolo, Firenze.* 12: 245–260.

Elmi, A.A. 1981. *Phenology, root growth and root carbohydrates of Lehmann lovegrass (Eragrostis lehmanniana) in response to grazing.* MS Thesis, University of Arizona, Tucson.

Fauck, R. 1977. Soil erosion in the Sahelian zone of Africa: Its control and its effect on agricultural production. In *Int. symp. on rainfed agriculture in semi-arid regions,* G.H. Cannell (ed.). Riverside: University of California.

FAO 1960. *Soil erosion by wind and measures for its control on agricultural lands.* FAO Agric. Dev. Pap. No. 71. Rome: FAO.

FAO 1965. *Soil erosion by water.* FAO Agr. Dev. Pap. No. 81. Rome: FAO.

Frasier, G.W., D.A. Woolhiser and J.R. Cox 1984. Emergence and seedling survival of two warm-season grasses as influenced by the timing of precipitation: A greenhouse study. *J. Range Manag.* 37(1): 7–11.

Gibbens, R.P. 1983. Grass and forb production on sprayed and nonsprayed mesquite (*Prosopis glandulosa* Torr.) dunelands in south-central New Mexico. *Proc. 14th Int. Grassland Congr.*, Lexington: Kentucky, June 15–24 1981, J.A. Smith and V.W. Hays (eds): 437–440. Boulder, Colorado: Westview Press.

Gibbens, R.P., J.M. Tromble, J.T. Hennessy and M. Cardenas 1983. Soil movement in mesquite dunelands and former grasslands of southern New Mexico from 1933 to 1980. *J. Range Manag.* 36(2): 145.

Gifford, G.F. and R.H. Hawkins 1978. Hydrologic impacts of grazing on infiltration ranges – a critical review. *Water Resour. Res.* 14: 305–313.

Gould, W.L. 1982. Wind erosion curtailed by controlling mesquite. *J. Range Manag.* 35(5): 563–566.

Green, C.R. and S.C. Martin 1967. *An evaluation of precipitation, vegetation and related factors on the Santa Rita Experimental Range*. Technical Reports on the meteorology and climatology of arid regions, No. 17. Tucson: Institute of Atmospheric Physics, The University of Arizona.

Haas, R.H. and J.D. Dodd 1972. Water stress patterns in honey mesquite. *Ecology* 53: 674–680.

Herbel, C.H., G.H. Abernathy, C.C. Yarbrough and D.K. Gardner 1973. Rootplowing and seeding arid rangelands in the southwest. *J. Range Manag.* 26: 193–197.

Hoffman, L., R.E. Ries and J.E. Gilley 1983. Relationship of runoff and soil loss to ground cover of native and reclaimed grazing land. *Agron. J.* 75: 599–602.

Hull, H.M., S.J. Shellhorn and R.E. Saunier 1971. Variations in creosotebush (*Larrea tridentata*) epidermis. *J. Ariz. Acad. Sci.* 6: 196–205.

Jenny, H. 1980. *The soil resource, origin and behaviour*. New York: Springer.

Jordan, G.L. 1981. *Range seeding and brush management*. Arizona Agric. Exp. Sta. Bull. No. T81121.

Leonard, J.N. 1973. *The first farmers*. New York: Time-Life Books.

Martin, S.C. and H.L. Morton 1980. Responses of false mesquite, native grasses and forbs and Lehmann lovegrass after spraying with picloram. *J. Range Manag.* 33(2): 104–106.

Orr, H.K. 1975. Recovery from soil compaction on bluegrass range in the black hills. *Trans. Am. Soc. Agric. Engin.* 18: 1076–1081.

Perry, R.A. 1978. Rangeland Resources: worldwide opportunities and challenges. In *Proc. First Int. Rangeland Congr.*, D.N. Hyder (ed.): 7–9. Denver: Soc. Range Manag.

Rauzi, F. 1960. Plant cover increases water intake rate on rangeland soils. *Crops Soils* 12: 30.

Rauzi, F. 1963. Water intake and plant composition as affected by differential grazing on rangeland. *J. Soil Water Conserv.* 18: 114–116.

Renard, K.G. 1980. Estimating erosion and sediment yield from rangeland. *Proc. Symposium on Watershed Manag.* July 21–23 1980: 164–165. Boise, Idaho: ASCE.

Sandanam, S. and C.C. Rasasingham 1982. Effects of mulching and cover crops on soil erosion and yield of young tea. *Tea Q.* 51(1): 21–26.

Sellers, W.D. and R.H. Hill 1974. *Arizona climate (1931–1972)*. Tucson: University of Arizona Press.

Skidmore, E.L. 1965. Assessing wind erosion forces: directions and relative magnitudes. *Soil Sci. Soc. Am. Proc.* 29: 587–590.

Skidmore, E.L. and F.H. Siddoway 1978. Crop residue requirements to control wind erosion. In *Crop Residue Management Systems*, Chapter 2, Am. Soc. Agron. Spec. Pub. No. 31.

Skidmore, E.L. and N.P. Woodruff 1968. *Wind erosion forces in the United States and their use in predicting soil loss*. USDA Agr. Handbook No. 346.

Smith, H.V. 1956. *The climate of Arizona.* Arizona Exp. Sta. Bull. No. 279.

Warskow, W.L. 1965. *Factors affecting stomatal opening of creosotebush (Larrea tridentata [DC] Cov.) and their combined effect of herbicide activity.* MS Thesis, Tucson: University of Arizona.

Wischmeier, W.H. and D.D. Smith 1960. A universal soil-loss equation to guide conservation farm planning. *Int. Congr. Soil Sci.,* Vol. II: 418–425.

Wood, M.K. and W.H. Blackburn 1981. Grazing systems: Their influence on infiltration rates in the rolling plains of Texas. *J. Range Manag.* 34: 331–335.

Woodruff, N.P., L. Lyles, F.H. Siddoway and D.W. Fryrear 1972. *How to control wind erosion.* ARS–USDA Agric. Inf. Bull. No. 354.

Woodruff, N.P. and F.H. Siddoway 1965. A wind erosion equation. *Soil Sci. Soc. Am. Proc.* 29: 602–608.

16 Nitrogen fixation in arid environments

J. I. Sprent

Department of Biological Sciences, The University, Dundee DD1 4HN, UK

Introduction: the pathways of nitrogen fixation

Plants have two main natural sources of combined nitrogen, nitrate and ammonium ions. In addition, many have access to urea, from either animals or fertilizer. Although it can be absorbed and metabolised by plants, in many soils urea is hydrolysed by the enzyme urease to ammonia and carbon dioxide. Urease is one of a number of enzymes found in a more or less free state in soil, following secretion by, or death of the organisms which produced them (Nannipieri *et al.* 1982, Nor 1982). Nitrate, ammonium and urea form essential parts of the familiar nitrogen cycle, as does the reduction of nitrogen gas (N_2, now often called dinitrogen) to ammonia. This latter process is carried out by an enzyme complex called nitrogenase which occurs only in certain prokaryotic organisms ('primitive' organisms lacking many of the components of cells of more advanced organisms (eukaryotes), i.e. plants, animals and fungi). Many of these nitrogen fixing prokaryotes associate more or less closely with eukaryotic organisms. The biology of the various systems has been described in recent books by Sprent (1979) and Postgate (1982). We shall be considering some examples of significance to arid zones in the next section.

The main stages from the reduction of nitrogen gas to the incorporation of reduced nitrogen into protein are summarised in Fig. 16.1, p. 216 (note that in cells, ammonia is protonated to ammonium which is the form which predominates in cell physiological reactions). For comparison, nitrate reduction and urea hydrolysis are also included. Ammonium assimilation can be seen to offer the simplest route. Additional comparisons amongst these sources of nitrogen will be made in later sections. Further details may be obtained from Postgate (1982) and Sprent (1984a).

CONDITIONS UNDER WHICH NITROGEN FIXATION MAY LIMIT PLANT GROWTH

The nitrogenase reaction is generally considered to be more costly in terms of photosynthate used than the assimilation of ammonium or nitrate (see Sprent *et al.* 1983). When internal physiological considerations, such as pH regulation, are taken

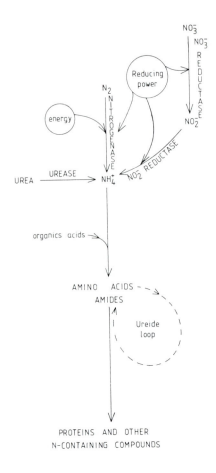

Figure 16.1 Outline of the processes whereby various nitrogen sources are incorporated into plant cells. Those from ammonium to protein are the same for all N sources, apart from the ureide loop which is largely confined to certain tropical legumes and then only when they are fixing nitrogen.

into account the differences may not be very great (Raven 1984). However, the nitrogen fixing reaction is much slower than the alternative processes and more expensive in terms of iron and molybdenum, which may be limiting in some soils (see Table in Sprent & Raven 1985). Thus in evolutionary terms it has been argued that selection pressures only favoured nitrogen fixation when supply of combined nitrogen was the major limiting factor (Sprent & Raven 1985). The same arguments can be applied to arid environments. Here water deficit is the major factor limiting plant growth. The supply of nutrients in the soil may or may not be adequate to support plant growth at the level permitted by the available water. As Felker (1981) and others have observed, many desert areas are dominated by higher plants which do not fix nitrogen. Felker suggested that where there were Caesal-pinioid legumes, e.g. *Parkinsonia* (= *Cercidium*) these may have primitive nodules not recognisable as such. It is, however, at least equally likely that there are suffic-ient nutrients in the soil to make an efficient scavenging root system (see also following sections) the best strategy. Lamont (1982) has given an excellent account of the ways in which plants of Mediterranean climates (hot, dry summers; cool, moist winters) may be adapted for nitrogen uptake. In selecting nitrogen fixing and other plants for arid areas the nutrient content of the soil should be con-sidered in conjunction with the amount and distribution of rainfall as well as soil type (for an introduction to these features, see Heathcote 1983). Some relevant points are summarised in Figure 16.2, p. 218 which is further discussed below.

Nitrogen fixing organisms in arid environments

CYANOBACTERIA

Cyanobacteria, formerly known as blue-green algae, are familiar in aquatic environ-ments (both marine and fresh water), but are also common in soils. All are photosynthetic; many also fix nitrogen. It has recently been shown that *Nostoc flagelliforme* from hot, arid areas of China can survive drought for two years and yet on rewetting it takes up water very rapidly (Scherer *et al.* 1984). It is thus able to recommence metabolic activity, including nitrogen fixation, during brief rainy periods. It may be equally important for its survival that it desiccates rapidly, thus going into a state of low metabolic activity; slow desiccation under hot conditions could result in metabolic imbalance and cell damage. Although the total amounts of nitrogen fixed by cyanobacteria may be small, in a water limited environment they could make a significant contribution.

CYCADS

All extant cycads which have been examined have endosymbiotic cyanobacteria in specialised coralloid roots. Although not significant on a global scale, in certain environments these cycads may form a dominant understory in forests. This is true

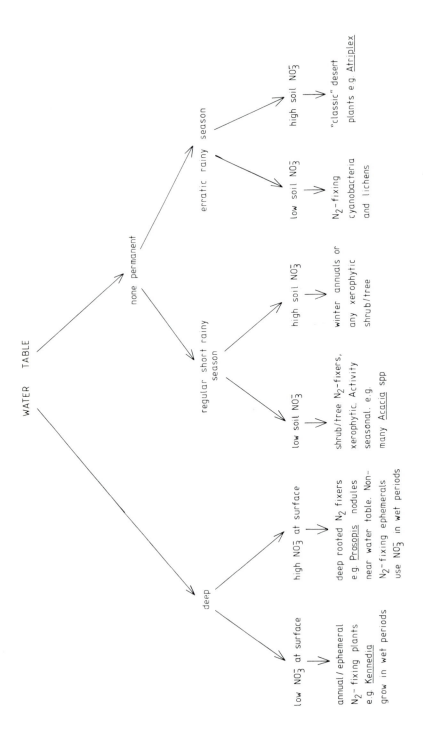

Figure 16.2 A possible scheme showing where and which nitrogen fixing organisms may be of significance in arid environments.

for example in certain *Eucalyptus* forests of arid, fire-prone areas of western Australia; here they contribute significantly to the nitrogen input (Grove *et al.* 1980). The relevant cycad, *Macrozamia riedlei*, has its symbiotic roots well buried beneath the soil surface; the bole of the plant is also protected and can regenerate new leaves following fire. Arid zones subjected to periodic wild- or managed fire are usually low in nitrogen and thus the ability to fix nitrogen is a definite advantage. Table 16.1 gives a nitrogen balance for such a system. Furthermore, cycads have leaves with sharp tips which may help them resist herbivores in the way that many legumes (and other plants) of arid zones use prickles. They are easily raised from seed and are worthy of more consideration for enriching the soil of certain arid areas.

Table 16.1 Possible contribution of nitrogen fixing cycad plants and seedling legumes to a *Eucalyptus marginata* forest in Western Australia following fire. Data and references from Grove *et al.* (1980).

Component	kg/N/ha
Litter in system (mainly from *E. marginata*) 6 years before fire	50
90% loss during fire	47.5
N remaining in litter after fire	2.5
N fixation by *Macrozamia riedlei* in 6 years	35
N fixation by seedling legumes at approximately 3 kg/ha/yr	18
N fixation by free living organisms in litter	9
Total N fixed in 6 years	62

LICHENS

Lichens may, but do not always contain a cyanobacterial component; when they do, they fix nitrogen. Nitrogen fixing lichens are found in varied habitats, e.g. on forest floors in Scandinavia, at the tops of tall trees in the Pacific North West of the USA, on the surfaces of rocks and on soils, including those of deserts. Like some free-living cyanobacteria, many lichens can withstand severe desiccation. On re-watering they may lose nitrogen compounds; the magnitude of and conditions for this are discussed by Millbank (1985). Rogers *et al.* (1966) concluded that the lichen *Collema coccophorus* fixed locally significant amounts of nitrogen. Considering specifically the arid areas, in southern central Australia, using the stable isotope ^{15}N, readers should consult the book edited by Bergersen (1980) for methods of measuring nitrogen fixation.

BACTERIA ASSOCIATED WITH PLANT ROOTS

Much publicity and much research has centred on these associations in recent years. The bacteria involved are described in Postgate (1982). In the context of arid zones, perhaps the most potential importance can be placed on those bacteria which associate with the roots of C_4 grasses, since such grasses are known to have high levels of water use efficiency. They include forage species such as *Digitaria decumbens, Brachiaria decumbens* and many others; also cereals such as maize. The bacterium involved is *Azospirillum lipoferum* (for references to recent work on *Azospirillum,* see Klingmuller 1983). There is no doubt that *Azospirillum* can enter plant roots, can fix nitrogen and that the fixed nitrogen can find its way into the 'host' plant. Whether this nitrogen passes to the host immediately after fixation, as in legume nodules, or only after death and breakdown of the bacterial cells as recently suggested by Okon *et al.* (1983) needs to be further examined. The whole question is complicated by the direct stimulation of plant growth, possibly as a result of hormone production by the bacteria, which is frequently observed. For a balanced assessment of the situation, see Brown (1982). As far as the farmer is concerned, it is the fact that plants grow better when associating with certain bacteria which is important, but if this fact is to be properly exploited, the means by which this better growth is achieved needs to be understood. At present consistent yield improvements are not obtained following inoculation with *Azospirillum* (see Klingmuller 1983).

ACTINORHIZAS

Scattered genera of dicotyledonous plants form root nodules with a filamentous bacterium (actinomycete) *Frankia.* The majority of these genera are temperate and may have colonised these regions following the retreat of glaciers (see Sprent & Raven 1985). However, there is one notable exception, namely the genus *Casuarina* which grows well in hot dry and saline environments where it may fix appreciable amounts of nitrogen as well as provide useful shade and fuelwood. The potential of this genus (now divided into three) is under intensive study at the present time. Recent volumes (Turnbull & Midgley 1983, National Research Council 1984) have reported on the current situation and so *Casuarina,* though important, will not be considered further here.

LEGUMES

As far as direct use is concerned, legumes are currently the most important nitrogen fixing systems, both as forage for stock feed and as a source of pulses.

Taxonomically, the family Leguminosae is generally divided into three subfamilies; Caesalpinioideae, Mimosoideae and Papilionoideae. Some of the relevant features of these subfamilies are summarised in Table 16.2. The classification adopted here is that given in Polhill and Raven (1981).

Table 16.2 Some features of the subfamilies of the Leguminosae relevant to growth in arid areas.

Feature	Caesalpinioideae	Mimosoideae	Papilionoideae
evolution	primitive	branch of Caesalpinioideae	? separate branch of Caesalpinioideae. Advanced
present distribution	mainly tropical few arid	many in arid areas	from equator to arctic
habit	mainly trees	mainly trees and shrubs	many herbaceous and annual spp.
nodulation	sporadic	very common	almost 100%
some nodulated genera important in arid areas	?	*Acacia* *Leucaena*	members of the tribe Mirbelieae, certain genotypes of many genera, e.g. *Arachis*

As far as arid regions are concerned, the Caesalpinioideae are comparatively unimportant. Many of the genera belong to tropical rain forests. Some are found in arid and semi-arid areas, for example the apparently non-nodulating genus *Parkinsonia* dominates much of the Coloradan and Sonoran deserts (Felker 1981). The Mimosoideae includes the very large genus *Acacia, Leucaena* and many other trees and shrubs, together with some annual and herbaceous forms. The plants are often extremely well adapted to arid areas. Nodulation is a more or less universal attribute and the nodules may be perennial. Normally nitrogen fixation is confined to wet periods and is related to shoot growth (Langkamp *et al.* 1982).

The third sub-family, Papilionoideae contains nearly all of the legumes currently exploited for grain and forage purposes. A few are trees, but most are annuals or short lived perennials. They normally nodulate freely and fix considerable quantities of nitrogen. They have radiated into temperate and even arctic areas from the tropics and on the whole are not very drought-tolerant. However, there is a range of tolerance and considerable scope for breeding more drought-tolerant lines. For example *Phaseolus acutifolius* (tepary bean) is more drought-tolerant than the much more widely grown *P. vulgaris* (Parsons & Howe 1984).

Ways of optimising nitrogen fixation by legumes under arid conditions

NODULE STRUCTURE AND DEVELOPMENT

Like leaves, nodules are adapted to optimise gaseous exchange with the environment. However, not only oxygen and carbon dioxide are involved but also nitrogen and hydrogen (Table 16.3). As with leaves, free exchange of gases frequently

Table 16.3 Gases which may diffuse into and out of nodules (see Sprent 1984a for further discussion) and which are involved in metabolic processes.

In	Out
oxygen	carbon dioxide
carbon dioxide	hydrogen
nitrogen	water vapour
? hydrogen	
? water vapour	

involves water loss. Although well vascularised, in dry soils nodules may lose water faster than the vascular system can supply it and hence suffer water stress. This may result in the collapse of cells near the surface, giving impaired diffusion which may, as with stomatal closure, reduce the adverse effects of drought (see Pankhurst & Sprent 1975; Ralston & Imsande 1982). Because nodules vary considerably in general structure, they also vary in their sensitivity to water stress. Three possible strategies are summarised in Table 16.4. Categories A and B well documented (see references in Sprent *et al.* 1983). Category C is more speculative, but is based upon macroscopic and microscopic observations on numerous arid zone plants by the author. Many of these produce suberised (corky) and in some cases lignified (woody) layers at or near the outside. These layers could act as diffusion barriers preventing water loss and also oxygen uptake. This is consistent with a lack of activity in the dry season. In wet periods, such nodules would develop aerating tissue.

ROOTING HABITS

As with any plant, the rooting habit of legumes is an essential part of drought survival. Deep rooting has an obvious advantage in getting nearer any low water table. *Prosopis* is the classic example of this (Felker & Clark 1982), nodules forming deep in soil, near the water table. Here they are less likely to suffer from either desiccation or heat stress. Deep rooting further assists in soil stabilisation as the use

Table 16.4 Structure and drought sensitivity of some types of legume nodule.

Nodule type and example	Structural features	Drought response
A – determinate: soybean *Phaseolus* spp. *Vigna* spp.	limited growth period. No apical meristem	can withstand limited stress (reversible) then die
B – indeterminate: *Trifolium* spp. *Vicia* spp.	longer growth period apical meristem	can withstand more stress than A because growth can recommence after rewatering
C – perennial corky nodules: many trees and shrubs	? gas impermeable corky tissue grows in dry season in place of aerating tissue in wet season	fixation only occurs in wet periods – may survive in inactive state for many months

of the tap rooted, nodulated *Lupinus arboreus* in sand dune reclamation shows (Sprent & Silvester 1973). The root:shoot ratio in legumes, as in other plants, is important in balancing water loss with water uptake.

Root hair production may be of particular importance to legumes because in many, but not all, rhizobial infection is via root hairs (Sprent & Raven 1985). Drought may enhance or depress root hair growth (Sprent 1984b) and stress, in this case salinity, may lead to production of root hairs in plants which do not normally produce them, e.g. *Arachis hypogea* (Sprent & McInroy 1984). Enhancement of root hair growth under nutrient stress, including nitrogen deficiency, has frequently been observed and this may be a way of using root dry weight to the best advantage (Robinson & Rorison 1983). These features can be of significance both in nitrogen fixation and general nutrient (including water) uptake.

USE OF PERENNIAL VERSUS ANNUAL SPECIES

Most grain legumes of economic significance are annuals. The major exception is *Cajanus cajan* which can be grown as a short-lived perennial. Most forage legumes on the other hand are perennial, the major exception being *Trifolium subterraneum*. Annual species are particularly useful where several crops can be grown in the same year, but whereas this is common in the humid tropics, it is generally not possible in arid zones without extensive irrigation. Perennial species have the advantage of minimising soil erosion and also a cover of leaves means that surface soil temperatures do not reach high values. However, the perennial forage species are not particularly drought-tolerant although some cultivars, for example of lucerne (*Medicago sativa*) are far more drought-tolerant than others (Aparicio-Tejo *et al.* 1980). For arid zones, shrubby perennial species may hold out the best prospects. Many legumes come into this category as has been reported elsewhere in this symposium. Normally in any legume, nodulation and nitrogen fixation are closely related to plant growth; the plant produces those nodules which it can support. With increasing age more nitrogen is recycled within the plant (see Sprent 1984c) and the proportion of nitrogen fixed compared with the total biomass falls. The proportion of photosynthetic tissue to total biomass also falls, but the relationship between nodules and photosynthetic area may remain more or less constant for the first few years (Langkamp *et al.* 1982). However, if such plants are harvested (by animal browsing, for green manure, food or any other reason), nitrogen rich material is removed and the nitrogen balance of the plant is upset. Although no direct evidence is available for shrubby species, the situation is analogous to grazing or cutting of forage legumes. This is followed by nodule degeneration, because there is insufficient photosynthate to support nitrogen fixation, and then regrowth. The phase of nodule degeneration results in release of nitrogen into soil which then becomes available for other plants including the legume. If this pattern of nodule loss and regrowth occurs in shrubby legumes then it is likely that regular harvesting will result in more nitrogen being fixed per unit area of land than if the same plants were not harvested. Some evidence that this may be so comes from the study of alley cropping of maize with *Leucaena* (Kang *et al.* 1981).

BREEDING FOR PHYSIOLOGICAL TOLERANCE TO DROUGHT

It must be recognised that, as a family, legumes are not noted for their water use efficiency (water used per unit of carbon dioxide fixed). The xerophytic ones are slow growing and the mesophytic ones drought-intolerant. No legume is known to have either C_4 photosynthesis or Crassulacean acid metabolism, the two major ways in which plants achieve maximum carbon gain for minimum water loss. Thus the span of potential for increasing tolerance to drought is limited to those possible for C_3 plants. If plants are to grow in arid regions outside the rainy season they must maintain high concentrations of solutes within their cells. Some of the options available are given in Table 16.5. All have their disadvantages, but this is true of drought tolerance mechanisms in all plants. Legumes can certainly vary their solute concentration in a way that is related to drought tolerance (Parsons & Howe 1984) and this should be taken into account in breeding programmes.

Table 16.5 Some solutes used by plants to maintain high internal levels of osmoticum to enable them to take up water during dry periods.

Substance	Comments
inorganic ions such as sodium, chloride	generally cheap to acquire, but need to be available. Only in vacuole, because may be toxic to cytoplasmic enzymes. Organic compounds (see below) thus needed for cytoplasm
sugars and related substances	good when excess photosynthate available. How often does this occur in N fixing organisms?
N compounds such as betaines, proline	if the environment is N-limited, these are of doubtful use

OXYGEN AND CARBOHYDRATE LIMITATIONS TO NITROGEN FIXATION

Drought may affect nitrogen fixation in nodules in various ways. In the short term reduced oxygen diffusion may be a problem and this is supplemented later by reduced supplies of photosynthate from the shoot (Sprent 1976). Recent work, (published in a series of papers by Sheehy, Minchin & Witty and discussed by Sprent & Minchin 1985) has suggested that nodules fall into two types, those which are more likely, under well watered conditions, to be oxygen limited and those more likely to be carbohydrate limited. These differences stem from fixed and variable resistances to oxygen diffusion. The problem arises from the fact that nitrogenase is inactivated by oxygen, but that oxygen is required for, and consumed in the processes producing energy for nitrogenase action from the products of photosynthesis. Thus the more photosynthate available, the more oxygen used and the more nitrogen can be fixed. However when less photosynthate is available, less oxygen will be used and if the oxygen diffusion rates cannot be adjusted, oxygen may inactivate nitrogenase, in spite of the oxygen flux control exerted by nodule haemoglobin (Appleby 1984). Therefore if photosynthate supply is likely to be very variable, a variable oxygen diffusion resistance is desirable. Not all nodules have this. The alternative appears to be a fixed diffusion resistance which is to some extent a compromise for 'average' supplies of photosynthate. The chances of oxygen inactivation are minimised, but the nodules are unable to utilise all the products of photosynthesis which may be available under good conditions. Is either of these strategies more or less suited to arid environments? It is difficult to answer this question at present as the nature and location of the diffusion barriers are not fully established. We do not know how (or if) the barriers described above relate to the collapse of aerating cells on the surface of some nodules when they are stressed (Pankhurst & Sprent 1975). Probably a variable resistance would be of most use to nodules with a fluctuating water supply (and hence presumably fluctuating photosynthate supply), whereas plants from a mediterranean climate, where rainfall is more predictable, may be able to manage with a fixed resistance set at a high value during the wet season. Until these factors are understood, breeding legumes for increased photosynthesis *per se* under stressed conditions may not lead to an increase in nitrogen fixing potential.

Comparison of nitrogen fixation with other forms of nitrogen nutrition

COMPARATIVE COSTS OF MAKING NODULES AND ROOTS

Much attention has been devoted to the costs of running nitrogen fixation, far less to the costs of establishing nodules. The options open to legumes are whether to make nodules to fix nitrogen, or to make roots to take up soil nitrogen. We have argued (Sprent & Raven 1985) that to make nodules is only a viable proposition if soil nitrogen is extremely low. There are two main reasons for this. First, that to make nodules instead of roots requires greater investment of nitrogen, because nodules have a higher content of nitrogen (4–7%) than roots (1–3%). Depending on the dimensions used, one can calculate that to grow a single nodule of the type found in *Phaseolus* (type A of Table 16.4) to the stage where nitrogen fixation begins, takes as much nitrogen as to make a piece of root suitable for ion uptake 5–10 mm in length (for assumptions and detailed calculations see Sprent & Raven 1985). Second, that if root tissue is made instead of nodule tissue, not only can soil nitrogen be acquired, but also all other nutrients and water.

COSTS OF RUNNING THE NITROGEN ASSIMILATING MACHINERY

These have generally been considered from the point of view of carbon utilisation. It has been argued that in plants where soil nitrate is transported to leaves for reduction, it is less costly since spare solar energy may be available, especially at low latitudes. However, other costs, notably for pH regulation, are incurred by leaf nitrate reduction which may offset the apparent savings. For an outline of the relevant processes see Sprent (1984a) and for detailed costing see Raven and Smith (1976).

For every CO_2 molecule fixed during photosynthesis there is a cost incurred in terms of water lost during gas exchange and used for carbohydrate production. In this sense, use of photosynthate for any purpose, including nitrogen fixation and ammonium assimilation, is linked to use of water. These costs may be more significant in low latitude, arid areas than carbon costs. Raven (1984 and in preparation) considers the 'water use' costs of different forms of nitrogen nutrition under various stress situations.

ADVANTAGES OF LEGUMES

In this chapter I have tried to put nitrogen fixation in perspective in terms of arid climates. To some extent this has involved stressing the disadvantages of legumes. However, there are two clear advantages which they possess. First, extreme versatility in nitrogen nutrition: they are able to use both nitrate and ammonium and are, indeed, good scavengers of soil nitrogen. Second, in the case of grain legumes in particular, they yield a high protein crop, and one which nutritionally balances that of cereals. The protein yield is virtually unaffected by N source. Thus for nutritional purposes it may be worth growing legumes even when their full nitrogen fixing potential is not expressed.

A further point in favour of certain tropical legumes (Sprent 1980) may be their ability to synthesise ureides (see Figure 16.1). This property is highly correlated with nitrogen fixation; plants assimilating combined nitrogen from the soil have low levels of ureides (see for example, Herridge 1982). The ureides allantoin and allantoic acid are produced in the nodules, transported to the leaves where they are broken down and the reduced nitrogen inserted into amino acids and amides. The advantage of ureides to plants is by no means clear; they are generally a waste product in animals and cannot be used as food. Thus it has been suggested (Sprent 1984a; Wilson & Stinner 1984) that ureide exporting nodules, e.g. *Vigna, Glycine, Phaseolus, Cajanus*, but not *Cicer* among tropical grain legumes, may confer some herbivore, especially insect resistance on nitrogen fixing plants.

General discussion and conclusions

Figure 16.2 attempts to put nitrogen fixing plants into context for various types of arid zone systems. For each situation, both high and low nitrate soils are considered. There are two points to be emphasised here. First, that dry soils often have high levels of soil nitrogen. Because they are aerobic this is usually in the form of nitrate although, shortly after input of organic nitrogen, ammonium may predominate since the bacteria which produce nitrate are more sensitive to drought than those which produce ammonium (Dommergues & Garcia 1980). Second, high levels of nitrate are frequently, but not always associated with high levels of chloride (Charley & McGarity 1964). Deep rooted nitrogen fixing plants such as *Prosopis* may contribute to high nitrate levels in upper layers of soil as a result of mineralisation of their litter.

In the absence of irrigation, legumes are likely to be a reliable crop only when rainfall is predictable, since they do not generally withstand droughts of several years duration. Under the latter conditions nitrogen fixing cyanobacteria and lichens may have a role to play in low nitrogen soils.

Acknowledgements

I should like to thank my colleagues, in particular Professor J.A. Raven, for helpful discussions and Miss S.M. McInroy for technical assistance.

References

Aparticio-Tejo, P.M., M.F. Sanchez-Diaz and J.I. Pena 1980. Nitrogen fixation, stomatal response and transpiration in *Medicago sativa, Trifolium repens* and *T. subterraneum* under water stress and recovery. *Physiol. Plantarum* 48: 1–4.

Appleby, C.A. 1984. Leghemoglobin and *Rhizobium* respiration. *Ann. Rev. Plant Physiol.* 35: 443–478.

Bergersen, F.J. (ed.) 1980. *Methods for evaluating biological nitrogen fixation.* Chichester: Wiley.

Brown, M.E. 1982. Nitrogen fixation by free living bacteria associated with plants – fact or fiction? In *Bacteria and plants*, M.E. Rhodes-Roberts and F.A. Skinner (eds): 25–41. London: Academic Press.

Charley, J.L. and J.W. McGarity 1964. High soil nitrate-levels in patterned saltbrush communities. *Nature* 201: 1351–1352.

Dommergues, Y. and J-L. Garcia 1980. Microbiological considerations. In *West African ecosystems*, T. Rosswall (ed.): 55–72. Stockholm: Royal Swedish Academy of Sciences.

Felker, P. 1981. Use of tree legumes in semi arid regions. *Econ. Bot.* 35: 174–186.

Felker, P. and P.F. Clark 1982. Position of mesquite (*Prosopis* spp.) nodulation and nitrogen fixation (acetylene reduction) in 3-m long phraetophytically simulated soil columns. *Plant Soil* 64: 297–305.

Grove, T.S., A.M. O'Connell and N. Malajczuk 1980. Effects of fire on the growth, nutrient content and rate of nitrogen fixation of the cycad *Macrozamia riedlei. Aust. J. Bot.* 28: 271–281.

Heathcote, R.L. 1983. *The arid lands: their use and abuse.* London: Longmans.

Herridge, D.F. 1982. Relative abundance of ureides ahd nitrates in plant tissues of soybean as a quantitative assay of nitrogen fixed. *Plant Physiol.* 70: 1–6.

Kang, B.T., G.F. Wilson and L. Sipkins 1981. Alley cropping maize (*Zea mays* L.) and leucaena (*Leucaena leucocephala* Lam.) in Southern Nigeria. *Plant Soil* 63: 167–179.

Klingmuller, W. (ed.) 1983. *Azospirillum II: genetics, physiology, ecology.* Basel: Birkhauser.

Lamont, B. 1982. Mechanisms for enhancing nutrient uptake in plants with particular reference to Mediterranean South Africa and Western Australia. *Bot. Rev.* 48: 597–689.

Langkamp, P.J., G.K. Farnell and M.J. Dalling 1982. Nutrient cycling in a stand of *Acacia holosericea* A. Cunn. ex G. Don. I. *Austr. J. Bot.* 30: 87–106.

Millbank, J.W. 1985. The role of lichens in natural ecosystems. *Proc. R. Soc. Edinb. B*, in press.

Nannipieri, P., B. Ceccanti, C. Conti and D. Bianchi 1982. Hydrolases extracted from soil: their properties and activities. *Soil Biol. Biochem.* 14: 257–263.

National Research Council 1984. *Casuarinas: nitrogen-fixing trees for adverse sites.* Washington, D.C.: National Academy Press.

Nor, Y.M. 1982. Soil urease activity and kinetics. *Soil Biol. Biochem.* 14: 63–65.

Okon, Y., P.G. Heytler and R.W.F. Hardy 1983. N_2 fixation by *Azospirillum brasilense* and its incorporation into host *Setaria italica. Appl. Environ. Microbiol.* 46: 694–697.

Pankhurst, C.E. and J.I. Sprent 1975. Surface features of soybean root nodules. *Protoplasma* 85: 85–98.

Parsons, L.R. and T.K. Howe 1984. Effects of water stress on the water relations of *Phaseolus vulgaris* and the drought resistant *Phaseolus acutifolius*. *Physiol. Plantarum* 60: 197–202.

Polhill, R.M. and P.H. Raven (eds) 1981. *Advances in legume systematics*. Kew: Royal Botanic Gardens.

Postgate, J.R. 1982. *The fundamentals of nitrogen fixation*. Cambridge: University Press.

Ralston, E.J. and J. Imsande 1982. Entry of oxygen and nitrogen into intact soybean nodules. *J. Exp. Bot.* 33: 208–214.

Raven, J.A. 1984. The role of membranes in pH regulation: implications for energetics and water use efficiency of higher plants with nitrate as a nitrogen source. *Ann. Proc. Phytochem. Soc. Eur.* 23 A. Boudet *et al.* (eds.) in press.

Raven, J.A. and F.A. Smith 1976. Nitrogen assimilation and transport in vascular land plants in relation to intracellular pH regulation. *New Phytol.* 76: 415–431.

Robinson, D. and I.H. Rorison 1983. Relationships between root morphology and nitrogen availability in a recent theoretical model describing nitrogen uptake from the soil. *Plant Cell Environ.* 6: 641–647.

Rogers, R.W., R.T. Lange and D.J.D. Nicholas 1966. Nitrogen fixation by lichens of arid soil crusts. *Nature* 209: 96–97.

Scherer, S., A. Ernst, T-W. Chen and P. Boger 1984. Rewetting of drought-resistant blue-green algae: time course of water uptake and reappearance of respiration, photosynthesis and nitrogen fixation. *Oecologia* 62: 418–423.

Sprent, J.I. 1976. Water deficits and nitrogen fixating root nodules. In *Water deficits and plant growth*, T.T. Kozloswki (ed.): 291-315. New York: Academic Press.

Sprent, J.I. 1979. *The biology of nitrogen fixing organisms*. Maidenhead: McGraw Hill.

Sprent, J.I. 1980. Root nodule anatomy, type of export product and evolutionary origin in some Leguminosae. *Plant Cell Environ.* 3: 35–43.

Sprent, J.I. 1984a. Nitrogen fixation. In *Advanced plant physiology*, M.B. Wilkins (ed.): 249–276. London: Pitman.

Sprent, J.I. 1984b. Effects of drought and salinity on heterotrophic nitrogen fixing bacteria and on infection of legumes by rhizobia. In *Advances in nitrogen fixation research*, C. Veeger and W.E. Newton (eds): 295–302. The Hague: Nijhoff.

Sprent, J.I. 1984c. Agricultural and horticultural systems: implications for forestry. In *Biological nitrogen fixation in forest ecosystems*, J.C. Gordon and C.T. Wheeler (eds): 213–232. The Hague: Nijhoff.

Sprent, J.I. and S.M. McInroy 1984. Effects of salinity on growth and nodulation of *Arachis hypogea*. In *Advances in nitrogen fixation research*, C. Veeger and W.E. Wheeler (eds): 546. The Hague: Nijhoff.

Sprent, J.I. and F.R. Minchin 1985. *Rhizobium*, nodulation and nitrogen fixation. In *Grain legume crops*, E.R. Roberts and R.J. Summerfield (eds). London: Granada Technical Books, in press.

Sprent, J.I. and J.A. Raven 1985. Evolution of nitrogen fixing symbioses. *Proc. Roy. Soc. Edinb.* B in press.

Sprent, J.I. and W.B. Silvester 1983. Nitrogen fixation by *Lupinus arboreus* grown in the open and under different aged stands of *Pinus radiata*. *New Phytol.* 73: 991–1004.

Sprent, J.I., F.R. Minchin and R.J. Thomas 1983. Environmental effects on the physiology of nodulation and nitrogen fixation. In *Temperate legumes*, D.G. Jones and D.R. Davies (eds): 269–317. London: Pitmans.

Turnbull, J. and S.J. Midgley (eds) 1983. *Proceedings of the Casuarina workshop*. Canberra: CSIRO, Division of Forest Research.

Wilson, K.G. and R.E. Stinner 1984. A potential influence of rhizobium activity on the availability of nitrogen to legume herbivores. *Oecologia* 61: 337–345.

17 The potential for the commercial utilisation of indigenous plants in Botswana

Frank W. Taylor

Veld Products Research, PO Box 2020, Gaborone, Botswana

Introduction

Botswana is a landlocked country lying on the tropic of Capricorn, and it covers an area of 582 000 km². Some 80 per cent of the country is Kalahari Sandveld with the Hardveld running roughly north/south along the eastern border with South Africa. The climate is semi-arid and much of the country is covered by open grass, shrub and tree savanna. The Okavango Delta, *c.* 15 000 km² in extent, is situated in northern Botswana and is an area of floodplains, papyrus sudd and deciduous tree savanna.

The population in 1984 is estimated 1 047 000, of whom 84 per cent live in the rural areas. The population growth rate is three per cent, one of the highest in the world. Less than 10 per cent of the population has cash employment in the country, with over 80 per cent of the paid employment being in the urban areas. People subsist from agriculture (when there is no drought) and from remittances from relatives working in towns or as migrant workers in South Africa. In 1974 it was noted that 46 per cent of the total population lived below the poverty datum line.

The Government of Botswana has been very concerned about the plight of the rural dwellers and the urgent need for income-generation in rural areas to improve the quality of life. As a direct result of earlier work undertaken by the author, the Government commissioned him to undertake a 20 month research project to identify indigenous plant products with potential commercial value.

The terms of reference of this project included a country-wide survey to identify indigenous plants with commercial potential, to recommend harvesting methods and rates and to give recommendations for monitoring any ecological impact caused by sustained harvesting. In addition, to develop simple, local, value-adding technologies to process the plant material, to research domestic and international marketing channels and to indicate potential economic impact on the gatherers. Nutritional analyses of wild foods were also to be undertaken and recommendations made concerning a long term monitoring system to measure the nutritional levels of the participating gatherers and their neighbours as wild foods become a commercial resource. Finally to recommend a plan for one or more pilot projects around the country to implement the findings of the survey.

Methodology

The areas around 83 villages spread throughout the country were selected for the research. The areas are representative of all the vegetation types and sub-types in Botswana. Data was obtained detailing density and geographical distribution of selected species which had indications of economic potential. The method used to obtain the data was a quantitative analysis of eight hectares by means of four belt transects each measuring 1000 X 20 meters.

Concurrent with the above survey, economically promising wild foods were harvested and experimentally processed, where necessary, to make them commercially acceptable. Records also were made of the yields harvested from each species.

Market research was undertaken in person on domestic and international markets for products which included wild foods, herbal medicines, and dried florist materials.

Some 25 people were involved in the research project of whom only the author and Helen Moss, a botanist, were involved for the full duration of the project. Four consultant botanists, one consultant entomologist and one food technologist provided short-term inputs.

The project started in April 1981, at the beginning of what has since become a fairly severe drought, thus in some of the areas investigated the majority of the annual species did not appear. The results have been published in three volumes, Taylor (1983) whose circulation is restricted, and Moss with Taylor (1983), the third volume being an appendix to volume 2 and contains maps showing the distribution and densities of 111 species at the 83 locations surveyed.

Selection of plants with economic potential

A wide range of plants were investigated for commercial potential but due to time constraints only the most promising were researched in depth. Species were categorised according to their potential use, i.e. food, medicine, etc. As the research progressed some plants were added to the list and others removed as investigation revealed their potential or lack of it.

For each species considered suitable for commercial harvesting, the following data were collected: local names; description of plant and parts to be used; habitat and distribution; current utilisation by humans, domestic livestock and wildlife; possible commercial uses; analyses, where relevant; yield; resource potential; harvest period; storage; potential impact of current use; impact of proposed use; recommendations for reducing the impact; and factors limiting commercial exploitation.

FOOD PLANTS

About 100 traditional food plants were investigated in the literature and/or in the field, and out of this large number only four were selected as having a clear commercial potential if easily accessible resources were to be exploited (Moss with Taylor 1983). A further 50 were identified as having some potential but each had certain constraints usually indicated by low plant populations, or the possibility of detrimental ecological disturbance, etc. The four food plants in the final selection are *Sclerocarya caffra, Vangueria infausta, Grewia retinervis* and *Citrullus lanatus.*

Sclerocarya caffra (morula)
Sclerocarya caffra is a medium to large tree; male and female flowers are usually borne on separate trees. Its fruits are extremely popular and are usually eaten freshly picked from the tree. The yield and composition of the fruit are shown in Tables 17.1 and 17.2, p.234.

Table 17.1 1982 Yields of *Sclerocarya caffra* in Gabane area (Moss with Taylor 1983).

Tree No.	Date started	Date ended	Total days	Average drop/day	Total fruits
1	12.2.82	29.3.82	46	1452	66 822
2	19.2.82	5.4.82	46	1369	62 991
3	16.2.82	2.4.82	46	1359	62 544
4	15.2.82	29.3.82	43	1074	46 213
5	22.2.82	19.4.82	57	671	38 280
6	15.2.82	8.4.82	52	671	34 905
7	15.2.82	26.3.82	40	552	22 084
8	15.2.82	15.3.82	29	760	22 058
9	15.2.82	1.3.82	15	1308	19 623
10	15.2.82	22.3.82	36	484	17 445
				Mean value	39 297

The date the fruit started to drop was not recorded in all cases but the accumulated fruit was counted.

Table 17.2 Composition of mature fruits of *Sclerocarya caffra* (Quin 1959).

Average weight per fruit	17.99 g
Peel per fruit	41.0 %
Nut per fruit	52.51%
Edible flesh per fruit	6.47%

A potent morula 'beer' is prepared by pounding the ripe fruit to remove the kernel and then adding 50 per cent water to the pulp. After 24 hours it is strained and ready to drink; potency increases with age. It is usually sold for 10 thebe per mug. According to Watt and Breyer-Brandwijk (1962) a potent spirit is distilled from the beer in Mozambique.

Some local housewives produce a delicious red jelly from the fruit, although this is not a traditional local practice.

When the fruiting season is over the children gather the kernels. Many are also harvested from the kraals, where the whole, clean kernel is passed through the digestive system of the animals. However, care is taken never to collect from the kraals after it has rained since the kernels then become tainted. The kernels are usually eaten fresh but are sometimes cooked with porridge. An analysis of the fruit and kernel is given in Table 17.3.

Table 17.3 Analysis of *Sclerocarya caffra* fruit and kernel/100 g edible portion (Wehmeyer 1966).

	Fruit	Kernel
Moisture	91.7 g	4.0 g
Protein	0.5 g	30.9 g
Fat	0.1 g	57.0 g
Ash	0.2 g	4.2 g
Fibre	0.5 g	2.4 g
Carbohydrate (by difference)	7.0 g	1.5 g
Calcium	6.2 mg	106.0 mg
Magnesium	10.5 mg	467.0 mg
Phosphorus	8.7 mg	836.0 mg
Iron	0.10 mg	0.42 mg
Copper	0.04 mg	1.99 mg
Sodium	trace	3.38 mg
Potassium	54.8 mg	637.0 mg
Thiamin	0.03 mg	0.04 mg
Riboflavin	0.05 mg	0.12 mg
Nicotinic acid	0.25 mg	0.71 mg
Vitamin C	67.9 mg	—

Fox (1966) and Wehmeyer (1980) both recorded far higher vitamin C levels, viz. 200 mg/100 g. The 67.9 mg recorded by Wehmeyer in 1966 may be of the pure juice only, excluding the skin and flesh.

The Zulu women are reported to boil the crushed kernels in water to produce a light non-drying oil which is used for cosmetic purposes (Watt & Breyer-Brandwijk 1962, Coates Palgrave & Drummond 1977). The Venda also use this oil to preserve meat, which is lightly steamed over water, gradually moistened with oil and then stored in a cool place. It is claimed that such meat can be preserved for up to a year. They also pound the kernels with lean, sinewless meat, which is then shaped into cakes and subsequently dried and stored (Palmer & Pitman 1972).

Although Fox and Norwood Young (1982) report that the Pedi cook the leaves as a relish, there is no evidence for this practice in Botswana.

The bark is widely used medicinally, especially for fever and diarrhoea (Watt & Breyer-Brandwijk 1962); it is also used as a preventative as well as a cure for malaria (Coates Palgrave & Drummond 1976).

The light, white wood is used for carving traditional wooden bowls and other craft items.

The possible commercial uses for the morula fruit are: fruit juice, dried fruit rolls, jams and jellies, kernels, fresh fruit and alcohol production.

Vangueria infausta (mmilo, wild medlar)

Vangueria infausta is a shrub or small tree. Its fruit may be eaten fresh from the tree or sun dried and eaten months later after soaking in cold water for 12 hours (warm water accelerates the process); the fruit tastes almost the same as when fresh. The average weight of the fruit is 1349 g; after shade drying for five to six days they were dry enough for bulk storage having lost 27.7 per cent by weight. After six months of open air storage the average weight of the fruit was 610 g, giving a total weight loss of 54.8 per cent. The maximum estimated number of fruits per tree was 1800, the lowest 3, with a mean value of 253. The composition and analysis of the fruit is shown in Table 17.4, p. 236.

Although there is no record of the Batswana using the fruit as a relish, the Pedi and Swazis are known to eat the pulp with porridge and the Pedi to drink it with milk (Quin 1959, Fox & Norwood Young 1982). In South Africa farmers' wives add sugar and milk to the pulp to make a substitute for apple sauce; the pulp can also be used for puddings (Coates Palgrave & Drummond 1977). In the western Transvaal a brandy is distilled from the fruits (Palmer & Pitman 1977).

The roots are used in folk medicine for treating chest complaints and menstrual problems. An infusion of the leaf is also used for toothache and a leaf poultice for swellings (Watt & Breyer-Brandwijk 1962).

The possible commercial uses of the fruit are as whole, dried fruit to be sold in the towns, flavouring agent in dried fruit rolls and for alcohol production. There is considerable superstition associated with the mmilo. Coates Palgrave and Drummond (1977) noted that many Africans will not use the wood even as firewood although they have no qualms about collecting the fruit. However enquiries have revealed that the growing of the tree in agricultural areas is not likely to give rise to superstitious concern amongst the majority of the inhabitants.

Table 17.4 Composition and analysis of *Vangueria infausta* fruit (Quin 1959).

Fresh fruit:	Peel per fruit	10.0%
	Stones per fruit	42.5%
	Pulp per fruit	47.5%
Pulp content:	Moisture	51.8%
	Crude protein (dry basis)	2.6%
	Citric acid	2.3%
	Vitamin C (dry basis)	7.6 mg/100 g

Grewia retinervis (motsotsojane)

The fruit of the multi-stemmed *Grewia retinervis* is extremely popular. It can be either eaten raw or dried and stored. The fruit can also be crushed and the resulting juice fermented to make an alcoholic drink. An analysis of the fruit is given in Table 17.5.

Table 17.5 Analysis of *Grewia retinervis* fruits/100 g.

	Fresh (1)	Fresh (2)	Dried (3)
Moisture	10.6 g	11.7 g	6.9 g
Ash	3.7 g	3.6 g	2.4 g
Protein	5.4 g	4.2 g	4.2 g
Fat	0.2 g	0.1 g	0.3 g
Fibre	12.6 g	24.7 g	18.7 g
Carbohydrate	67.5 g	55.7 g	67.5 g
Calcium	157 mg	133 mg	137 mg
Magnesium	172 mg	229 mg	166 mg
Iron	4.7 mg	4.26 mg	2.86 mg
Copper	0.4 mg	1.29 mg	0.25 mg
Sodium	31.0 mg	4.8 mg	4.48 mg
Potassium	655 mg	1061 mg	853 mg
Phosphorus	–	60 mg	54 mg
Zinc	1.6 mg	1.42 mg	0.2 mg
β carotene	–	–	–
Thiamin	–	0.029 mg	trace
Riboflavin	–	0.042 mg	0.433 mg
Nicotinic acid	–	1.916 mg	1.305 mg
Vitamin C	–	–	–
Energy (kJ)	293	1010	290

There is a substantial demand in the urban areas for both the fresh and dried fruit.

Citrullus lanatus (tsamma, marotse or mokatse)

There are both wild and cultivated forms of *Citrullus lanatus* growing in Botswana. The wild form, known as tsamma, has small, round fruits *c.* 15 cm in diameter, with a white flesh which has a high water content. At times the tsamma is the sole means of survival for the Bushmen of the Central Kalahari, who not only use it as a source of water but also collect, crush and cook the seeds as a porridge. Wild life in the Kalahari also depend on the tsamma.

The cultivated forms are larger, with orange, yellow or white flesh. The marotse are eaten by both the Batswana and their livestock whereas the mokatse is cultivated solely for the livestock. An analysis of both the wild and cultivated forms is given in Table 17.6. It is not clear whether the analysis of the seeds refers to wild or cultivated forms.

Table 17.6 Analyses of *Citrullus lanatus* fruit and seeds/100g (Wehmeyer pers. comm. 1982*, Wehmeyer 1968**, Fox 1966***).

	Melon* cult. Zululand	Melon** Tsamma	Seeds***
Moisture	86.9 g	97.9 g	4.7 g
Ash	1.1 g	0.5 g	–
Protein	1.3 g	0.1 g	22.1 g
Fat	0.1 g	0.02 g	31.5 g
Fibre	2.3 g	0.6 g	–
Carbohydrate	–	0.9 g	10.7 g
Zinc	0.14 mg	–	–
Calcium	17.20 mg	18.9 mg	180 mg
Iron	0.30 mg	0.19 mg	18 mg
Magnesium	37.2 mg	9.1 mg	–
Sodium	9.7 mg	0.72 mg	–
Potassium	286 mg	68.6 mg	–
Copper	6.30 mg	0.06 mg	–
Phosphorus	12.60 mg	–	480 mg
Thiamin	0.22 mg	0.20 mg	–
Nicotinic acid	0.36 mg	0.05 mg	–
Vitamin C	21.7 mg	3.8 mg	–
Energy (kJ)	–	4.0	415

A number of named cultivars are recognised by the Batswana. Some are cooked and eaten or cut into slices and dried (lengagale), to be cooked and eaten later in the year. The dried slices are especially sweet. The small, white skinned mmamonwaana melon with very sweet flesh is the only one to be eaten both raw and cooked. The seeds of those cultivars with plump seeds are roasted and eaten, those with thinner seeds are ground and cooked to be eaten with other foods. Sesowane is a small melon with a dark green skin, of which only the very soft seeds are eaten after roasting.

The cultivated melons generally produce 10–15 melons/plant; melons can average 10 kg. No figures are available for the tsamma but casual observation suggests about five melons per plant.

The commercial possibilities of the marotse melon are for jams, preserves, dried fruit rools, lengangale, melon juice and melon seed meal/snacks.

It is interesting to note that the *Citrullus lanatus* is a traditionally cultivated melon which is the same species as the wild melon. This melon has been selectively grown over a long period in Africa and now is a distinctive cultivar. Of the above food plants, probably the *Sclerocarya caffra* and *Citrullus lanatus* could be of the most interest to the international researchers due to their high yield even in drought years.

There are two notable exceptions to the above list, these being *Ricinodendron rautanenii* (mongongo) and *Tylosema esculentum* (marama). Both of these produce fruit of a high nutritional value, but as the dense plant populations of these two species generally are far from habitation, they could not be economically exploited.

It was considered that most of the wild foods could be absorbed on the local urban markets, but the dried fruit rolls of the *Sclerocarya caffra* and *Citrullus lanatus* were shown to have very strong prospects on the export market in the health food shops in particular. The employment potential for a pilot project could be several hundred people harvesting on a piecework basis.

MEDICINAL PLANTS

Botswana is endowed with a wide variety of traditional medicinal plants but in the final analysis it was recommended that only one, *Harpagophytum procumbens* (sengaparile or grapple) be commercially exploited, subject to certain very strict controls. After discussions with traditional herbalists and officials of the Natal Parks Board in South Africa, it was decided not to recommend that the lucrative traditional medicine market be developed in Botswana. In spite of supposedly strict controls in South Africa many species of plants have totally disappeared from certain areas because their roots or bark were highly valued for the medicinal properties. Botswana should not follow suit.

H. procumbens, which grows almost exclusively in the Kalahari Sandveld, has a vertical primary storage root from which radiates horizontal secondary roots, on each of which one or more enlarged storage roots may be found. Each of these can weigh up to four kg. It is intended that the woody primary root is left intact in the ground to regenerate and maintain populations, while only the secondary storage roots are harvested. These are cut up, dried and exported mainly to Germany where it is processed into tablets, capsules, etc., and re-exported around the world. It is reputed to be one of the world's most effective remedies for arthritis, as well as diabetes, gall bladder, liver and kidney complaints, and is highly researched. It is a registered drug in several European countries.

The commercial harvesting of this plant started in Namibia over 25 years ago and in recent years up to 1000 tons fresh secondary roots have been dug up annually. Due to diminishing supplies caused by serious over-harvesting, Namibia buyers encouraged harvesting across the border into Botswana. In order to protect the resource the Botswana authorities have instituted strict controls, mainly through the use of permits and a limited harvesting season. Considerable interest was experienced on the international market for this product.

The survey has revealed that the plant occurs throughout most of the Sandveld, seldom in great concentrations, but certainly in economic quantities. The employment potential for a pilot project to develop this resource could be *c.* 200 more than the 500 or so people presently involved in harvesting.

FLORIST MATERIALS

A wide range of dried florist material of potential economic value was collected. Altogether some 45 species were found to have dried components of some aesthetic interest, these usually were pods, fruits, stems with dried inflorescences, etc.

The most important export markets were found to be in western Europe and the USA which require container-loads, mostly at marginal prices.

The employment potential for the pilot project could be 500–1000 people in harvesting.

MISCELLANEOUS CRAFTS

Crafts generally depend on the use of indigenous raw materials; the most important craft in Botswana is basketmaking, which utilises the leaves of the *Hyphaene ventricosa* for weaving, and *Berchemia discolor* (mutsintsila or bird plum) and *Euclea divinorum* (motlhakola or magic quarri) roots to dye the leaves. About 4000 people in the Okavango Delta area are very dependent on this export-orientated craft but the severe pressure on the trees has caused them to become depleting resources. A government Development Officer requested an investigation and two experts were brought in to study the problem. Their report included a number of recommendations regarding management of the resources through improved harvesting methods and controls, propagation, utilisation of alternative species, etc.

The only other major craft which provides income on any scale is woodcarving, involving some 300–400 men in eastern Botswana. In this case the mature *Colophospermum mopane* (mopane) trees were under increasing pressure: ten years ago the fairly extensive mopane forests in the woodcarving area had good stands of mature trees but by 1982 woodcarvers were having to cycle up to 60 km to find suitable trees to cut down. Although mopane coppices extensively, it is a slow grower.

Alternative crafts were investigated in order to take the pressure off the threatened species. These included papyrus 'parchment' from *Cyperus papyrus,* and engraved gourds from Cucurbitaceae.

Markets for crafts were not investigated as this activity is undertaken by a parastatal agency.

OTHER WOOD USES

The potential for charcoal production, which is not a traditional activity, was also investigated. In certain areas of low human population there are substantial quantities of dead trees which are almost totally unutilised. In other areas where there is severe bush encroachment by *Dichrostachys cinerea* and *Acacia tortilis,* it may be possible to aid recovery of the grazing through the removal of these invasive species for commercial charcoal production.

The market for charcoal would be both domestic and export to South Africa. The employment potential for the pilot could be about 50 people.

Discussion

The over-riding concern of both Government and those involved in the research was that whatever steps were to be taken to commercially exploit Botswana's indigenous plant resources, these should include sufficient practical controls to ensure sustained harvesting without unduly disturbing the ecology or adversely affecting the social, health or nutrition of the people in the affected areas. It was acknowledged that there are very strong pressures for full exploitation of resources which could lead to their eventual demise.

Current conservation methods, both traditional and legislative were examined, and specific recommendations made in those areas which would be affected by the harvesting of recommended plant products, i.e. a system of controls and monitoring. The controls would centre upon quotas of products from areas, harvesting methods, etc., while the monitoring would be an on-going operation involving both the ecology and the socio-economic impact on the people. The monitoring would be undertaken by ecologists and health/social welfare people.

It was proposed that overall control and co-ordination of the commercial utilisation of indigenous plant products should be by an independent body, an 'Eco Committee', with its members drawn from relevant government departments and one or two representatives from the implementing agency(s). One ecologist should work full-time on the monitoring and allied research, and be directly responsible to the Committee.

It was recognised that the duration of the research project was so short (20 months) that many of the results were of an indicative nature only. There are many variables to be considered such as drought, erratic rainfall, plant metabolic cycles, etc. Considerable research will be necessary for many years to come before relatively firm conclusions may be reached concerning the commercial utilisation of many indigenous plants. Therefore, it was recommended that a system of adaptive management be used to facilitate commercial utilisation. This entails commercial harvesting which is monitored as an experimental research programme, thereby establishing a data base from which a sound long-term management plan could be developed.

Government has undertaken to ensure that the indigenous plant resources will be protected from over-exploitation and have prepared legislation with regard to the harvesting and marketing of *Harpagophytum procumbens* roots as recommended in the report. In addition new areas will not be opened up for the harvesting of these roots until more is known about both propagation of the plant and the regeneration of the primary roots after the secondary roots have been harvested.

Government is planning a national workshop to discuss the implementation of the project's findings. In the meantime two organisations have started to adopt some of the proposed projects.

Conclusion

The constraints and dangers to wide-scale commercial utilisation of indigenous plant products are considerable but it is considered that, with due caution and careful oversight by the proposed Eco-Committee, the original objective can be achieved. This could provide cash incomes to substantial numbers of rural people and at the same time protect their long-term interests by the wise utilisation of the plant resources.

References

Coates Palgrave, K. and R.B. Drummond 1977. *Trees of Southern Africa.* Cape Town: Struik.

Fox, F.W. 1966. *Studies on the chemical composition of food commonly used in Southern Africa.* Johannesburg: South African Institute for Medical Research.

Fox, F.W. and M.E. Norwood Young 1982. *Food from the veld – edible wild plants of Southern Africa.* Johannesburg: Delta Books.

Moss, H. with F.W. Taylor 1983. *The potential for commercial utilisation of indigenous plants and insects in Botswana. Vol. II: The resources and its management.* Gaborone, Botswana: Ministry of Commerce and Industry.

Palmer, E. and N. Pitman 1972. *The trees of Southern Africa.* 3 vols. Cape Town: Balkema.

Quin, P.J. 1959. *Food and feeding habits of the Pedi.* Johannesburg: Witwatersrand University Press.

Taylor, F.W. 1983. *The potential for commercial utilisation of indigenous plants and insects in Botswana. Vol. I: The resource and its commercial utilisation.* Gaborone, Botswana: Ministry of Commerce and Industry.

Watt, J.M. and M.G. Breyer-Brandwijk 1962. *The medicinal and poisonous plants of Southern and Eastern Africa,* 2nd edn. Edinburgh: Livingstone.

Wehmeyer, A.A. 1966. The nutrient composition of some edible wild fruit found in the Transvaal. *S. Afr. Med. J.* 40: 1102–1104.

Wehmeyer, A.A. 1980. *Some Botswana veld foods which could possibly be used on a wider scale. Food processing opportunities in Botswana.* Gaborone, Botswana: Ministry of Commerce and Industry.

18 Ecodevelopment of arid lands in India with non-agricultural economic plants – a holistic approach

T. N. Khoshoo and G. V. Subrahmanyam

Department of Environment, Bikaner House, Shahjahan Road, New Delhi 110011, India

Introduction

Nearly 14 per cent of the earth's population (*c.* 630 million people) live in arid and semi-arid areas, which constitute about one third of the earth's surface (Eckholm & Brown 1977). At one time in their history, many of these regions supported magnificent civilizations with prosperous towns. Today these have retrogressed to levels which are well below subsistence and their emaciated children wander around aimlessly, suffering from disease and malnutrition. The basic question before this Conference is whether plant scientists can do something to ameliorate the situation by husbanding these lands, making them more productive, where possible, with species that are traditional and indigenous to the region as well as by introducing species that are exotic yet are relevant to a particular situation.

A very large majority of the rural poor in arid regions meet their daily household needs through biomass or biomass-based products such as food, fuel (firewood, cowdung, crop wastes), fodder, fertilizer (organic manure, forest litter, leaf mulch), building materials (poles, thatch), medicinal herbs, etc. Water, another crucial product for survival, is not biomass by itself, but its availability is closely related to the level of biomass available in the surrounding environment and is thus biomass-dependent.

Biomass resources also provide a range of raw materials for traditional occupations and crafts and are, thus, a major source of employment. Reduction in the available biomass and biomass-related products for household needs, e.g. fuel, fodder and water, has meant considerably increased physical labour for women and more and more child labour, which negates the objectives of family planning (Agarwal 1984). Biomass scarcity reduces nutrition to livestock. Thus there is increasing numbers but not quality of livestock. The result is a situation where there is less than subsistence living, which often becomes a vicious cycle and one that is very difficult to break. Furthermore, it is not only the type of biomass that is important for meeting basic household needs but also the diversity necessary to meet the people's equally varied requirements. Thus many different kinds of plants are necessary to increase the overall bioproduction. Such plants need to be integrated into the Arid Land Ecosystem of which human beings are an integral part. The challenge before plant scientists lies in devising land-use plans and bioproduction systems which will meet the total and integral needs of fuel, fodder, food and other basic requirements of the diverse sections of a rural population. Thus, it requires a determined holistic concept, rather than a compartmentalised or sectorial approach, which alone can bring redress to the arid regions.

Arid lands

Nearly one third of the earth's surface or *c.* 49 million km², excluding polar regions, has been classified as arid land according to their soil types (acidic, leached soils of the tropics and subtropics; neutral to alkaline, calcareous soils, etc.), vegetation (drought escaping ephemerals, drought evading, drought resistant) and climate (low and highly variable precipitation, high evaporation, high temperatures and lower humidity) (Dregne 1983, Dalsted & Myers 1983). Arid land can be subdivided into three categories: semi-arid, arid and extremely arid. In a semi-arid region, precipitation is sufficient in most seasons but droughts occur so frequently that the practice of rainfed farming is a somewhat hazardous venture. An arid zone is one in which rainfed farming is entirely marginal, successful in some years and hopelessly unsuccessful in others. However, an extremely arid region is one in which average annual precipitation is insufficient to sustain any agricultural crops, and hence ordinarily rainfed farming is impossible (Hillel 1982).

ARID LANDS IN INDIA

Arid lands in India are characterised by low and highly variable precipitation ranging from 150–500 mm/year, high evaporation and extremes of temperatures. These regions can be grouped into two categories: hot arid and cold arid. The hot arid regions of India cover *c.* 300 000 km² which include many portions of western Rajasthan and Gujarat, south-western portions of the Punjab and Haryana, and parts of Karnataka and Andhra Pradesh. The cold arid areas of India cover *c.* 100 000 km² in Ladakh area of Jammu & Kashmir State in Northwest India (Das Gupta 1977).

The dynamics of arid land ecosystems is controlled by the low availability of renewable resources, e.g. water, soil, flora and fauna, which is responsible for their low primary productivity. This reduction in productivity or degradation of arid land environments has been termed 'desertification' or 'desertization'. The major causes of desertification are: overgrazing, lack of erosion control, improper water management, cutting of trees for fuel, crusting of soil surface and burning of rangeland.

In spite of the adverse prevailing environment, the arid region maintains a diverse flora which is well adapted to the xeric conditions and composed of as many as 700 plant species, including 107 grasses (Mann & Dhir 1984). The vegetation of by far the largest portion of the hot semi-arid area consists of species of grasses such as *Dichanthium, Lasiurus, Phragmites* and *Saccharum* while that of the cold semi-arid region is covered by species of *Themeda* and *Arundinella* and temperate alpine cover types.

The major crops of hot semi-arid regions are cotton, groundnut, sugarcane, maize, bajra (pearl millet), jowar (sorghum), wheat, rice, millets, pulses, oil seeds and fruit crops, e.g. wild jujube (*Ziziphus nummularia*).

Large amounts of top-feed from plants like *Ziziphus nummularia, Prosopis cineraria, Calligonum polygonoides, Capparis decidua* and *Salvadora oleoides* are exploited by domesticated animals (Saxena 1977). According to 1966 livestock census, India has 175 million head of cattle, 52 million buffalo, 64 million goats and 42 million sheep. This amounts to nearly 11 per cent of the total livestock population of the world (Singh & Joshi 1979). Livestock population in arid areas of the country, is 22.89 million, of which cattle, buffaloes, sheep, goats, camels and other species comprise 7.27, 2.67, 6.90, 5.14, 0.65 and 0.23 millions respectively (Ahuja 1977).

Among the fairly abundant large wild mammals are the Indian gazelle (*Gazella gazella bennetti*), blue bull (*Boselaphus tragocamelus*) and the Indian black buck (*Antilope cervicapra rajputanae*). Those which were formerly abundant but are now either restricted to certain favourable pockets or are fast disappearing, include the caracal (*Lynx caracal*), wild ass (*Equus hemionus khur*), tiger (*Panthera tigris*), leopard (*Panthera pardus*), chital (*Cervus axis*), etc.

Among the world's arid regions, the Indian arid zones are characterised by a long history of human settlement and by high population pressure. The population in the arid tracts of India is over 19 millions, with an average density of 61 persons/km² as against 3 persons/km² in most other deserts of the world (Mann & Dhir 1984).

In recent years world attention has been increasingly drawn to the arid lands because of the exceptional riches beneath their sands, and also on account of the ecological deterioration that has culminated in human disaster. Due to over-grazing by livestock and the activities of rodent pests, the pastures and browse shrubs have reached a stage of severe degradation where the plant diversity has dropped from 112–414 spp/ha to 6–18 spp/ha (Mann 1977). Soils have also been simultaneously degraded. To check over-exploitation, it is essential to first prepare an inventory of the natural resources of the arid and semi-arid lands. Soil erosion in desert areas can be checked by planting wind breaks, creating shelter belts and covering and stabilising sandy tracts and dunes with adapted grasses and shrubs, e.g. *Lasiurus, Ziziphus, Crotalaria,* etc. The technology of sand dune stabilization is a contributing factor to efficient land management including the planting of photo-insensitive, aridity-resistant and remunerative tree species, e.g. *Acacia* and other genera.

The fuel requirement of the relatively dense population of the Indian arid zone is usually met from the locally available brushwood and trees. Most of the fuelwood comes from *Prosopis cineraria, P. juliflora, Acacia* spp. and some local shrubs.

In order to stabilise the ecological conditions and ameliorate the lot of people who live in such inhospitable habitats, a systematic survey of possible species which could meet the escalating demands for the growing human and livestock population with regard to food, fodder, fuel, fertilizer, fibre, shelter, medicare, etc. needs to be undertaken. For such a survey both local species and exotics, which can be grown without any adverse environmental impact, need to be considered.

Non-agricultural economic plants

A few definite examples of non-agricultural economic plants which have been fully researched in our country with reference to their utilisation as food, multipurpose trees and medicinal and industrial plants are dealt with here. The present paper consolidates information on such species regarding current level of knowledge, socio-economic potential, together with cultivation practices and future research needs.

FOOD PLANTS

Obtaining enough high-quality protein in their diet is a problem for many of the world's people, especially those who rarely eat meat and animal products and must, therefore, get most of their nourishment from grains and vegetables. Even when their calorie-supply is adequate, these people may suffer from malnutrition because cereal proteins do not supply all the amino acids that the body needs. One way to attack this problem is to breed new strains of cereals containing proteins of enhanced nutritive value. Another approach is to look for alternative nutritious food crops which are not being exploited at present as a food source but have potential for development. The amaranths are among such potential food plants that have been neglected. These may be used for three principal purposes, namely, protein-rich grain, carotene-rich leaves and as ornamentals.

Amaranthus (Amaranthaceae)

Distribution

The grain species (*Amaranthus* section *Amaranthus*) are native to Central and South America. Although these were grown in historical times in Mexico, India is the only country where grain amaranths are now cultivated on a wide scale. Grain amaranths are mainly used as subsidiary food and are usually cultivated in small patches throughout India. In Hamachal Pradesh and other areas in the Himalayan belt, they are grown as a regular crop and constitute a staple food in place of bread-wheat. In India the most commonly cultivated species is *A. hypochondriacus.*

The vegetable types (*Amaranthus* section *Blitopsis*), whose leaves are very rich in carotene content, are essentially native to India and south-east Asia.

Botany

Typical grain amaranths, e.g. *A. hypochondriacus, A. cruentus* and *A. caudatus* are sturdy plants with erect habit, well developed main stems and large terminal inflorescences. They are pale or white seeded, although black seeded types also occur.

Typical vegetable amaranths (*A. tricolor* and several weedy species) are characterised by a succulent spinach-like habit; young plants have short, tender stems and broad leaves with a high moisture content. The inflorescences are relatively small and the seed production is rather low (200–500 kg/ha).

Economic Importance

Both grain and vegetable amaranths are a source of unusually high quality protein. Amaranth leaves are a good source of carotene, iron, calcium, vitamin C, folic acid and other micro-nutrients, these levels are very high when compared with tomato and cucumber (Table 18.1).

Table 18.1 Content of essential nutrients per 100 grammes of edible product of tomato, cucumber and amaranth leaves; and total annual nutrient production in kg/ha, based on 65.5 tonnes/ha of edible product per year (FAO 1968).

	Tomato		Cucumber		Amaranth	
	Content	kg/ha	Content	kg/ha	Content	kg/ha
Dry matter	6.5 g	5200	4.9 g	3920	16.0 g	12 800
Carotene	0.5 mg	0.4	trace	trace	5.7 mg	4.6
Iron	0.6 mg	0.5	0.5 mg	0.4	8.9 mg	7.1
Calcium	10 mg	8	13 mg	10	410 mg	330
Vitamin C	26 mg	21	16 mg	13	64 mg	51
Protein	1.0 g	800	0.8 g	640	4.6 g	3680

The grains are commonly sold in markets throughout India and are used in a number of ways, as flour for 'chapatis', puffed grain for porridge and sweetmeat balls, gruel, substitute for opium poppyseed, and in confectionary, etc. The protein content is 15–16 per cent. The special value of grain amaranths apart from good baking and organoleptic quality, is their favourable amino acid composition. The lysine level scores high (5.0%) and so do sulphur-containing amino acids (4.4%).

The proportion of essential amino acids like lysine is much higher than in corn, wheat and rice (Table 18.2). However, it is limited in another amino acid, leucine, but with an amaranth-corn mixture, this deficiency can be restored (Kaufmann & Hass 1983).

Table 18.2 Comparison of amino acid content of grain amaranth and common grains (g amino acid/100 g grain) (Rodale Research Center Supplement 1983).

Amino acid	Amaranth	Corn	Whole wheat	Rice	FAO/WHO Daily recommendation
Isoleucine	.48−.624	.350−.46	.403−.577	.300−.352	1.68
Leucine	.752−.919	1.3−1.192	.756−.892	.574−.648	2.94
Lysine	.701−.908	.19−.265	.288−.374	.266−.299	2.31
Methionene/Cystine	.591−.752	.290−.329	.420−.528	.21−.267	1.47
Phenylalanine/Tyrosine	.96−1.253	.827−1.06	.902−1.154	.637−.720	2.52
Threonine	.40−.577	.342−.40	.324−.383	.259−.307	1.68
Tryptophan	.137−.224	.056−.67	.142−.164	.07−.98	.42
Valine	.576−.72	.461−510	.516−.616	.433−.524	2.10

Some reasons for amaranth being regarded as an important potential crop for arid regions are their relatively low water requirement and photosynthetic efficiency, being C_4 plants which grow rapidly during the warm season. There is a potential for these rapidly maturing crops which fit into multiple cropping systems. These species can play a special role in relieving protein malnutrition because they have a higher percentage of lysine than other cereals. Thus amaranth is a prime candidate for cultivation by the small farmers of the less developed countries where protein deficiencies are often major nutritional problems.

The advantages are: (a) protein content averages 16 per cent; (b) high lysine content which is nearly three times that of wheat and corn; (c) higher fibre content than wheat, corn, rice or soya bean; hardy and adaptable to many different areas; (e) can be eaten without elaborate preparation and processing; and (f) yield potential of grain amaranth is comparable to average world yields for the seven grains. Similarly, the vegetable amaranths have wide adaptability and are also rich in vitamins A and C, riboflavin, iron and folic acid (Anon. 1983a).

Cultural practices

The main cultivation method of vegetable amaranths is either direct sowing in rows by broadcasting or by transplanting. Grain amaranths are mostly cultivated as secondary crops in mixed croppings with maize, sorghum, millet, castor, egg-plant, hot peppers or other food crops. However, one limitation is that the domesticated strains can hybridize with their ubiquitous weedy relatives and may thus produce seed of low quality.

Research needs

There is considerable genetic diversity in amaranths which can be exploited by: (a) widening their adaptability to arid habitats; (b) enhance their productivity/area/ season by increasing yield and/or seed size; (c) using appropriate cultivation practices to suit arid regions; and (d) socio-economic benefits including the possibility of making amaranths a cash crop.

MULTIPURPOSE TREES

Growing multipurpose trees and shrubs in the arid regions is vitally important for firewood, fodder, industrial raw materials, fruit and timber. The species that are relevant to the arid zones in India are *Acacia* species such as *Acacia albida, A. nilotica* (= *A. arabica*) (including gum-exudate), *A. auriculiformis, A. tortilis* (sand dune stabilization), *A. catechu* (tannin) and *A. senegal* (gum-exudate), *Albizia lebbek* and *A. procera, Azadirachta indica* (non-edible oil, pesticide, fertilizer), *Cassia siamea, Leucaena leucocephala* (paper and pulp), *Prosopis juliflora* and *P. cineraria, Sesbania grandiflora*, (paper and pulp), *Phyllanthus emblica* (fruit), *Ziziphus mauritiana* (fruit), etc.

Advantages

The leaves, pods and young shoots of several legumes produce the main browse for wildlife and have excellent palatability for domestic livestock. The great promise of forage acacias is for dry regions where pasture grow poorly or only seasonally. Most acacias grow vigorously, coppice readily and withstand heavy browsing. The nitrogen fixing rhizobium present in the root nodules of legumes improves the soil fertility and its physical properties.

Acacias are also grown to control erosion, to stabilize sand dunes and to reclaim land lost to unpalatable grasses. For example, over the past 15 years *A. tortilis* has proved to be the most useful tree for growing in the arid and semi-arid regions of the State of Rajasthan, and have been a boon to the people of the desert who suffer from shortage of fuel, fodder and timber. It is also excellent for soil stabilization, and in Rajasthan over 800 ha of shifting sand dunes have been stabilized by planting *A. tortilis* (National Academy of Sciences 1979).

Acacia senegal is the main source of gum-arabic which is commercially the most important natural gum used in food, beverages, pharmaceutical preparations and confectionary and it has a wide range of industrial applications. *A. auriculiformis* is a promising tree for fuel and as a source of paper and pulp.

Of all the tropical legumes, *Leucaena* offers the widest assortment of uses. It is an excellent tree for reforestation of denuded areas and is highly palatable and nutritious, suitable as forage for ruminants such as cattle, goat, etc. *Leucaena* wood makes excellent firewood and charcoal with calorific value of 4000–4600 kcal/kg. The specific gravity of the wood is 0.5 to 0.6. The average timber yield is 30–40 m³/ha/year. *Leucaena* is also a major source for pulp and paper, roundwood and construction materials (National Academy of Sciences 1977a).

Species of *Prosopis* such as *P. juliflora* and *P. cineraria* are remarkable trees which grow in the hot, dry salt desert. They are useful as fuelwood, timber, forage for sheep and goats, stabilizing sand dunes, as shade plants and windbreaks. Some species, e.g. *P. juliflora* are relevant to apiary.

Sesbania grandiflora is a very fast growing tree, which is useful for firewood and fodder. In India it is managed in rotation of 5–15 years or more, and produces about 5 m³/ha/year.

Phyllanthus emblica (= *Emblica officinalis*) is useful for its fruit and medicinal properties; it also has a potential for small timber and firewood. The acid-tasting fruit is one of the richest known natural sources of vitamin C and they are used fresh, dried or pickled in cooking preserves, relishes and candies.

Ziziphus mauritiana is a poor man's fruit. The fruits are eaten either fresh or pickled and the juice can be made into a refreshing drink. It is also an excellent fuel wood tree and makes a good charcoal with a heat content of 4900 kcal/kg (National Academy of Sciences 1980). Cytologically the species is very variable with a ploidy ranging from 2x to 8x.

Limitations

The vigour and adaptability of acacias make them potential weeds; many *Acacia* and *Prosopis* spp. have long and spiny thorns or hooked prickles and are not readily eaten by livestock; *Leucaena* forage contains mimosine, an uncommon amino acid, an excess of which causes cattle to produce reduced quantities of the thyroid hormone, thyroxine, resulting in the loss of tail and rump hair and in extreme cases becoming completely debilitated with goitre.

Research needs

The following lines of research are suggested for the development of multipurpose trees in arid regions: (a) nutritional recycling in relation to short rotation and high density; (b) mycorrhizal and rhizobial research; (c) germplasm collection; (d) genetic system; (e) breeding methodology; (f) parameters of selection – fast growing, high productivity, hardy, mimosineless (*Leucaena*) and spineless (*Prosopis* species), coppicing ability, etc.; (g) multiplication of elite lines; and (h) development of suitable practices (spacing, fertilizers and irrigation etc.).

MEDICINAL PLANTS

Most of the drugs from higher plants which have become important in modern medicine invariably had a folklore origin and are traditional in systems of medicine such as Ayurvedic Pharmacopoeia; *Charaka Samhita,* one of the earliest treatises of Ayurveda, 600 BC, lists a total of 341 plants and plant products. Thus, Ayurvedic drugs represent a treasure trove waiting to be explored by all the modern techniques. Some medicinal plants relevant to the arid regions with proven pharmocologically uses are discussed here.

Commiphora wightii (= *C. mukul*) (Burseraceae)

C. wightii is a much branched, spiny shrub or a small tree 2–4 m high with crooked and knotty branches ending in sharp spines. The gum-oleoresin of *C. wightii,* Indian bedellium, is the main source of guggulu. The oleo-gum-resin is present in resin ducts in the soft stem bark and also in larger veins of the leaf.

Distribution

C. wightii is one of the most widely distributed species of *Commiphora* which occurs in the extremely arid and semi-arid tracts of Rajasthan, Gujarat, Maharashtra and in parts of Karnataka and Madhya Pradesh. In Rajasthan it is very widely distributed including the Aravallis and desert areas, while in Gujarat it is distributed throughout the Kutch division.

Medicinal importance

Guggulu is one of the important drugs of Ayurveda with a wide range of therapeutic uses. The classical Ayurvedic literature claims its efficacy in a variety of metabolic disorders, especially rheumatoid arthritis and obesity. It also provides a good source for a number of useful steroids. It was found that two steroids, named Z- and E-guggulsterone 11, 12 are responsible for the hypolipaemic/hypocholesterolemic activity of guggulu. Investigations at Central Drugs Research Institute, Lucknow and Regional Research Laboratory, Jammu show that Guggulipid should reach the Indian market as a hypocholesterolemic drug by 1984–85 (Sukh Dev 1983).

Cultivation

Guggulu plants can be easily raised from stem or branch cuttings, preferably *c.* 30 cm long and 1–3 cm in diameter. The best results are obtained from cuttings planted during June.

Tapping and collection

Healthy plants which are over 5 years of age are selected for tapping, which is carried out from November to January and the collection of the oleo-gum continues until May or June. About 200–500 gms of dry guggulu is generally obtained from a plant in one season (Raghunathan 1982).

Limitations and special requirements

In recent years gum-guggulu is becoming increasingly scarce due to indiscriminate felling and exploitation of the plants by drug companies, together with the adoption of faulty tapping techniques. Improved tapping methods need to be developed that will help to maintain an optimum yield of oleo-resin while conserving the plant population.

Research needs

Research into the following topics is required: (a) survey and screening for better oleo-gum-resin yield; (b) genetic upgrading through selection and breeding; (c) development of cultural methods for large scale propagation; (d) experiments need to be conducted to develop tapping processes that can help in optimum oleo-resin yield and plant conservation.

Azadirachta indica (Meliaceae)

Azadirachta indica is commonly known as neem tree or margosa tree. It is a deep-rooted, medium sized, evergreen tree with rather rough, greyish or brownish bark, and delicate foliage. Its short bole produces wide spreading branches forming a round or oval crown and the flowers appear from March until May and fruits ripen from June until August.

Distribution

It is native to the dry forest areas of India including the forests of Deccan. It is cultivated all over India but thrives best in drier climates of the north-western parts where normal rainfall ranges between 450–1000 mm and maximum temperature may be as high as 48°C.

Economic importance

The neem tree has great potential for agricultural, industrial and commercial exploitation because of its multiple uses such as firewood, timber and in pharmaceutical, entomological preparations, etc. Furthermore, it holds great scope for developing village industries, generating employment and thus boosting the economy. It can also provide a series of products which are highly competitive in price and quality with petroleum-based synthetics.

Almost every part of the tree from its roots, trunk, bark, leaves, flowers to fruits and seeds is known to have some use. Neem tree and its constituents have a wide range of uses in the arid zone countries.

Neem seed contains about 20% of an oil which in turn contains about 20% of active compounds with a promising potential in the manufacture of pharmaceutical and insect-repellent preparations. Further processing of this oil yields several industrial products such as myristic and lauric acids and it is probably that the soap produced from it will have better lathering detergent properties than the soap derived from the major edible oils (Mitra 1963, Ketkar 1976). The use of non-edible neem oil in soap manufacture may result in the replacement of edible oils (palm, coconut and groundnut oils) used for this purpose at present. Consequently, large quantities of the latter could become available for human consumption.

Neem fruit pulp is promising for methane gas generation and it may also serve as a carbohydrate rich base for other industrial fermentations.

Neem seed cake, the residue remaining after extraction of the oil, represents 80 per cent by weight of the whole seed; its chemical composition is 3.56 % N, 0.83% P, 1.67% K, 0.77% Ca and 0.75% Mg (Ketkar 1976). It is likely that many degraded and leached soil types in arid regions could be enriched by the application of the cake, which would also have a beneficial effect on the soil structure. The cake contains a relatively high proportion of sulphur, and its slow decomposition in the soil may result in the release of carbon disulphide/or other compounds, which could act as volatile fumigants; in addition to this probable biocidal effect, these compounds may repel some soil insect pests. Neem cake should also be used to raise the pH and thus improve nutrient balance in acid soils in both arid and humid regions.

Practically all the parts of the neem tree find use in Ayurveda and the tree is considered to be a village dispensary by itself. The twigs are still used as toothbrush in rural India. The oil is used as medicinal hair oil and also in curing rheumatism and leprosy (Mitra 1963, Ketkar 1976).

Neem seed is considered to be entomologically useful. Neem seed extracts are effective in the control of several insect pests (Pradhan *et al.* 1962, Gill & Lewis 1971, Anon. 1983b, Butterworth & Morgan 1984). Unlike most insecticides available, they are non-poisonous to humans or animals, non-polluting to the environment and have a systematic action when applied to some crops.

Neem tree is also used as firewood; its wood having a very high calorific value. The wood is relatively heavy with specific gravity varying from 0.56 to 0.85 (average 0.68).

Cultivation

Both direct seeding and use of nursery transplants are used in establishing neem plantations. The choice depends on the soil type, climate and systems of planting. In natural habitats with well drained soils, direct seeding has been successful in afforestation.

Limitations and special requirements

Neem seed has a very low viability and has to be sown within two or three weeks after harvest. The seedlings are killed by frost or fire. A number of strains will have to be tested to find the forms best adapted to local conditions. The neem tree is un-demanding and grows well on most soils, including dry, stony, clay and shallow soils. It will not grow on seasonally waterlogged soils or in deep dry sands where the dry-season water-table lies below 18 m. The optimum pH should be 6.2 or above, although neem will grow at pH 5, bringing the surface soil to neutral pH by its leaf litter. It does not grow well on saline soils.

Research needs

For further development research should be conducted into the following topics: (a) coordinated multidisciplinary research and development programme leading to agricultural, industrial and commercial utilisation of neem tree; (b) survey and screening for the elite types for different uses; (c) studies on the autecology of neem; (d) standardization of silvicultural practices; (e) collection, harvesting, storage and processing of neem seed; (f) assessment of the contribution of the neem cake to soil organic matter and its effect on soil physical properties; (g) investiga-tion of pesticidal properties of neem extracts in tropical and temperate field crops and in stored farm produce; (h) economic feasibility studies, such as market research for neem products, costs in relation to economic and social benefits with an overall appraisal of the role of neem tree in agriculture, industry and commerce.

SEED GUMS

Several million rupees worth of foreign exchange is earned through export of plant gums, which has benefited the rural sector. However, in order to cope with the increasing demand both in the national and international markets, newer and cheaper sources of gums need to be discovered. Apart from the guar gum, *Plantago ovata* and *Sesbania bispinosa* are important sources (Farooqi 1976). While the culti-vation of former species can be extended, the latter species may prove to be a good source of gum that could be introduced in the world market.

Plantago ovata (Plantagineae)

Plantago ovata, commonly known as psyllium, is an annual with an unbranched stem and narrow leaves. It is native of Iran but was introduced into commercial cultivation in India over a century ago. Currently, India holds the monopoly both in production and world trade of psyllium husk and seed. The crop is mainly grown in Mahasana and Banaskantha districts of north Gujarat but can be extended into Rajasthan.

Cultivation

Psyllium is a 110–120 days cold weather crop, sown any time from mid November to mid December and harvested from the last week of March to early April. The crop needs continuous dry weather from February to April, and at that stage, rain or dew can ruin the standing crop.

The seed of *P. ovata* is treated with thiram or some other mercurial fungicide and is then sown by broadcasting at a rate of 4 kg seed/ha. It receives from five to seven light irrigations. Fertiliser requirements are low and if a legume crop precedes it, the nitrogen requirement of psyllium is met, otherwise 30 kg N and 25 kg P_2O_5 per hectare is enough to raise a good crop. The estimated yield ranges from 800–1000 kg/ha.

Economic importance

Psyllium seed is known for its medicinal properties. Several pharmaceutical preparations of the husk are used as bulk laxative. Indian psyllium has almost replaced other psyllium, including French psyllium, and in the year ending March 1975, India exported about 10 million dollars worth of husk. A new use of the husk is as cervix dilator (isaptent) for the termination of pregnancy. This is prepared from granulated *P. ovata* (isapgol) seed husk (Khanna *et al.* 1979). Husk is also used in textile, food and cosmetic industries.

Sesbania bispinosa (= *S. aculeata*) (Leguminosae)

Sesbania bispinosa commonly called 'dhaincha' is known throughout India as an important leguminous green manure as well as a fodder crop. It also yields a cheap fibre, paper pulp and rural fuel. However, in recent years, a detailed chemical investigation of the seeds of dhaincha revealed that the seeds may also be a possible source of commercial gum for sizing and stabilising purposes and the seed meal could be useful as poultry and cattle feed (Farooqi & Sharma 1972). In view of the varied uses, especially as a potential source of commercial gum, and the fact that it can be easily and cheaply grown on alkaline-saline soils, dhaincha assumes added importance. Its large scale cultivation could, therefore, be a very vital and welcome step towards the development and utilization of usar (alkaline) or semi-usar land.

Botany

Sesbania bispinosa is an erect, suffruticose, multibranched annual shrub growing 1.0–2.50 m in height. The papilionaceous flowers, borne on racemes, bloom from September to December. It grows wild as a weed throughout India.

Cultivation

Dhaincha is commonly propagated by seeds. The best time for sowing the seeds is the second week of June. About 20–25 kg seed are required for one hectare of land.

Economic Importance

Dhaincha plant is used for fibre, fodder, fuel and pulp. Dhaincha seed has been found to be a potential source of cheap galactomannan gum. A sizeable amount of highly proteinaceous seed meal for cattle feed is obtained as a byproduct. Seed gums are important vegetable products with varied uses in paper, textile, pharmaceutical, cosmetic, food and other industries for sizing textiles and paper products and for thickening and stabilizing solutions. The production of seed gum in India, has, therefore, come to be an important industry. More than 50 000 tonnes of various seed gums were exported during the year 1977–78 earning foreign exchange amounting to *c.* 250 million rupees (Farooqi 1976, Chandra & Farooqi 1979). In view of their ever increasing uses, it is desirable to explore new economic sources of these gums. These new sources may not necessarily be used as substitutes for the already known ones, but could be utilised by totally new industries, depending upon their properties, and thus meet the increasing demand for gums in the world market. If efforts are made towards organising large scale cultivation and collection of dhaincha seeds thus assuring a regular supply of seeds at a reasonable rate, some manufacturers of guar seed gum may also take up the production of dhaincha seed gum. Seed yield is 1000 kg/ha and in addition 10 000 kg/ha of stems are available as fuel.

Limitations and special requirements

The plant grows well in areas with annual rainfall between 550–1100 mm. Frost and cold weather are unfavourable for the development of the crop, and although sandy loam soil is the best suited for dhaincha, it grows satisfactorily in clay-loam as well as black cotton soils. It does not require irrigation, fertiliser or even weeding.

Research into the following topics is required:— (a) genetic upgrading of *Plantago ovata* and *Sesbania bispinosa*; (b) standardisation of appropriate cultural practices; (c) survey for newer and cheaper sources of seed gums to cope with the increasing demand of international market; (d) need for a well planned and scientifically organised collection of plant exudates of commercial importance; (c) need for production of gum-derivatives for use in specific industries.

OTHER PLANTS OF SOME MEDICINAL POTENTIAL

Some other plants relevant to the arid regions with proven pharmacological uses are *Achyranthus aspera* (diuretic), *Butea monosperma* (= *B. frondosa*) (anthelmintic), *Boerhaavia diffusa* (antiinflamatory), *Syzygium cumini* (= *Eugenia jambolana*) (hypoglycenic), *Euphorbia thymifolia* (antiasthmatic) and *Sida rhombifolia* (anabolic) (Sukh Dev 1983). These need to be investigated systematically and suitable agricultural practices established.

INDUSTRIAL PLANTS

Over the ages, the arid and semi-arid regions of the world have supplied mankind with a rich variety of raw materials of plant origin for a wide variety of uses and purposes. New products and recently developed technological processes may be capable of changing formerly uneconomic plants into viable industrial crops. Recent examples are production of a high quality, resin-free, natural rubber from the guayule plant and oil from jojoba as a substitute for sperm whale oil. The successful commercial exploitation of both these plant species is about to be realised, and it seems reasonable to believe that many others of the known potential industrial plant species could be similarly exploited within a short period of time.

Parthenium argentatum – guayule (Compositae)

The principal source of natural rubber is *Hevea brasiliensis,* a large tree, which meets nearly one-third of the world's rubber demand, while the remainder is met by synthetic products from petroleum-based chemicals (National Academy of Sciences 1977b). However, following the petroleum crisis, the price of synthetic rubbers is less competitive on account of its high cost and non-availability of the petrochemicals. There has been an ever-increasing demand for natural rubber on account of its elasticity, resilience, tackiness and low heat build up (National Academy of Sciences 1977b). By the 1990's, the projected international demand for natural rubber is expected to outstrip the *Hevea* rubber production, leading to a worldwide shortage. In India there is expected to be a short-fall of 176 000 tonnes by 1990 (Patel *et al.* 1981). This shortage cannot be met by *Hevea* rubber because of the limitations in extending its plantation. World attention has been drawn to a search for alternative sources of natural rubber and guayule is currently being considered as the best contender for this purpose, since it can be grown in dry, arid, semi-arid and non-agricultural lands.

Distribution

Guayule is a native to north-central Mexico and south-west USA, where it grows in arid, semi-arid, poor desert land. It occurs in drylands over an area of 276 700 km² and occupies favourable sites on limestone hillsides at an altitude of 1 200—2 100 m in regions with less than 400 mm rainfall.

Cultivation

The work on guayule at National Botanical Research Institute, Lucknow has established that it can be successfully grown in India. The plant has been grown from seed for three successive generations.

Seeds are obtained from April to October but the best seeds are available during the dry and hot months from mid-April to mid-June. They are sown outdoors in raised nursery beds during October to January. About one month old seedlings in polythene bags or earthern pots are transplanted directly in the ground. The seedlings are best transplanted during the winter months (November—February).

Guayule can also be propagated by shoot cuttings. Vegetative propagation by apical shoot cuttings has been successful when undertaken in July or August.

Botany

Plants start flower production in March when they are 3—4 months old and continues until November. Flowering and growth cycle in old plants start again in March.

Genetic system

The species is genetically very variable. Broadly speaking it falls in three groups — the diploids (2n=36), triploids (2n=54) and tetraploids (2n=72), based on x=18. However, individual plants have been found at penta-, hexa- and even at octoploid levels. The species has sporophytic incompatibility. The diploid alone is sexual and is localized in Eastern Durango in Mexico which may be the centre of the origin of the species. All the higher ploids are pseudogamous facultative apomicts and apomixis is through diplospory (Powers & Rollins 1945, Esau 1946).

Guayule reproduces sexually as well as apomictically, but when it reproduces by facultative apomixis a certain amount of reduction takes place. It is through these chance reductions with subsequent fertilization that new combinations can be obtained. The degree of apomixis plays an important role in breeding and selection. The higher the frequency of sexual off-types the greater is the possibility of obtaining favourable combinations, whereas low frequency of sexual off-types is more desirable for greater uniformity. Guayule, therefore, presents an unusually fine opportunity for breeding and selection for higher rubber yields.

Thus guayule is an agamic complex with versatility in breeding system and forms a poly-aneuploid system ranging from diploid (2n=36) to octoploid (2n=144). The genetic system in guayule has potentialities to conserve and preserve heterozygosity through agamospermy, provided the genotype in question has an adaptive value (Khoshoo 1982).

Research needs

Considerable work must be done in order to place guayule among plants of great commercial importance. This can be accomplished by improving yield potential, cultural and harvesting procedures, and developing new processes for rapid extraction and purification of rubber.

The following research thrusts could significantly enhance the commercial potential of guayule:– (a) collection of wild and cultivated germplasm of guayule; (b) study of adaptability of the species to different agro-climates and soil types particularly in marginal, sub-arid land and identify suitable areas for its extensive cultivation; (c) cytogenetical analysis of different strains of guayule in order to see whether there is any correlation between polyploidy and rubber content; (d) standardisation of breeding methodology; (e) regulating high rubber content in standing crop through use of growth bioregulators; (f) studies of seed production, dormancy, storage and germination; (g) adoption of appropriate cultural practices and protection of plants from diseases and pests; (h) acceptability of the product and economic feasibility of cultivation (i) development of a method for quick evaluation of individual plants using microgram quantities or morphological markers for high rubber content.

Jatropha curcas (Euphorbiaceae)

With the energy demand in developing countries increasing sharply, research on alternative sources has found that *Jatropha curcas,* an oil plant which grows in the tropical areas of the world, can be a suitable source for biomass energy especially in rural areas. Its oil can be substituted for diesel engine fuel. It is native to tropical America, but is now widespread throughout India and in the Andaman Islands.

Botany

It is a large shrub, 3–4 m high, fruits are 2.5 cm long, ovoid, black, breaking into three 2-valved cocci; seeds ovoid-oblong, dull brownish black. It flowers in hot and rainy seasons, and sets seed in winter when it is leafless.

Economic importance

High quality oil with a calorific value of 9470 kcal/kg can be extracted from the seeds of *J. curcas,* which compares favourably with 10 600 kcal/kg for gasoline and 6400 kcal/kg for ethanol (Anon. 1959). The oil is readily soluble in diesel oil and gasoline and can be used in fuel mixtures for gasoline engines. Substitution of diesel oil is of prime importance to developing countries and a diesel engine performance test (the Yanmer SA 70–L diesel engine) indicates that the engine performance and fuel consumption are very similar for *J. curcas* and diesel oil when the same engine is used (Takeda 1983). If a *J. curcas* oil industry is developed, it would essentially be an agro-industry because vegetable oil seeds would be used as the raw material for an alternative fuel.

Other uses of this oil are as an illuminant, for making soaps and candles, and it also has medicinal properties. Seed cake can be used as manure and the leaves are used as feed for raising eri-silk worm.

Cultivation

The plant can be grown from the seed, or from stem cuttings 45–100 cm long which when buried in the soil soon develop roots. The mean annual yield of air-dried seeds from a 5 year old tree is 2–3 kg.

Advantages

J. curcas has the following advantages:— (a) the plant is adaptable to a wide range of soil types, including those of arid regions; (b) it grows quickly and is easily culti-vated; (c) the processing for oil extraction is very simple; (d) the energy balance appears to be favourable; (e) it is available for direct practical use in the rural areas; (f) no engine structure improvements are necessary and (g) it can be grown under conditions which offer no competition to food or animal feed production.

Research needs

Research needs to be undertaken on the following:— (a) development of high oil yielding varieties; (b) development of suitable cultural practices; (c) technological research and development in all phases of oil-seed processing, including extraction and refining, but especially for the development of small-scale technology which can be used in rural areas.

Strategies for development and management of arid lands

The ever-increasing human and livestock populations and associated increases have put tremendous pressure on the semi-arid ecosystems which are already water stressed and fragile. Large scale deforestation coupled with grazing results in forests turning into grass lands. Heavy grazing pressure decreases the plant cover and carrying capacity of these grass lands, and encourages invasion by less palatable plants with low conservation value. This, together with increasing soil loss due to erosion, results in a loss of perennial plant cover moving the land towards desert conditions. This calls for immediate attention for evolving rational management development strategies.

One of the very important vocations of arid zone people is animal husbandry particularly sheep and goat farming. This provides self-employment to the people inhabiting the area and protects them from the vagaries of drought and famine. For example the animal wealth in Rajasthan State alone accounts for 12 per cent of the total revenue that accrues to that state. Besides, a large portion of the population in the area is also engaged in animal by-product industries, e.g. skins, hides, hair, wool, milk and milk products (Kanodia & Patil 1983).

The main emphasis in the desert development should be in the improvement of range lands through regeneration of natural vegetation and reseeding, coupled with other soil conservation practices and afforestation. The silvi-agri-pastoral system of development of desert areas would result in the availability of sufficient grazing resources which can be judiciously utilized to maintain a large sheep and goat population, which in turn would improve the multifold socio-economic condition of the rural mass of arid and semi-arid areas (Acharya & Patnaik 1977).

In the desert zone the limiting factors for production are set by conditions of water stress, harshness of climate and salinity of soils. Researches into techniques for improving crop yields under dry-farming conditions can contribute greatly. Furthermore, it is essential that suitable fodder trees and shrubs are planted in the range lands to provide shade and top feed for the animals. The fodder trees that can be propagated may include *Prosopis cineraria, Acacia* spp., *Ailanthus excelsa,* etc. These fodder trees in addition to increasing carrying capacity of the pasture land could also provide employment to the rural people through large scale collection of tree leaves which can be preserved as dry leaves or in meal form for use as supplementary feed during lean periods.

Natural regeneration of range lands needs to be undertaken through proper protection from the biotic factors and/or reseeding them with appropriate grass species where natural grasses are of poor quality or regeneration is not possible. Since the grasses are low in protein except during their early stages of growth, it will be essential that some perennial legumes, e.g. *Lablab purpureus* (= *Dolichos lablab*) and *Macroptilion atropurpureus* (= *Phaseolus atropurpureus*), are incorporated in the regenerated or reseeded range lands. In addition to the perennial species of grasses and legumes used for development of permanent pastures, certain annual fodders can be grown under dryland conditions and conserved for use during the dry season. Legumes are considered as a useful, dependable and the cheapest source of nitrogen in pastures and thus provide a quality of protein at all stages of their growth for animal feed.

Industrial plant species, e.g. medicinal plants, seed gum and mucilages, hydrocarbon plants, paper and pulp, can contribute to the economic development of arid zones and at the same time aid in stabilising soil and preventing erosion. Many industrially valuable trees and shrubs can also provide firewood for local population. In addition to supplying valuable industrial materials, these desert species also have the potential to be utilized for the rehabilitation of desertified areas. This could enable establishment of industries based upon these plant materials and this type of agro-industrialization may become, in time, an important element in the development of arid regions.

At present, most of the under-exploited economic plants are either used from wild or semi-wild populations in nature, and have not been domesticated. There is a need to assess their domestication potential which would take into account: breeding cycle, ease of harvesting, accessibility of the economic part, distribution and ease of propagation, handling and storage qualities.

Not all taxa are suitable for immediate domestication, even when they possess a number of positive characteristics. The potential of such taxa is not completely lost because the option still exists to maximise their utilization in the wild, particularly in areas where they are abundant. However, the possibility of domestication should not be totally ignored, until ways of overcoming their apparent negative qualities have been studied.

The following research needs have been identified to hasten the process of domestication:– (a) knowledge about composition of economic products (edible food plants, fodder, fuel, industrial materials, etc.); (b) understanding of the genetic systems; (c) standardising breeding methodology; (d) evolving better agricultural practices; (e) multiplication of the elite types through seeds, vegetative means and tissue culture.

GREENING THE ARID LANDS

Only 3.93 per cent (Bains 1977) of the total area in the arid zone of India is under forests as against 33 per cent required under the National Forest Policy for maintaining the ecological balance. A large measure of desertification has been caused by depletion of woodland and shrubs in those regions where only wood is available as fuel. The most important step to improve the arid land ecosystem would, therefore, be revegetating the arid lands with firewood and fodder species. This would result in whole range of social, economic and environmental benefits:– (a) conservation of soil and water; (b) better microclimate; (c) resolving energy and fodder crisis; (d) conservation of biological diversity; (e) creation of aesthetic land-

Figure 18.1   Photosynthetic model of Development; the balance as it should be.

scape; (f) employment generation; (g) better health; (h) better quality of life; (i) halting influx of rural population into urban areas; and (j) decentralised economy.

SILVI-AGRI-PASTORAL SYSTEM

Silvi-agri-pastoral system of development of arid lands would result in a self-sufficient system where all the primary needs of the rural family will be met by growing food, pasture and cash crops, rearing dairy or beef animals, sheep, pigs or poultry or combination of these and selling of surplus produce or animals.

The model of development emerging for these regions has to be holistic, involving forestry, agriculture, horticulture and animal husbandry, which needs to be properly integrated so as to help the under privileged people of arid regions. In essence this is what may be called the Photosynthetic Model of Development. This is necessary on account of the fact that the chief vocation of these regions at least in India, is animal husbandry, which contributes materially to the economy of Rajasthan State.

The hidden riches (minerals and oil) being found under these regions if explored without due consideration of the environment may change the face of these regions, and will ecologically further degrade the area. Hence, the two models of development (the Photosynthetic and the Industrial) need to be balanced properly. In fact, there needs to be given a small but a perceptible tilt in favour of the Photosynthetic Model (Fig. 18.1).

Once a holistic approach to development is accepted, mere identification of new economic plants such as multipurpose trees, grass land species (including perennial legumes) and industrial species for the arid regions is not enough. It is necessary to integrate these plants into a system of the rotational cycle to make these acceptable to the people of these regions. If this is not done, the new plants though very useful, would not be accepted by the people, who on account of the vagaries of nature do not want to take risks through innovative changes, unless their practical value can be demonstrated.

References

Acharya, R.M. and B.C. Patnaik 1977. Role of sheep in the desert ecosystem and drought proofing through improved sheep production. In *Desert ecosystem and its improvement*, H.S. Mann (ed.): 350–370. Jodhpur: Central Arid Zone Research Institute.

Agarwal, A. 1984. Beyond pretty trees and tigers: the role of ecological destruction in the emerging patterns of poverty and peoples' protests. *Fifth Vikram Sarabhai Memorial Lecture*. New Delhi: Indian Council of Social Science Research.

Ahuja, L.D. 1977. Grassland and range management. In *Desert ecosystem and its improvement*, H.S. Mann (ed.): 296–322. Jodhpur: Central Arid Zone Research Institute.

Anonymous 1959. *Wealth of India (raw materials)*. Vol. 5: 293–295. New Delhi: Publication and Information Directorate, Council for Scientific and Industrial Research.

Anonymous 1983a. *Amaranth: a nutritious crop for the Developing World*. Emmaus, Penn.: Rodale Press.

Anonymous 1983b. Neem in agriculture. *Res. Bull.* 40. New Delhi: Indian Agriculture Research Institute.

Bains, S.S. 1977. Role of consumers in the functioning of desert ecosystem. In *Desert ecosystem and its improvement*, H.S. Mann (ed.): 335–340. Jodhpur: Central Arid Zone Research Institute.

Butterworth, J.H. and F.D. Morgan 1968. Isolation of a substance that suppresses feeding in locusts. *Chem. Communications* 11: 23–24.

Chandra, V. and M.I.H. Farooqi 1979. 'Dhaincha' for seed gum. *Extension Bulletin* No. 1. Lucknow: Economic Botany Information Service, National Botanical Research Institute.

Das Gupta, K. 1977. Ground water resources. In *Desert ecosystem and its improvement*, H.S. Mann (ed.): 26–30. Jodhpur: Central Arid Zone Research Institute.

Dasted, K.J. and V.I. Myers 1983. Remote sensing to detect desertification process. In *Natural resources and development in arid regions*, E. Campos-Lopez and R.J. Anderson (eds): 63–84. Boulder, Colorado: Westview Press.

Dregne, H.E. 1983. Soils of semi-arid regions. In *Natural resources and development in arid regions*, E. Campos-Lopez and R.J. Anderson (eds): 53–62. Boulder, Colorado: Westview Press.

Eckholm, E. and L.R. Brown 1977. *Spreading deserts – the hand of man*. Worldwatch Paper 13. Washington, D.C.: Worldwatch Institute.

Esau, K. 1946. Morphology of reproduction in guayule and certain other species of *Parthenium. Hilgardia* 17: 61–120.

FAO 1968. *Food composition table for use in Africa*. Rome: FAO.

Farooqi, M.I.H. 1976. Plant and seed gums for the development of rural economy. *Khadi Gramodyog* 22: 377–388.

Farooqi, M.I.H. and V.N. Sharma 1972. *Sesbania aculeata* Pers. seeds – a new source of gum. *Research & Industry* 17: 94–95.

Gill, J.S. and C.T. Lewis 1971. Systematic action of an insect feeding deterrent. *Nature* 232: 402–403.

Hillel, D. 1982. *Negev. Land, water and life in a desert environment.* New York: Praeger.

Kanodia, K.C. and B.D. Patil 1983. Pasture development. In *Desert resources and technology* Vol. 1, Alam Singh (ed.): 263–305. Jodhpur: Scientific Publishers and Geotech-Academia.

Kaufmann, C.S. and P. Wagoner Hass 1983. *Grain amaranth: an overview of research and production methods.* Emmaus, Penn.: Rodale Press.

Ketkar, C.M. 1976. Utilization of neem (*Azadirachta indica* Juss.) and its byproducts. *Final technical report of non-edible oils and soap industry.* Hyderabad: Khadi and Village Industries Commission.

Khanna, N.M., J.P.S. Sarin, R.C. Nandi, S. Singh, B.S. Setty and V.P. Kamboj 1979. Isaptent – a new aid for cervical dilation. *Internat. Planned Parenthood Federation. Med. Bull.* 13.

Khoshoo, T.N. 1982. Energy from plants: problems and prospects: Presidential address. *Proceedings of the 69th Session of Indian Science Congress.* Mysore (separate).

Mann, H.S. 1977. Desert ecosystem and its improvement. In *Desert ecosystem and its improvement,* H.S. Mann (ed.): 1–7. Jodhpur: Central Arid Zone Research Institute.

Mann, H.S. and R.P. Dhir 1984. Critical research problems in arid zones: insights from twenty-five years of the Jodhpur Institute in India. In *Ecology in practice Part I: Ecosystem management,* F. Dicastri, F.W.G. Baker and M. Hadley (eds): 243–261. Paris: UNESCO.

Mitra, C.R. 1963. *Neem.* Hyderabad: Indian Central Oil Seeds Committee.

National Academy of Sciences 1975. *Underexploited tropical plants with promising economic value.* Washington, D.C.: National Academy of Sciences.

National Academy of Sciences 1977a. *Leucaena: promising forage and tree crop for the tropics.* Washington, D.C.: National Academy of Sciences.

National Academy of Sciences 1977b. *Guayule: an alternative source of natural rubber.* Washington, D.C.: National Academy of Sciences.

National Academy of Sciences 1979. *Tropical Legumes: resources for the future.* Washington, D.C.: National Academy of Sciences.

National Academy of Sciences 1980. *Firewood crops: shrubs and tree species for energy production.* Washington, D.C.: National Academy of Sciences.

Patel, S., N.S. Sharma and A. Sarabhai 1981. *Commercialization of guayule in India.* Ahmedabad, India: ve-Par, Shahibagh House.

Powers, L. and R.C. Rollins 1945. Reproduction and pollination studies on guayule (*Parthenium argentatum* Gray) and *P. incanum* HBK. *Am. Soc. Agron. J.* 37: 96–112.

Pradhan, S., M.G. Jotwani and B.K. Rai 1962. The neem seed deterrent to locusts. *Indian Farming* 12: 7–11.

Raghunathan, K. (ed.) 1982. Guggulu and its cultivation. *News Letter* 4. New Delhi: CCRAS.

Saxena, S.K. 1977. Vegetation of Indian Desert biome. In *Desert ecosystem and its improvement,* H.S. Mann (ed.): 144–165. Jodhpur: Central Arid Zone Research Institute.

Singh, J.S. and M.C. Joshi 1979. Ecology of the semi-arid regions of India with emphasis on land-use. In *Management of semi-arid ecosystems,* B.H. Walker (ed.): 243–273. New York: Elsevier.

Sukh, Dev 1983. Natural products in medicine – present status and future prospects. *Curr. Sci.* 52: 947–956.

Takeda, Y. 1983. *Jatropha curcas* – a substitute for diesel oil. *Sci. Service* 2: 3–4.

19 Some indigenous economic plants of the Sultanate of Oman

R. M. Lawton

formerly Land Resources Development Centre, Tolworth Tower, Surbiton, Surrey KT6 7DY, UK
now 46 High Park Avenue, East Horsley, Leatherhead, Surrey KT24 5DD, UK

Introduction

The Sultanate of Oman covers an area of approximately 300 000 km² of south-eastern Arabia, between 16°40′–26°20′N and 51°50′–59°50′E. The arid climate of the Sultanate falls into two major zones.

Northern and eastern Oman, dominated by a range of limestone mountains reaching an altitude of nearly 3000 m, forms the first major climatic zone. Here the climate is of the Mediterranean type with an erratic and unreliable winter rainfall from December to March. In the mountains the annual rainfall varies from 300–500 mm, some of which may fall in July due to the influence of the southwestern monsoon, whereas on the coastal plain it is only *c.* 50–100 mm. During the winter months the day temperature on the mountains is *c.* 15°–20°C and may fall to below 0°C at night; during summer the mountain day temperature will rise to between 25°–30°C and fall to about 15°C at night. There is a variation in the range of temperatures along the coastal plain within the range of 20°–35°C for winter day temperatures, falling to below 10°–15°C at night (Whitcombe 1982). During summer the day temperatures on the coastal plain rise to almost 40°C and may exceed 45°C, at night they fall to 25°–32°C. The relative humidity on the coastal plain during the day varies from 25–80 per cent; at night the relative humidity is higher and frequently in the region of *c.* 80 per cent.

The second major climatic zone is the southern region of Dhofar. A range of limestone mountains runs in an easterly direction from the South Yemeni border. To the south of this mountain range there is a narrow crescent-shaped coastal plain; to the north the desert stretches into Saudi Arabia. The mountains rise steeply from the plain to a height of *c.* 900 m, with a number of higher peaks and a 10–15 km wide, undulating, central plateau.

Dhofar has a monsoon climate with moisture-laden winds of the southwestern monsoon blowing in from the Arabian Sea and bringing heavy cloud cover and precipitation mainly in the form of drizzle, or mist, from late June to September. The drizzle is precipitated by broad-leaved woodlands that cover the south-facing escarpment and steep wadi (valley) sides of the mountains, and by the tall grasses on the plateau. Cyclonic storms may occur at any time of the year. There are no adequate rainfall records for the mountains, but it is probably in the region of 500 mm per annum. On the coastal plain the rainfall is lower, varying from 100–500 mm per annum depending upon the frequency of cyclonic storms. Destructive flash floods occur about once every decade or so. The temperatures are much more equable in Dhofar; they range from *c.* 27°C to a maximum of 35°C during the day and fall to 16–28°C at night, throughout the year. Relative humidity is high, usually around 70 per cent and over 90 per cent during the monsoon.

The two major climatic zones are separated by the central Desert of Oman, where the wide range of day and night temperatures gives rise to radiation at night, and this results in heavy falls of dew for part of the year.

Indirect economic attributes

Vegetation in arid environments is an important attribute of indirect economic value, e.g. trees and shrubs provide shade and shelter for human settlement, live-stock and agriculture. Shelterbelts of trees protect agricultural crops from desic-cating winds. Some trees and shrubs stabilize sand dunes and so protect agricultural crops and settlements from dune encroachment.

In Oman the following trees and shrubs are common stabilizers of sand dunes:— *Prosopis cineraria* (ghāf), *Acacia nilotica* subsp. *adstringens* (= *A. nilotica* subsp. *adansonii*) (karat), *Calligonum crinatum* subsp. *arabicum* (arta) and *Tamarix* spp. (athel).

The broad-leaved woodlands that cover the escarpments and steep wadi sides in the mountains of Dhofar, precipitate the monsoon mists, or drizzle. This moisture re-charges the water-table that sustains the irrigated agricultural crops grown on the coastal plain. Without the natural woodland cover much of the mist would be dissipated over the desert.

The large trees on the summit and upper slopes of Jabal Akhdar and the other northern mountain ranges, in particular *Juniperus macropoda* (al'alān) and *Olea europaea* subsp. *africana* ('itm), intercept the rain and check surface run-off, thus enabling the water to penetrate the mountain soil. This water eventually reaches the water-table that supports the irrigated agricultural crops growing along the Batinah coastal plain.

On the central desert region, wildlife rarely has access to surface water; it is dependent for its water requirement upon the dew that is trapped or condensed by the vegetation during the night. The gazelle (*Gazella gazella arabica*) obtains its moisture when it is browsing shoots of *Acacia ehrenbergiana* (salam), *A. tortilis* (samr) and other vegetation. The desert grasses that are grazed by the Arabian oryx (*Oryx leucoryx*) are wet with dew during the night and provide the oryx with water. Other wildlife species, including the plants themselves, obtain their water from the same source.

The economic value of browse

Trees and shrubs are perennial sources of browse for livestock, in particular goats and camels, the two main browsing domestic animals in the area. In 1984 a drought occurred in northern Oman, and there was no grass growth and very little herb and forb growth, but the trees and shrubs came into new leaf in spring (March–April) and flowered and set fruit, despite the severe lack of rain. Some of the main browse species in the rangelands of Oman are: *Prosopis cineraria* (ghāf), *Acacia tortilis* (samr), *A. ehrenbergiana* (salam), *A. senegal* (thumōr), *Anogeissus dhofarica* (sughat), *Maerua crassifolia* (sarh), *Olea europaea* subsp. *africana* ('itm) *Ziziphus spina-christi* (sidr), *Jaubertia aucheri* (khurmān) and *Pteropyrum scoparium* (sidāf).

The coppice growth around the base of the tree is the browsing zone for goats and gazelles while the upper crown is browsed by camels. The browsing animals tend to select the young shoots before the thorns have hardened, although camels are able to cope with thorny mature growth.

Analyses of the common browse species is given in Table 19.1.

It is surprising that the young leaves of *Acacia tortilis* have a low crude protein content of 7%, particularly as young leaves and shoots usually have a higher crude protein content than mature foliage.

Table 19.1 Analysis of common browse species in Oman.

Species	Material	Ash %	C.P. % (N × 6.25)	P%	K%	DOMD* %	Source of data
Acacia ehrenbergiana	young shoots and leaves	7.34	10.62	1.03	0.54	58.2	TSAU 1984**
Acacia senegal		15.3	21.25	0.11	0.9		GRM 1982
	old leaves	11.56	17.43	0.12	0.82	59.2	TSAU 1984
Acacia tortilis		8.4	17.50	0.14	0.5		GRM 1982
	young leaves	7.71	7.06	0.07	0.66	41.2	TSAU 1984
Anogeissus dhofarica	old leaves	8.84	11.81	0.14	1.10	54.4	TSAU 1984
Maerua crassifolia		23.4	21.25	0.07	2.8		GRM 1982
Prosopis cineraria		9.8	13.13	0.09	0.7		GRM 1982
	young leaves	6.9	14.31	0.36	1.92	25.7	TSAU 1984
	mature pods		14.25	0.17	2.06		TSAU 1984
Jaubertia aucheri		12.1	8.13	0.08	0.50		GRM 1982
Pteropyrum scoparium		12.2	10.00	0.09	2.20		GRM 1982
Ziziphus spina-christi	old leaves	5.78	13.25	0.22	1.61	43.1	TSAU 1984

* % digestibility of organic matter in dry matter (DOMD) determined according to method of Tilley and Terry 1963.

**TSAU – analyses carried out by the Tropical Soils Analysis Unit, Land Resources Development Unit, Tolworth Tower or by the Ministry of Agriculture, Fisheries and Food, UK at the request of TSAU, on plant material collected by the author.

Figure 19.1   *Prosopis cineraria* (ghāf) woodlands in eastern Oman.

There are extensive areas of *Prosopis cineraria* (ghāf) woodlands on the edge of the sand dunes in eastern Oman (Fig. 19.1). These woodlands are lopped to provide browse for herds of camels and goats owned by nomadic pastoralists. It is proposed to control and manage the exploitation of this woodland resource in order to conserve it and provide a continuous yield of browse.

Camels, goats and wildlife eat the fallen ripe and unripe pods of *P. cineraria* (ghāf) which contain a pulp rich in crude protein (*c.* 18%) and carbohydrates (*c.* 50%) (Gupta *et al.* 1974). The fallen flowers and pods of *Acacia tortilis* (samr) are nutritious and form part of the rangeland resource.

Anogeissus dhofarica (sughat) is an important source of camel browse in the escarpment woodlands of Dhofar. The outer branches of the trees are partially cut through and bent over to bring the leaves within reach of the camels. The branches are not killed and the leaves sprout again. *Anogeissus* remains in leaf for three or four months after the monsoon and responds quickly to cyclonic storms at any time of the year by coming into leaf.

On the central plateau the lichen *Ramalina duriaei* occurs on the dead lower branches of the crowns of *Acacia tortilis* (samr). The dew-soaked lichen is soft and leafy during the night and remains so until about 0.900 hours in the morning; it is eaten by gazelles. Although the crude protein content at 7.4% is low, digestible carbohydrates content at 28.7% is high (Hawksworth *et al.* 1984). The lichen is therefore a source of energy for gazelles. This is of particular interest because the caribou in the cold northern deserts also get some of their energy needs from lichen.

A rare leguminous tree, *Ceratonia oreothauma* (tiyū) that grows on one or two of the eastern limestone mountains is a browse species and yields a nutritious pod. *Ceratonia* is potentially a valuable browse species and should be grown in the northern limestone mountains.

Honey

Indigenous trees and shrubs are a source of forage for the little bee *Apis florea* and the local strain of honey bee *A. mellifera.* Local honey is highly esteemed and highly priced, a 450 g jar or bottle may cost over 20 Rials Omani (*c.* £41 sterling).

In the spring of 1984 a heavy flowering of the following bee forage trees in northern Oman was observed by the writer: *Acacia ehrenbergiana* (salem), *A. tortilis* (samr), *Prosopis cineraria* (ghāf), and *Ziziphus spina-christi* (sidr).

In the escarpment woodlands of Dhofar the main bee forage trees were: *Acacia nilotica* subsp. *kraussiana, A. senegal sensu lato* (thumōr), *Anogeissus dhofarica* (sughat) and *Ziziphus spina-christi* (sidr).

On the Dhofar coastal plain, apart from cultivated plants, small areas of mangrove, *Avicennia marina* and the low woody shrub *Limonium* sp. are a source of bee forage.

Apiculturists are working with traditional bee keepers in northern Oman and in Dhofar in order to introduce modern methods of bee keeping and to increase the yield of local honey.

Resins, gums, latex and oil

Frankincense is a resin obtained from *Boswellia sacra* (mughur) that grows in the dry wadis of Dhofar. It occurs both north and south of the mountains. Dhofar has been famous for frankincense, locally known as lubān, or olibanum in the trade, for thousands of years. In the days of King Solomon frankincense from Dhofar was considered to be the most precious. An exhaustive history of the trade in frankincense and myrrh has been written by Groom (1981).

Incisions are made in the main stem and branches of the tree with a special chisel-like tool and the lubān oozes from the phloem; if the flow is copious it may fall to the ground and dry, otherwise it coagulates on the tree and forms tears. The highest quality lubān is white or clear, the lower grades are reddish or brown. The

Figure 19.2   A flow of lūban that will coagulate to form 'tears' of frankincense.

lūban is collected every 7–14 days and is spread out to dry in small caves. At the same time the edge of the incision is re-cut to stimulate the flow again. Figure 19.2.

Collection and drying can only be done during dry weather, so trees growing south of the mountains, in the wadis of the coastal plain, are not exploited during the monsoon (June–September) when humidity is high, but trees growing north of the mountains beyond the effects of the monsoon can be tapped for most of the year, although they are not tapped during the cold weather from November to about the end of February.

Local opinion maintains that the best quality lūban is obtained from the dry northern wadis, and the lūban from the coastal plain wadis is of inferior quality. Samples of lūban are now being collected from different localities at different seasons of the year, to determine whether there are differences in quality between localities and whether quality varies with the time of collection.

High quality lūban is used in the manufacture of perfume. A new perfume called 'Amourage' is being made in Oman using Dhofar frankincense. The price of a 50 ml bottle is over £500 sterling. This high price is partly due to the cost of the silver casket containing the bottle of perfume; re-fills are a little cheaper.

A number of *Commiphora* spp. occur in Dhofar; some of them have a scented resin that may be of value in the perfumery trade. It is suggested that this resource should be evaluated.

The large *Juniperus macropoda* ('al 'ālan) of the northern mountains yields a resin which is burnt like frankincense and an oil that is used locally as a cosmetic or to rub in the beard.

An oil is derived from the large seeds of *Moringa peregrina* (shūa'), a tree that grows on rocky mountain slopes. The oil was used for cooking and may still be so used by some of the poorer bedu who still follow the traditional nomadic life.

In the escarpment woodlands of Dhofar grows *Acacia senegal sensu lato* (thumōr), and this yields a clear gum, which comes within the complex of 'gum arabic' although it contains some slightly atypical morphological features (Anderson *et al.* 1983). The gum was used in the past to make a glue. It is proposed to use this tree to plant up gaps in the natural woodland, mainly because it is a good browse species and bee forage; it may also be possible to exploit the gum in future.

A latex is collected from the shrub *Euphorbia balsamifera* subsp. *adenensis* (tishaq'), that grows on the plateau and northern slopes of the Dhofar mountains. This latex is dried and sold as a chewing gum. It has no flavour, but perhaps a flavour could be added enabling a small local industry to be established.

Fruit trees

The most widespread indigenous fruit tree that grows in wadis throughout Oman is *Ziziphus spina-christi* (sidr). The small brown fruits (nabaq) of this tree are sold in the suk. An improved variety with large fruits is cultivated.

On the northern mountains *Sideroxylon mascatense** (= *Monotheca buxifolia*) (būt) produces a heavy crop of sweet black berries in about July. Sometimes the fruit is sold in the sūk for one Rial Omani per kilo. The fruit can be dried and stored.

Another small tree of the mountains that has a sweet red berry is *Sageretia spiciflora* (nimt). A breeding programme to select and develop varieties with larger fruit of both these species would be well worth while.

Tamarindus indica is another indigenous fruit tree that is sometimes cultivated.

---

*Sideroxylon mascatense (A.DC.) Penn. *comb. nov.*; basionym: *Monotheca mascatensis* A.DC. in DC., *Prodr.* 8: 152 (1844) & in Delessert, *Icon. Select.* 5: 15, t.35 (1846). For full synonymy see Friis in Kew Bull. 33, 1: 94–96 (1978) (T.D. Pennington).

Herbal tea

The leaves of a species of *Origanum,* (? *O. syriacum* L.) known as za'atar, makes an excellent herbal tea. Za'atar is a woody perennial herb which grows on the eastern slopes of Jabal Akhdar, the main part of the northern mountain range, and on some other northern and eastern limestone mountains. The leaves are sun-dried and sold together with petioles and pieces of twig.

It is proposed to carry out a feasibility study to determine the natural distribution of za'atar and to investigate methods of improving processing and marketing. It should be possible to establish a small industry to market za'atar in tea-bags. Za'atar is said to have medicinal properties and is particularly good for colds and coughs. It can also be used as a condiment and can be blended with other herbs such as mint.

Other plants

The pods of *Acacia nilotica* subsp. *adstringens* (karat) are a source of tannin.

The dried plants of *Euphorbia larica* (isbaq) are used to fire pottery kilns. This species is rich in hydrocarbons and produces a fierce burn. *E. larica* is very common in many parts of Oman and it is suggested that this plant could be grown in other desert regions and exploited for the production of fuel or hydrocarbons.

Many indigenous plants are used as traditional medicines. Some information on the medicinal and nutritional uses of Dhofar plants is being collected (M. Morris pers. comm.), but there is scope for further studies.

It is clear from this brief account that the Sultanate of Oman is rich in indigenous plants of economic value.

Acknowledgements

I wish to thank His Excellency Engineer Abdul Hafiz Bin Salim Rajab, the Minister of Agriculture and Fisheries, for permission to submit this paper. I thank my colleagues in the Ministry for their help, in particular Dr Abdul-Mumin Bin Mohammed Bin Ali Al Mjeni, the head of Agriculture Research for suggesting the inclusion of the Arabic names of plants after the botanical names, and my counterpart Engineer Said Bin Hamed Bin Khamis Al Alawi.

I thank Mr Richard Baker of the Tropical Soils Analysis Unit for the analysis of browse material and Mr A.J. Smyth, Director of the Land Resources Development Centre, for arranging for me to attend the Kew International Conference on Economic Plants for Arid Lands.

References

Anderson, D.W.W., M.M.E. Bridgeman, J.G.K. Farquhar and C.G.A. McNab 1983. The chemical characterization of the test article used in toxicological studies of gum arabic (*Acacia senegal* (L.) Willd.). *Int. Tree Crops J.* 2: 245–254.

G.R.M. 1982. *Range and livestock survey. Final report by G.R.M. International Pty Ltd to the Ministry of Agriculture and Fisheries, Sultanate of Oman.* Brisbane: G.R.M. Pty Ltd.

Groom, N. 1981. *Frankincense and myrrh. A study of the Arabian incense trade.* London: Longman.

Gupta, M.C., B.M. Gandhi and B.N. Tandon 1974. An unconventional legume – *Prosopis cineraria. Am. J. Clinical Nutr.*: 1035–1036 (27 October 1974).

Hawksworth, D.L., R.M. Lawton, P.G. Martin and K. Stanley-Price 1984. Nutritive value of *Ramalina duriaei* grazed by gazelles in Oman. *Lichenologist* 16: 93–94.

Tilley, J.M.A. and R.A. Terry 1963. A two-stage technique for the *in vitro* digestion of forage crops. *J. Brit. Grassl. Soc.* 18, 2: 104–111.

Whitcombe, R.P. 1982. *Preliminary Report Vol. 4. Climate No. 1. Meteorological conditions in the Batina Region of Oman.* Durham: Durham University Khabura Development Project, Dept. of Geography, University of Durham.

20   The ecological role of plant resources in the arid regions of China

Ren Jizhou, Hu Zizhi and Fu Yikun

Gansu Grassland Ecological Research Institute, Gansu Agricultural University, Lanzhou, China

Dry regions of China

The continentality and aridity of the climate in China sharply increases from the coast inland owing to the decreasing effect of the monsoon from the Pacific Ocean. The dry country covers about 40 per cent of the total area of China. According to the aridity, and their hydrothermic pattern, three categories may be recognized in China (Table 20.1 & Figs 20.1 & 2). They are distributed mainly in the northwest part of China. Every hydrothermic type has its own specific vegetation and species components.

Table 20.1 Arid regions in China.

	K value*	Landscape
Sub-arid	0.86−1.18	Tall steppe
Arid	0.28−0.85	Semi-desert
Super-arid	⟨ 0.28	Desert

$$* \, K \text{ value} = \frac{\text{annual precipitation}}{\text{accumulated temperature above } 0° \times 0.1}$$

The xerophytes in China and their ecological effects

According to their ecological characters, the xerophytes, can be divided into four categories: true xerophytes, super-xerophytes, extra-xerophytes and pseudo-xerophytes.

TRUE XEROPHYTES

The true xerophytes are the plants distributed on typical steppe with its regional vegetation. The typical steppe in China is situated in the vast area from the middle

Figure 20.1   The arid regions of China.

part of Shuang-Lio plain, north-eastern China, by way of the Inner Mongolia Plateau, to the south-west Plateau. Its north-west border merges into semi-desert. Its south-west edge extends to the typical steppe of Qinghai and Tibet Plateau.

The rainfall in the northern part of the steppe is 200–300 mm while that in the southern part is 300–450 mm. The K value is 0.86–1.18, and the soil is rich in organic matter.

The most typical true xerophytes are *Stipa grandis, S. krylovii, S. bungeana, S. capillata, Festuca valesiaca* subsp. *sulcata, F. ovina, Cleistogenes squarrosa, Agropyron cristatum, Koeleria micrantha (= K. cristata), Artemisia frigida, A. gmelinii, A. giraldii*, etc. All these are the principal constituents of the vegetation on China's steppe. Except for *Thymus mongolicus, Artemisia frigida, A. gmelini* and *A. giraldii*, the others are members of the Gramineae (grasses).

Their evapotranspiration varies greatly. For example, on the steppe of Hulumbeier, the evapotranspiration is 1:2000, but the ratio drops to 1:675–780 when the

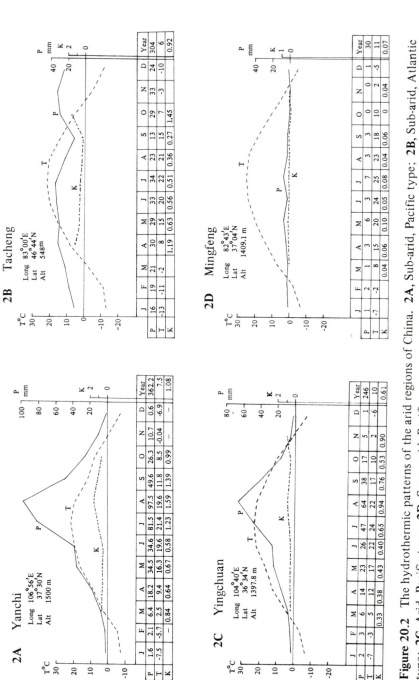

2A Yanchi
Long 106°56′E
Lat 37°30′N
Alt 1500 m

	J	F	M	A	M	J	J	A	S	O	N	D	Year
P	1.6	2.1	6.4	18.2	34.5	34.6	81.5	97.5	49.6	26.3	10.7	0.6	362.2
T	-7.5	-5.7	2.5	9.4	16.3	19.6	21.4	19.6	11.8	8.5	-0.04	-6.9	7.5
K	–	0.84	0.67	0.64	0.58	1.23	1.59	1.39	0.99	–	–	1.08	

2B Tacheng
Long 83°00′E
Lat 46°44′N
Alt 548m

	J	F	M	A	M	J	J	A	S	O	N	D	Year
P	16	19	21	30	29	33	34	23	13	29	33	24	304
T	-13	-11	-2	8	15	20	22	21	15	7	-3	-10	6
K				1.19	0.63	0.56	0.51	0.36	0.27	1.45			0.92

2C Yingchuan
Long 104°40′E
Lat 36°34′N
Alt 1397.8 m

	J	F	M	A	M	J	J	A	S	O	N	D	Year
P	2	3	6	14	23	26	47	64	38	17	5	1	246
T	-7	-3	5	12	17	22	24	22	17	10	2	-6	10
K			0.33	0.38	0.43	0.40	0.65	0.94	0.76	0.53	0.90		0.61

2D Mingfeng
Long 82°43′E
Lat 37°04′N
Alt 1409.1 m

	J	F	M	A	M	J	J	A	S	O	N	D	Year
P	1	2	8	3	6	3	7	3	3	0	0	1	30
T	-7	-2	8	15	20	24	25	23	18	10	2	-5	11
K			0.04	0.06	0.10	0.05	0.08	0.04	0.06		0.04		0.07

Figure 20.2 The hydrothermic patterns of the arid regions of China. **2A**, Sub-arid, Pacific type; **2B**, Sub-arid, Atlantic type; **2C**, Arid, Pacific type; **2D**, Super-arid, Pacific type.

rainfall is plentiful. It shows that the productivity of these grasses can be sharply increased with an improvement in the humidity.

The grass on the typical steppe in North China has the following characters in relation to forage production: (a) There is only one rest period throughout the year, which comes in winter, and none during the summer growing season. (b) There is only one peak of production in summer. The largest amount of dry matter forms rather late, approximately one month before the end of growing season, i.e. from the end of July to the beginning of September. (c) In autumn the grasses grow weakly and the amount produced is insignificant. The aerial parts of the plants wither before winter comes.

In such areas of xerophyte distribution in China, as in Inner-Mongolia, the temperature and humidity is insufficient for cultivated plants. So it is a traditional nomadic region, and good local, domestic breeds have been raised such as Mongolian horse, Mongolian cattle and Mongolian sheep. Grazing and browsing of domestic and wild animals is an essential factor for the maintenance of the grassland ecosystem and rational grazing is necessary for the maintenance of good pasture. Overgrazing is now a problem and causes the grassland to become depleted, followed by soil erosion and desertification.

SUPER-XEROPHYTES

Further west the environment becomes drier (in the north rainfall is 150–250 mm, while that in the south is 200–300 mm, K value is 0.28–0.85). Because of the drought, soil salinity increases and desertification becomes more severe. Super-xerophytic plants have taken the place of ordinary xerophytic plants and have become the main constituent of the vegetation of the semi-desert, the typical soil of which is brown earth.

Typical species of super-xerophytes forming the vegetation are *Stipa gobica, S. breviflora, S. glareosa, S. orientalis, S. caucasica, S. klemenzii, S. stapfii, Cleistogenes songorica, Allium polyrhizum, Ajania achilleoides* a small subshrub, *Artemisia dalai-lamae, A. salsoloides* var. *wellbyi, Hippolytia trifida,* etc.

The environmental conditions of the super-xerophylous plants, of which the water factor is particularly important, are harsher than those of normal xerophytes, and this fact is undoubtedly reflected by the morphological characters and the developmental rhythm of these species. For example, *Stipa gobica* has the most widespread distribution in China and plays an important role in the formation of desert-steppe vegetaion, is shorter than the typical steppe *Stipa,* usually only 10 to 20 cm high, more densely clustered, having a well developed fibrous root system and twisted thread-like leaves. Its vegetative renewal is closely related to the precipitation. Only in the period of August to September, when the soil moisture is relatively higher, tillering buds are formed to start the renewal process.

Among the super-xerophylous forbs, *Allium polyrhizum,* a caespitose bulbous plant, forms clumps 5 to 10 cm wide and 15 to 20 cm high, containing many fibrous-sheathed bulbs. In those years when there is a higher rainfall in the spring, it sprouts early, comes into head and blooms in July, with seed ripening in August. But when the drought occurs, it can remain in a dormant state into the late spring then only short and thin leaves emerge, which will soon become withered and yellow at the upper part, and the further development will be halted for the remaining months of the growing season.

EXTRA-XEROPHYTES

In the north-west of China, including the whole of the Xingjiang Autonomous Region, the Chaidamu Basin in Qinghai Province, the west section of the Hexi Corridor in Gansu Province, and the west and south part of the Inner Mongolia Autonomous Region, a vast area dominated by extra-xerophylous vegetation, covers about 2.2 million km^2, more than one fifth of the total area of China. This area has an extensive desert and gobi of about 1 million km^2, its centre lies more than 2000 or 3000 km in any direction from the sea. The precipitation varies between 50 and 200 mm, mostly less than 100 mm; various desert soils are the major soil types found in this area.

About 1000 species have been identified as extra-xerophylous plants in the true desert habitat of this area. Among those which form the main desert are as follows: *Haloxylon ammodendron, H. persicum*, Ephedra przewalskii, Zygophyllum xanthoxylum, Nitraria sphaerocarpa, N. roborowskii, Gymnocarpos przewalskii, Calligonum roborowskii, C. mongolicum, C. leucocladum, C. rubicundum, Ammopiptanthus mongolicus, Potaninia mongolica, Tetraena mongolica, Helianthemum soongoricum, Caragana korshinskii, C. tibetica, Ammodendron argenteum, Opuntia vulgaris (= O. monacantha), Reaumuria soongorica, Ceratoides latens, Salsola passerina, S. abrotanoides, Sympegma regelii, Iljinia regelii, Nanophyton erinaceum, Anabasis aphylla, A. salsa, A. brevifolia, Halostachys belangeriana, Halocnemum strobilaceum, Atriplex cana, Suaeda physophora, S. microphylla, Kalidium schrenkianum, K. cuspidatum, Artemisia sphaerocephala, A. arenaria, A. santolina, A. terrae-albae, A. borotalensis, A. kaschgarica, A. parvula, Brachanthemum gobicum, Asterothamnus centrali-asiaticus.*

In the long evolution process through the natural selection by the harsh environmental factors, especially by drought, the super-xerophylous and extra-xerophylous plants have developed various physiological functions and morphological structures to adapt themselves to waterstress. Some examples are as follows:

(a) The leaf blade (transpiration surface) has become smaller or degenerated, and the process of photosynthesis is carried out by green branchlets or stems, e.g. *Anabasis aphylla, Haloxylon ammodendron* and *Ephedra przewalskii.*

(b) The leaf blade or branch has well developed protective tissues (cuticle, wax layer, hairs and specific stoma structure and the way in which it acts) or has a protective colour, greyish white, e.g. *Calligonum* spp., *Ammodendron argenteum, Ceratoides latens, Artemisia* spp. and *Nanophyton erinaceum.*

*see footnote p. 129

(c) Plants having fleshy leaves or stems and supported by the water store in these tissues are enabled by the reduced transpiration to survive the drought period and the intense heat; such examples are *Opuntia vulgaris, Nitraria sphaerocarpa, Salsola passerina* and *Sympegma regelii.*

(d) The cell sap has a high salt content which maintains a high osmotic pressure to enable the plant to extract moisture from dry or saline soil, e.g. *Kalidium* spp., *Suaeda* spp., *Atriplex cana* and *Halostachys belangeriana.*

(e) Plants possessing deep and wide-spreading root systems to absorb water from a huge volume far below the ground surface (most of these are super-xerophylous or extra-xerophylous plants of salt-avoiding species).

(f) Plants having a dormant state, e.g. *Artemisia terrae-albae, A. borotalensis* and *Haloxylon ammodendron* or a partial defoliation (in most of the cases), or some branchlets fall off, e.g. *Haloxylon ammodendron* and *Calligonum* spp. during the extreme dry and hot period.

(g) Plants having a high tolerance to intense heat and dehydration, e.g. *Sympegma regelii, Anabasis brevifolia,* including some annuals, such as *Salsola lanata.*

(h) Plants sensitive to the rainfall and quickly respond to it, such as *Potaninia mongolica* and *Salsola passerina.* When they receive rainwater in the growing season, they come into bloom and form seed; this can be repeated in the same plant several times during the year.

Because of the uncongenial water and heat factors, the coarse parent soil material, the high salt content and the serious wind erosion, the desert ecosystem is very depauperate both in numbers of species and biomass. The conditions are so severe in this environment that the critical factors for survival of plant life are only just met. Nevertheless, the extra-xerophylous plants are able to tenaciously cling to life and form the principal substrate of the energy cycle in the ecosystem, thereby supporting some livestock production and slowing or halting the further deterioration of the desert environment. Therefore, from a global view, the importance and beneficial effect of the extra-xerophylous plant must not be underestimated.

PSEUDO-XEROPHYTES

In the typical steppe, semi-desert or desert, there are some other species not identified as any of the normal, super or extra-xerophylous plants. They are distributed in some relatively humid niches or periods which take place in the above areas. In fact, they are mesophytes, here temporarily called pseudo-xerophytes and divided into the following groups:

(a) Sand plants. In a typical steppe area dominated by normal xerophytes, some places are covered by a thick deposit of sand which can retain moisture underneath and mesophytes occur in these places, with such characteristic species as *Pinus sylvestris* subsp. *kalundensis (= P. sylvestris* subsp. *mongolica), Ulmus pumila,* etc.

(b) Meadow plants. In dry conditions, mesophytes can be found in some places with a relatively high underground water table, e.g. *Phragmites australis (= P. communis), Aneurolepidium dasystachys, Iris lactea, Glycyrrhiza inflata, Alhagi sparsifolia, Trachomitum lancifolium,* etc.

(c) Ephemeral plants. The desert region in the west of Xinjiang receives its rainfall mainly in the spring. Therefore, mesophylous ephemerals and ephermeral-like species have a chance to complete their life cycle in this humid period and to form thriving meadows in some areas. When the rainy season of 1 to 2.5 months ends, the life cycle of the ephemeral plants will also be completed and their remains quickly decomposed, and the normal desert features will be restored to the habitat once again. These ephemerals include *Papaver pavoninum, Trigonella arcuata, Alyssum desertorum, Chorispora soongorica, Erodium hoefftianum, Poa bulbosa* subsp. *viviparum, Carex physodes, C. stenophylla (= C. pachystylis), Tulipa iliensis,* etc.

The existence of the pseudo-xerophytes is evidence of the effect produced by the humid niches and humid periods in an arid region. It indicates that within the limit of the space and the period, the energy and water flows have been enlarged and converted into organic matter in the most economical way. As a result, the biomass has acquired a substantial increase which could make a great contribution to local pastoral farming and forestry production. Therefore, research into the humid niches and the periods of rainfall in arid regions would have great significance for an increase in the ecological elasticity and productivity of the region. It is sad indeed that not only is no attention paid to this aspect, but the most serious damage takes place in those areas. The sand lands have been deforested, the meadows and the ephemeral vegetation have been overgrazed and destroyed to such an extent that development has been halted or largely extinguished. This is a great waste of the precious water resources available and potential in the ecological-economic system of an arid region.

The situation for xerophytes in China

Because of various ecological environments, great and significant differences exist in the situation of Chinese xerophytes.

The first group of xerophytes have lived and adapted in this extreme climate for centuries and are very short and difficult to use. Such true xerophytes as *Stipa gobica, S. breviflora, S. glareosa, S. orientalis, S. caucasica,* etc., and subshrubs such as *Ajania achilleoides, Hippolytia trifida, Artemisia dalai-lamae, Helianthemum soongoricum* grow in an ecological environment with low humidity, high entropy value and low vegetation cover and density. There is very little biomass and a great reduction in the animals and microbes who live on the primary producers; it is an unsaturated environment. But xerophytes living in deserts can maintain an adequate water balance within the environment because of their sparse distribution, large stem:root ratio and morphological characters for drought resistance. They are protected naturally and they can complete their life cycle without any interruption because their low productivity makes them useless for grazing and fuel. These plants seem to have been forgotten by man and form the most stable xerophytic subsystem in grassland ecosystems today; they could be called 'inert plants'.

The second group of xerophytes have an important position in the food chain of secondary production. They are continuously struggling between exhaustion and recovery; because they are the most unstable they could be called 'active plants'. For example, the natural vegetation is destroyed after the ploughing of grassland, but it will recover naturally through a series of successions. Normally, in the first and second years after cultivation has ceased, a number of annuals appear and they are unstable. If cultivated or grazed again without any protection, it would result in serious desertification. If it is left, three or four years later, rhizomic grasses will predominate and this is called 'the stage of rhizome grasses'. In this stage, the grasses are suitable for cutting for maximum feeding value and herbage yield. But if no artificial methods of management are adopted within 7–8 years, the soil would begin to cap and the rhizome grasses would be replaced by xerophytic tussock grasses and some annual xerophytes, and this is called 'the stage of tussock grasses'. After ten or more years, subshrubs begin to invade and weeds spread, to change the relative stable original vegetation.

The third group is the medium type of xerophytes intermediate between the 'inert plants' and 'active plants'. However, the majority of species between the two groups are essentially xerophytes.

The first group of xerophytes contains two sub-groups. One of them is poisonous and harmful to animals, their hard hairs, awns and thorns can injure animals mechanically, or they are inedible due to their bad smell or taste. Such plants are not or seldom eaten by animals. Basically, this group does not take part in the grassland process of secondary production, and they can propagate and develop steadily. For instance, in over-grazed grasslands or areas surrounding drinking water or herdsmen's houses, good pastures suffer from repeated grazing and trampling, and become very weak and eventually die out, creating a niche for *Stellera chamaejasme, Calystegia hederacea* and *Thermopsis lanceolata,* which are inedible and poisonous, and propagate so successfully that they become dominant in the community. This group of xerophytes can be referred to as inert plants in the ecosystem.

Inert plants play an important role in eco-production.

(a) As an indicator of change in ecological environment, caused by bad conditions of soil moisture, salinization and the pH value.

(b) They reduce the area available for good pasture species of higher feeding value and compete with them for water and nutrients. Primary production of grassland, both in quality and quantity, is reduced and the first link of the food chain in grass-animal-man ecosystem becomes weak and fragile.

(c) They reduce biomass from secondary production. The more inert plants increase, the less edible plants will be present and this would cause reduction in animal production. During the serious drought of 1976, for example, many good pastures in the south of Igzhao prefecture of Inner Mongolia, were extremely poor, but *Oxytropis glabra* grew vigorously and attracted the animals to eat more. As a result, 11 per cent of the animals in Mushenzhao commune died from *Oxytropis glabra* poisoning; 45 per cent were horses. Another example was shown by the Sanjuchen sheep farm in 1954; their sheep flock was injured by the needlegrass seeds when returning to the farm after dipping, causing a serious economic loss. In Huang Chentan sheep farm in Gansu, over 100 hybrid sheep died for the same reason.

(d) This inert plant group plays an important role in checking wind, controlling sand movement and stopping the desertification. To provide favourable conditions for good pastures, inert plants normally need to be controlled by means of chemical, biological and ecological methods; those of value in soil conservation and Chinese herbal medicine should be preserved.

Xerophytic grass species of high quality play an important part in energy flow of the ecosystem. This group of xerophytes are referred to as active plants and to a certain extent they can be counted as the key component or mainstay in the arid ecological environment.

Table 20.2 Herbage Productivity in Arid Grasslands.

Types of dry grassland	Grass yield kg/ha	Grasses %	Legumes %	Sedges %	Forbs %	Bushes %
Sub-arid	1500–4000	50–80	3–9	2–4	15–45	0–10
Arid	800–1500	32	–	–	12	56
Alpine-arid	300–1500	40–50	5–15	0.5	15–20	10–15
Super-arid$_a$	500–1000	28	–	–	–	72
Super-arid$_b$	400–800	0–1	–	–	–	99–100

Active plants can be used by animals as their food resource in secondary production. The productivity and stocking rate of Chinese xerophytes are gradually reduced from east to west with the increasing of aridity. The area of grassland required per sheep unit is about 1 ha for sub-arid grassland, 1.6 ha for arid grasslands and 2.6–4 ha for super-arid and Alpine-arid grasslands.

From the nutritional point of view, the content of crude protein in the arid tussock grasses can be about 10 per cent during the growing season. It it worth noting that the drier the climate or the more arid the environment, the higher the protein content. However, the more arid the ecological environment the lower the carbohydrate content. The protein content of some grass groups growing in super-arid grassland can be over 14 per cent, but N-free extracts only 29.5 per cent.

There is a close relationship between the type of xerophytic community and animal breeds in a specific sub-subsystem. For example, Mongolian sheep, Mongolian horse and Kazak sheep are adaptable to the grassland where arid plants predominate, while goats, camels and lamb-skin breeds are adaptable to grassland composed of super-arid plant species.

Xerophytes react differently to situations brought about by the increasing number of ecosystems interrupted by human activities. Under the conditions of overgrazing, frequent cutting, burning and ploughing, xerophytes decline rapidly resulting in soil erosion and desertification so that plants would be destroyed and the ecosystem would be seriously out of balance.

Table 20.3 The Succession Series by Grazing in Arid Grasslands.

Normal grazing	*Caragana microphylla, Stipa krylovii*
Little heavy grazing	*Caragana microphylla, Stipa krylovii, Artemisia frigida*
Heavy grazing	*Artemisia frigida, Stipa krylovii, Heteropappus altaicus*
Extreme over-grazing	*Heteropappus altaicus, Artemisia frigida, Artemisia scoparia*

The proper management of the ecosystem by man, such as by rotational grazing, improvement and reseeding with good quality grasses, has caused a wide range of active plants to be protected and grown well, and grassland productivity has been greatly improved on a large scale.

Active plants are mainly: *Aneurolepidium chinense, Agropyron cristatum, A. desertorum, A. mongolicum, Koeleria macrantha, Semiarundinaria shapoensis*; many species of *Stipa*; Legumes – *Trigonella ruthenica, Astragalus melitotoides, A. scaberrimus, Oxytropis squamulosa, Lespedeza dahurica* and *Caragana* spp. There are also other plants such as *Kochia scoparia, Ceratoides latens, Artemisia frigida, A. ordosica, Allium mongolicum,* etc. Some super arid-shrubs and subshrubs can be used not only as forage but also as windbreaks, such as: *Haloxylon persicum, H. ammodendron, Salsola passerina, S. arbuscula, S. rigida (= S. orientalis), Artemisia borotalensis, A. argyi, A. kaschgarica, A. terrae-albae, A. schischkinii, Calligonum mongolicum, C. caput-medusae, Hedysarum scoparium,* etc.

The main inert plants are: *Achnatherum inebrians, Clematis aethusaefolia, Chelidonium majus, Hyoscyamus niger, Astragalus variabilis, Oxytropis glabra, Thermopsis lanceolata, Stellera chamaejasme,* etc.

Conclusions

Drought resistance of plants has attracted the attention of botanists for many years. This adaptability for drought stress always shows active response to biotic and abiotic factors. The activity of arid land plants is very sensitive to humidity and some give peak production and flowering after every rainfall. They contribute the main constituent of the arid grassland ecosystem's primary production and provide pastures for the animals. The biomass of the active xerophytes can increase and decrease rapidly. They play an important role in both pasture and soil conservation.

Unlike the active plants, some plants always seem more 'inert' in any environment, whether it is favourable or not. They are very persistent but they are hardly changed by improved circumstances. They contribute little to the pastures but may be very important for soil conservation.

Table 20.4 Activity of Plants.

	Inert	Medium	Active
Persistence in poor sites	+++	++	+
Influence by the rainfall increase	+	++	+++
Palatibility	−	++	+++
Contribute to the ecosystem biomass	+	++	+++

The main characteristics of different activities of plants are summarised in Table 20.4. The persistence of the inert plants is much better than the active ones, but the response to the rainfall and grazing is much less. The active species form the key component of the vegetation in dry regions. We should pay much greater attention to them for land conservation and forage production.

21 Plants of the Australian arid zone – an undeveloped potential

The late J. R. Maconochie

Arid Zone Research Institute, Conservation Commission of the Northern Territory, PO Box 1046, Alice Springs, NT 5750, Australia

Introduction

Until more recently, the plants of the Australian arid zone have had very little exploitive pressure applied. It is only in the last 100–150 years has there been a grazing pressure of any significance from rabbits, horses, donkeys, cattle, camels and in some parts, sheep and goats. These are all animals introduced by European man to the Australian continent.

Prior to European man, the aboriginal people were hunters and gatherers, harvesting the seed, fruits and tubers when conditions were favourable and hunting the marsupial kangaroo, euro and wallaby or the goanna lizard as an important source of protein. These animals were hunted with spears or trapped and harvested through the use of bushfires. It would appear that the aboriginal social customs of designating certain waterholes and surrounding areas as sacred, acted as a very effective means of conserving and maintaining the marsupial population, and hence their own protein supply, during periods of drought when plant resources were low.

Summaries of the knowledge of the biology of arid zone shrubs have been presented by McKell *et al.* (1972). The report by the United States National Academy of Sciences (Anon. 1975) has given a more overall picture of under-exploited tropical plants with economical potential, this report also incorporates select arid zone species.

McKell (1975) has reviewed the role and value of shrubs in the arid land environment. In his review, he critically re-assessed many of the premises and misconceptions that are often presented, and in particular, the often quoted statement that shrubs are worthless and replace the more palatable herbs and grasses.

In Australia, there has been limited work undertaken on assessing the potential of plants of the arid zone, except for those plants which are utilized by the grazing industry. Everist (1959), Cunningham (1971), and Askew and Mitchell (1978) have summarized the information on the fodder trees and shrubs of the inland regions of the different Australian states.

This report presents information of a selection of Australian arid zone trees, shrubs and grasses which have considerable potential for revegetating and stabilizing degraded and denuded parts of the world's arid lands.

Plants for arid lands   © *Royal Botanic Gardens, Kew, 1985*

Forage grasses

Table 21.1 presents information on a number of native grasses and two introduced species, one of which is used in soil reclamation work in central and northern Australia. The table gives details of habit, soil type and the mean annual temperature and rainfall ranges for the natural distribution of the species in central and northern Australia.

In general, these grasses do not significantly differ in nutrient content from the two introduced grasses *Cenchrus ciliaris* and *Rhynchelytrum repens. Cenchrus* has been used extensively for soil and land conservation and reclamation while *Rhynchelytrum* has become an adventive weed along road verges.

The native grasses are generally more palatable to cattle than *Cenchrus* and this is one of the reasons for the latter's success in land restoration. There has been little or no breeding or selection work undertaken on the native grasses of arid Australia, and it is probable that if the same effort and research was spent with these species as has been done for *Cenchrus,* then suitable cultivars would be available for different uses.

Forage trees and shrubs

Table 21.2, p.294 presents a selection of trees and shrubs useful for forage, fuel and amenity purposes. These species are listed on a multi-purpose basis as this presents a better evaluation of their use to the arid zone communities of developing countries. These communities obviously need plants of a multi-use function, namely, forage — fuel — soil stability if the degradation and desertification is to be hindered or controlled. The *Acacia* species as well as providing stock forage, fuel and honey also improve the nutrient status of the soil (Lawrie 1981).

The eucalypts, which do not rate high as stock fodder because of their high oil and wax content of the foliage, provide fuel for cooking and in the case of the mallee species, readily coppice from the ligno-tuber after cutting.

The ligno-tuber gives stability to the soil surface, acting as a litter and sand trap against wind and water erosion. The eucalypt flowers provide large quantities of pollen and nectar for bee forage and thus can enhance local village industries.

Of the other species listed, the members of the Proteaceae (*Hakea* and *Grevillea* spp.) are prolific in their seasonal flowering and these flowers often drip nectar, again providing ample bee forage for honey production. *Atalaya, Capparis, Santalum* and *Ventilago* are all highly palatable species to cattle and sheep, probably providing important nutrients when the native grasses have dried off. The fruits of *Santalum* and *Capparis* in addition are suitable for human consumption.

Table 21.1 Summary of the habit, habitat, chemical composition and palatability of a selection of forage grasses from central Australia

A = annual, P = perennial, x–xxx = increasing palatability

Species	Life cycle	Height (cm)	Soil type	Annual rainfall (mm)	Average temperature range °C	Ca(%)	Mg(%)	N(%)	P(%)	K(%)	Na(%)	DMD	Palatability
Astrebla pectinata	P	30–120	clay	250–500	3–38	0.51 0.16	0.12 0.17	0.55 0.94	0.13 0.11	0.52 0.78	0.02 0.02	31.8 37.3	x
Bothriochloa ewartiana	P	60–90	alluvial water courses red earths	125–500	5–38	0.18 0.14	0.10 0.06	0.59 0.67	0.10 0.16	0.43 0.85	0.01 0.01	33.3 45.5	xxx
Cenchrus ciliaris	P	60–100	alluvials; red earths; clays	250 +	–	0.38 0.21	0.13 0.10	1.81 0.69	0.17 0.22	1.96 1.95	1.40 0.03	56.9 31.2	x
Dichanthium sericeum	P	50–70	red & yellow earths; clays	200–750	3–38	0.32	0.05	0.35	0.11	0.76	0.01	38.1	xxx
Digitaria brownii	P	30–45	red earths	125–500	5–38	0.11	0.13	0.58	0.21	0.61	0.01	45.4	xxx
Digitaria coenicola	P	30–45	alluvials; sandy red earths	200–250	3–37	0.22	0.10	0.73	0.14	1.28	0.01	39.8	xxx
Enneapogon avenaceus	A	15–25	calcareous & alluvial sands	125–175	3–38	0.44	0.10	1.28	0.14	0.84	0.03	43.4	xxx
Enneapogon polyphyllus	A	20–45	alluvials; sandy clays	125–400	3–38	0.95	0.07	0.66	0.07	0.78	0.01	41.9	xxx

Table 21.1 – continued

Species	Life cycle	Height (cm)	Soil type	Annual rainfall (mm)	Average temperature range °C	Ca(%)	Mg(%)	N(%)	P(%)	K(%)	Na(%)	DMD	Palatability
Enteropogon acicularis	P	45–100	alluvials	175–275	3–38	0.25	0.10	0.84	0.13	1.00	0.02	34.8	xxx
Eragrostis setifolia	P	22–45	alluvials; red earths; clays	125–400	3–38	0.12 0.23 0.23	0.16 0.07 0.06	0.57 0.54 0.87	0.10 0.09 0.11	0.46 0.31 0.68	0.01 0.02 0.03	20.9 25.0 40.0	xx
Eulalia fulva	P	50–100	alluvial water course; red earths	150–750	3–38	0.29	0.03	0.33	0.05	0.43	0.01	25.0	xx
Monachather paradoxa	A/P	30–60	red earths	125–250	3–38	0.16	0.05	0.96	0.09	0.97	0.01	42.0	xxx
Panicum decompositum	A/P	45–100	alluvials; clays	200–750	3–38	0.24	0.23	0.71	0.10	1.0	0.02	40.2	xxx
Rhynchelytrum repens	A	30–50	disturbed road sides	250 +	–	0.23	0.08	0.73	0.19	1.88	0.02	44.6	x
Thyridolepis mitchelliana	P	20–30	red earths; clayey sands	125–275	5–38	0.09	0.05	0.60	0.06	0.48	0.01	39.8	xxx
Triraphis mollis	A/P	30–45	alluvials; red earths	125–300	3–38	0.14	0.07	1.57	0.21	0.86	0.02	54.5	x

Psoralea walkingtonii and *P. australasica* are two highly nutritious and palatable species to cattle and sheep. *P. australasica* is an annual to about 1 m high and occurs along floodouts in sandy alluvial soil. Under natural rainfall this species yielded about 3.5 tonnes/ha of dry matter or about 530 kg protein/ha, its potential under irrigation has yet to be assessed. *Psoralea walkingtonii* is an erect perennial to 3 m high and is found only on alluvial floodouts in a few localities. It is highly preferred by cattle and they literally break it down. If developed this plant has considerable potential as a forage species, particularly in the arid sub tropics.

Erosion control and land stabilization

In Table 21.2, p.294, a number of the trees and shrubs are rated for their value for soil stabilization and amenity or shade/wind break planting.

Acacia ligulata and *Allocasuarina decaisneana* are found associated with the sand-dune system of central Australia and, in conjunction with the 'spinifexes' or 'hummock' grasses *Triodia* and *Plectrachne,* give considerable stability to these dune fields by ameliorating the effects of wind and water erosion. Both species are low in palatability to cattle but provide fuel for cooking and improve the nitrogen level of the soil, and thus provide a more favourable environment for the establishment of forage grasses.

There are three grasses in central Australia which are important sand binders, namely *Amphipogon caricinus, Paraneurachne muelleri* and *Zygochloa paradoxa. Amphipogon caricinus* is a perennial bunch or tussock grass which can form relatively thick swards on open sand plains. *Paraneurachne muelleri* is a perennial tussock grass found on the flat sand plains and spreads by stolons which may extend several metres. *Zygochloa paradoxa,* 'sandhill cane grass', is a round but divaricately branched perennial tussock grass which grows mostly on the sides and crests of sand-dunes. In gross morphology it is very similar in appearance to *Panicum turgidum* of the Afro-Arabian-Indian region, but this latter species is found in a wider array of habits than *Zygochloa.* These two taxa, which grown in similar environments, appear to exhibit parallel evolution of habit, stem and leaf morphology.

The other important group of plants which should be considered for desert sand plain/sand-dune reclamation are the genera *Triodia* and *Plectrachne.* These perennial hummock grasses dominate the arid lands of Australia, they are low in palatability, extremely drought resistant and able to reproduce by stolons or seed. They may well provide the basis for re-vegetating and stabilizing the margins of the Sahara and other very mobile deserts. In Australia, the 'spinifex' deserts are considered to be of no value because cattle cannot survive on the forage available, but in de-vegetated zones of the other parts of the arid zone, these plants may provide the stability for re-colonization of more palatable species. They have a high lipid content

Table 21.2 Summary of habit, habitat, usage and chemical composition of select top feed or browse species from central Australia
T = tree, S = shrub, A = annual, + = bee forage, x–xxx = increasing potential.

Species	Habit (m)	Growth rate	Regeneration	Soil type	Annual rainfall (mm)	Average temperature range °C	Food	Forage	Fuel	Soil stability	Amenity	Ca	Mg	N	P	K	DMD
Acacia aneura	T/S 1–4	slow	seed	red earths	125–300	3–38	x	x	x	x	x	1.85	0.15	1.87	0.11	0.85	40.4
Acacia calcicola	S 1–2	slow	seed	red earths over limestone	125–300	3–36			x	x	x	na	na	na	na	na	na
Acacia cowleana	S 1–3	fast	seed	skeletal; alluvials	250–375	5–38	x		x		x	na	na	na	na	na	36.4
Acacia cuthbertsonii	S 1–2	slow	seed	skeletal; rocky	125–375	5–38	x		x		x	na	na	na	na	na	na
Acacia estrophiolata	T 4–10	slow	seed	alluvials; sands	125–275	3–38		xx	x	x	x	1.92	0.20	1.78	0.11	1.01	39.1
Acacia kempeana	S 1–3	mod–fast	seed	alluvials; red earths over limestone	125–300	3–38	x	+xx	x	x	x	1.56	0.20	1.61	0.09	0.56	36.3
Acacia ligulata	S 1–3	fast	seed	skeletal, rocky, sands	125–500	3–38		x	x	x	x	2.10	1.60	1.61	0.04	na	48.3
Acacia murrayana	S 1–3	fast	suckers, seed	alluvials; red earths; sands	125–275	3–38	x	+x	x	x	x	2.0	0.5	2.51	0.08	na	31.2

Table 21.2 – continued

Species	Habit (m)	Growth rate	Regeneration	Soil type	Annual rainfall (mm)	Average temperature range °C	Food	Forage	Fuel	Soil stability	Amenity	Ca	Mg	N	P	K	DMD
Acacia victoriae	S 1–3	fast	seed	alluvials; clays	125–500	3–38	x	+xx	x	x	x	2.30	0.30	1.78	0.09	0.64	37.4
Allocasuarina decaisneana	T 4–10	slow	seed	dune sands	125–200	3–38			x	x	x	1.5	0.50	0.8	0.02	na	na
Atalaya hemiglauca	S/T 2–5	fast	seed suckers	alluvials	175–750	5–38		xxx	x	x	x	2.39	0.45	1.69	0.17	1.10	33.3
Capparis nummularia	S 1	fast	seed	alluvials	200–250	3–38	x	xxx			x	2.3	0.37	4.55	0.40	3.49	70.7
Eucalyptus camaldulensis	T 4–10	fast	coppice, seed	alluvials; riparian	125–750	3–38		+	x		x	2.31 2.23 2.52	0.27 0.22 0.24	1.15 1.20 0.87	0.16 0.08 0.07	0.74 0.97 0.99	32.7 32.6 31.0
Eucalyptus gamophylla	S 1–3	fast	coppice, seed	sands	125–300	3–38		+	x	x	x	1.06 1.01	0.28 0.27	1.04 1.04	0.07 0.10	0.57 0.92	32.4 34.2
Eucalyptus intertexta	T 4–8	mod.	coppice, seed	alluvials	125–300	3–38		+x	x		x	1.60 3.58	0.48 0.37	0.88 0.99	0.06 0.08	1.01 0.89	24.2 28.6
Eucalyptus microtheca	T 4–8	mod.	coppice, seed	alluvials	125–750	3–38		+x	x		x	1.23	0.30	1.25	0.01	1.34	34.0
Eucalyptus oxymitra	S 1–3	mod.	coppice, seed	skeletal; sands	125–250	3–38		+	x	x	x	1.60	0.50	1.60	0.06	na	na

Table 21.2 – continued

Species	Habit (m)	Growth rate	Regeneration	Soil type	Annual rainfall (mm)	Average temperature range °C	Food	Forage	Fuel	Soil stability	Amenity	Ca	Mg	N	P	K	DMD
Eucalyptus pachyphylla	S 1–2	mod.	coppice, seed	sands	125–350	3–38		+	x	x	x	0.90	0.80	0.80	0.06	na	na
Eucalyptus socialis	S 1–3	mod.	coppice, seed	skeletal	125–275	3–38		+	x	x	x	1.34	0.14	0.79	0.08	0.79	37.4
Eucalyptus terminalis	T 3–5	mod.	coppice, seed	alluvials; red earths	125–750	3–38		x +	x		x	1.83 1.05 1.36	0.32 0.20 0.29	0.64 1.33 0.89	0.28 0.15 0.09	0.47 1.42 1.10	20.3 19.8 23.3
Grevillea juncifolia	S 2–4	mod.	seed	sands	125–300	3–38		+x	x	x	x	3.5	1.1	0.94	0.05	na	na
Grevillea stenobotrya	S 2–3	mod.	seed	sands	125–300	3–38		x +	x	x	x						
Hakea suberea	T 3–5	slow	seed	alluvials	125–275	3–38		x +	x	x	x	0.84 0.73 1.18	0.11 0.13 0.25	0.71 0.62 0.84	0.05 0.05 0.05	1.01 0.66 0.57	18.8 20.5 18.3
Psoralea australasica	A 1	fast	seed	alluvials; sands; red earths	125–375	3–38		xxx		x		1.53	0.22	2.46	0.23	1.96	69.1
Psoralea wal-kingtonii	S 1–2	fast	seed	alluvials	275–300	5–38		xxx			x	1.1 1.2	0.42 0.25	3.2 3.2	0.22 0.21	2.2 2.7	49.8 60.7

Table 21.2 – continued

Species	Habit (m)	Growth rate	Regeneration	Soil type	Annual rainfall (mm)	Average temperature range °C	Food	Forage	Fuel	Soil stability	Amenity	Ca	Mg	N	P	K	DMD
Santalum acuminatum	S 1–3	mod.	seed root parasite	sands	125–275	3–38		x	xx		x	3.2	1.2	1.86	0.06	na	na
Santalum lanceolatum	S 1–2	mod.	seed root parasite	alluvials	125–750	3–38	x	xxx				1.03	0.19	1.38	0.29	2.84	59.4
Ventilago viminalis	T 2–4	fast	seed, suckers	alluvials; red earths; clays	250–1000	5–38	x	xxx	x	x	x	3.18	0.2	1.59	0.17	0.95	39.0
												3.64	0.23	1.42	0.13	1.14	

(13.7–17.3% dry weight) and may have the potential for use as a compact pellet fuel, an area of fuel technology not as yet considered. Spinifex deserts are naturally burnt as a result of lightning strikes or man made fires, there is a need for an assessment of its use in soil stabilization.

Food plants for human consumption

The Australian aboriginal who inhabited the arid inland was basically a hunter and gatherer, and the population was in an ecological balance with the resource available.

They had no beast of burden nor did they till the soil. The plant foods which they collected were rich in nutrients but low in bulk. Latz (1982) has prepared a compendium on the 140 food plants used by the Australian aborigines of this region and several examples will be given here of those with economic potential.

Solanum centrale: one of several species which have edible fruit. The fruit has a raisin or sultana appearance and contains between 2.9–8.4% protein, 5.5% fat, 53.9% carbohydrate and 25.3 mg/100 gm ascorbic acid (Latz 1982, Peterson 1979).

Vigna lanceolata: a perennial spreading sub-shrub producing bimorphic seed which are probably edible, and underground tubers which have 2.8% protein and 17.5% carbohydrate (Dadswell 1934). Both these species have the potential for providing food to village communities having limited water resources.

Acacia species: There are a number of *Acacia* species which have edible seed and these seeds are usually cooked on the fire by aboriginal people before being consumed. Weder (1978, 1981) has determined the proteinase inhibitor levels of some of these seed and it is likely that the heating of the ground seed meal destroys these inhibitors. Table 21.3 presents some data on nutrient content and seed yield of several *Acacia* species.

Table 21.3 Selected *Acacia* seeds, their nutrient content (Latz 1982) and yield

Species	Protein	Fat g/100g	Carbo- hydrate	Seed yield kg/tree
A. aneura	27.7	5.8	53.8	1.5
A. cowleana	23.5	8.9	55.3	0.23
A. cuthbertsonii	na	na	na	1.44
A. dictyophleba	23.1	8.7	54.1	na
A. murryana	18.4	4.0	62.5	na
A. victoriae	16.1	2.8	68.5	3.3

Thus, the *Acacia* species are of considerable importance in any revegetation and re-afforestation program, as these provide food, stock forage, fuel, as well as improving the nutrient status of the soil through nitrogen fixation and a build up of organic material below the canopy. Further research is needed here to assess their potential.

Medicinal plants

Knowledge of the medicinal plants used by the people of arid Australia is still rather scanty, the health workers, pharmacists and doctors in the Northern Territory Department of Health and others, Henshall *et al.* (1982) and Latz (1982), are slowly accumulating data, but there are considerable difficulties in separating the religious or psychological effects from the chemo-therapeutic role of the aqueous extracts of these plants.

Two groups of plants that may have value are the *Eremophila* spp. (Myoporaceae) and members of the Asteraceae.

Eremophila alternifolia is an aromatic bushy shrub to 2 m high and the plant is held in high esteem by the old men of the Pitjantjatjara people. The flowers and leaves are dried and ground up into a paste and applied to the body or a small amount may be added to water and the infusion drunk. No analyses have been made for tannin, alkaloid or saponin content of this plant so its commercial potential is as yet unknown.

Pterocaulon sphacelatum and *Streptoglossa odora* are both viscid and aromatic composites which are used as decongestants by grinding finely and inhaling, or as a pillow or an infusion made from the material and the liquid rubbed on the chest.

Southwell and Maconochie (1977) analysed a related species, *Streptoglossa bubakii* (= *Pterigeron bubakii*), for essential oils as it is used as a fly repellent by aboriginal women, but the results showed no significant repellency against the house fly.

There are a number of aromatic herbs and sub-shrubs used by the aboriginal people (Latz 1981) and it is possible that these plants may have economic potential once the cost of synthetic compounds reaches a critical point. Until that time there is still the need to continuously document the usage of such plants by the communities so that the information is not lost as tribal elders die.

Amenity planting

The use of trees and shrubs for purposes other than producing fuel or fodder should not be neglected in developing countries. Amenity planting for beautification and shade may create a more favourable situation in a hostile environment for a potentially successful program of better land management. Central Australia has a number of trees and shrubs which are unpalatable to stock, but while growing in a hostile environment, do have attractive shapes, forms and flowers. Table 21.2, p.294 indicates the amenity usage of a number of fodder trees, but as well as these there are genera in the Chloanthaceae (*Dicrastylis* and *Newcastelia*), Papilionoideae (*Clianthus, Indigofera* and *Templetonia*), Mimosoideae (*Acacia*), Proteaceae (*Grevillea* and *Hakea*), Myrtaceae (*Eucalyptus, Callistemon* and *Melaleuca*) and other families all with potential for development in the Amenity/Horticulture field. Very little is known of the biology of most of these species so that there are probably many drought survival mechanisms which need to be assessed before successful horticultural development can be achieved.

Fuel

Table 21.2 gives a brief assessment of the fuel value of some central Australian trees and shrubs. There are more *Acacia* and *Eucalyptus* species with potential, but those listed in Table 21.2 should be given priority. Boland and Turnbull (1981) have reviewed a selection of Australian trees other than eucalypts for this purpose.

The mallees eucalypts with their ligno tubers and numerous short trunks (2–5 m) have a considerable potential for providing fuel and windbreaks for village communities. There may also be the potential for eucalyptus oil extraction at the village level if the most suitable species are chosen.

Summary

This paper presents a list of central Australian plants with potential economic value to developing countries with a semi-arid or arid climate and having land in need of re-vegetation and re-afforestation. To date only mulga (*Acacia aneura*), river gum (*Eucalyptus camaldulensis*) and old man saltbush (*Atriplex nummularia*) have been used in any large scale trials, thus the Australian arid flora has a tremendous potential yet to be tapped.

Acknowledgements

I am very grateful to P.K. Latz for the use of unpublished data and to D.M.R. Newman and I. Hamdorf for digestibility and chemical values respectively.

References

Anon. 1975. *Underexploited tropical plants with promising economic value.* Washington, D.C.: National Academy of Sciences.

Askew, K. and A.S. Mitchell 1978. *The fodder trees and shrubs of the Northern Territory.* Darwin: Division of Primary Industry, N.T. Government.

Boland, D.J. and J.W. Turnbull 1981. Selection of Australian trees other than eucalypts for trials as fuelwood species in developing countries. *Austr. Forestry* 44(4): 235–246.

Cunningham, G.M. 1971. Recognizing western fodder trees. *J. Soil Cons. Serv. N.S.W.* 27(1): 25–61.

Dadswell, I.W. 1934. The chemical composition of some plants used by Australian aborigines as food. *Aust. J. Exptl. Biol. and Medical Sci.* 12: 13–18.

Everist, S.L. 1969. *Use of fodder trees and shrubs.* Qld. Dept. Primary Production. Advisory Leaflet No. 1024.

Henshall, T.S., D. Jampijinpa, J.N. Spencer and F.J. Kelly 1982. *Ngurrju maninja kurlangu yapa nyurnukurlangu: Bush medicine.* Yuendumu: Warlpiri Literature Production Centre.

Latz, P.K. 1982. *Bushfires and bushtucker: Aborigines and plants in Central Australia.* University of New England: M.A. Thesis.

Lawrie, A.C. 1981. Nitrogen fixation by native Australian legumes. *Aust. J. Bot.* 29: 143–157.

McKell, C.M. 1975. Shrubs – A neglected resource of arid lands. *Science* 187: 803–809.

McKell, C.M., J.P. Blaisdell and J.R. Goodin (eds) 1972. *Wildland shrubs – their biology and utilization. An International Symposium Utah State University, Logan, Utah July 1971.* USDA Forest Service General Technical Report INT–1.

Peterson, N. 1979. Aboriginal uses of Australian Solanaceae. In *The biology and taxonomy of the Solanaceae,* J.G. Hawkes, R.N. Lester and A.D. Skelding (eds): 171–189. New York: Academic Press.

Southwell, I.A. and J.R. Maconochie 1977. The essential oil of the fly-repellant shrub, *Pterigeron bubakii. J. Proc. R. Soc. N.S.W.* 110: 93–94.

Weder, J.P.K. 1978. Occurrence of proteinase inhibitors in *Mimosoideae. Z. Pflanzenphysiol.* 90: 285–291.

Weder, J.P.K. 1981. Protease inhibitors in the *Leguminosae.* In *Advances on Legume Systematics,* R.M. Polhill and P.H. Raven (eds), vol. 2: 533–560. Kew: Royal Botanic Gardens.

22 Wild and semi-cultivated legumes as potential sources of resistance to bruchid beetles for crop breeder: a study of *Vigna/Phaseolus*

N. Birch[1], B. J. Southgate[2] and L. E. Fellows[3]

[1]*formerly Jodrell Laboratory, Royal Botanic Gardens, Kew, Richmond, Surrey TW9 3AB, now Zoology Department, Scottish Crop Research Institute, Invergowrie, Scotland, UK*

[2]*10 Spinney Way, off Deerleap Way, New Milton, Hants BH25 5DN, UK*

[3]*Jodrell Laboratory, Royal Botanic Gardens, Kew, Richmond, Surrey TW9 3AB, UK*

Introduction

According to FAO estimates (Poleman 1975) protein deficiency affects at least 500 million people and is most serious in arid regions where crop failures are common. In 'Tropical Legumes: resources for the future' (National Academy of Sciences 1979) the cultivation of existing or new legume crops was cited as being the most promising way to quickly provide sufficient quantities of vegetable protein to supplement cereals in meat-deficient diets, and the development of varieties with improved resistance to pests, diseases and drought is a research priority. Insects in particular, which can attack grain legumes as growing plants as well as after harvest, are a major problem.

Bruchids (seed bettles) are one of the most important insect groups attacking both grain and tree legumes (Southgate 1978, Johnson 1983) in the arid tropics. Most of the 1300+ species of bruchids living on wild plants have a requirement for developing pods for egg laying and usually have one generation a year. The bruchids of pest status however (10 major species to date) have evolved the additional ability to lay eggs in or on the stored mature seed (Fig. 22.1, p. 304), and can continue their life cycle over several generations a year. Adult bruchids mate within 24 hours of emergence, after which females lay up to 80 eggs on or in the stored seed over 7–10 days. Larvae penetrate the seed testa and cotyledon from the eggs and develop through five larval instars before pupating just below the seed testa. Adult emergence time depends on the host seed and storage conditions (temperature, humidity) but is usually within 20–30 days.

As a result of storage and international transport of legume seeds, pest bruchids are now distributed throughout the Old and New World tropics and some species have also spread into temperate N. America and Europe (Table 22.1, p. 304). Greatest losses are caused by *Callosobruchus maculatus*) (cowpea seed beetle) and *C. chinensis* (adzuki seed beetle) of Old World origin and *Zabrotes subfasciatus* (Mexican bean seed beetle) and *Acanthoscelides obtectus* (bean seed beetle), both of New World origin. Host ranges of these pest bruchids are shown in Table 22.2, p. 305. The net result of bruchid attack is a loss of food quantity, quality and

Figure 22.1 Bruchids laying eggs on surface of cowpeas. Note also emergence holes of the adult beetles (photograph MAFF, Slough Laboratory. British Crown Copyright (C)).

Table 22.1 Present known world distribution of important stored product Bruchidae: + indigenous species, o introduced species.

	Africa	Asia	Australasia	Europe	America
Callosobruchus maculatus	+	+	o	o	o
Callosobruchus chinensis	+	+			o
Acanthoscelides obtectus	o	o		o	+
Zabrotes subfasciatus	o	o			+

germination viability (Gupta *et al.* 1981). For example, in arid areas of Nigeria (Booker 1967) and the Republic of Niger (Alzouma 1983) bruchids can destroy up to 100 per cent of the semi-arid staple crop the cowpea (*Vigna unguiculata*) in six months storage, even though initial infestation of pods in the field is low. Losses in Nigeria alone have amounted to 1.6 million U.S. dollars per year (Caswell 1970).

The need for new sources of bruchid resistance

Crop plants have been primarily selected for high yields and nutritional value together with low anti-mammalian chemical factors. This has severely disrupted the

Table 22.2 Host specificity of bruchid species. + indigenous (field and store), o introduced species (new host).

Bruchid	Cajanus cajan	Cicer arietinum	Glycine max	Lablab purpureus	Lens culinaris	Macrotyloma uniflorum	Pisum sativum	Phaseolus lunatus	Phaseolus vulgaris	Vigna aconitifolia	Vigna angularis	Vigna mungo	Vigna radiata	Vigna subterranea	Vigna unguiculata
Callosobruchus maculatus	+	+	o	+	+		o			+	o	+	+	+	+
Callosobruchus chinensis	+	+			+	+	+			o	+	+	+	o	+
Acanthoscelides obtectus								+	+						
Zabrotes subfasciatus								+	+					o	o

coevolutionary processes between plants and insects (Feeny 1976) such that very few cultivated species have retained the insect resistance of their wild relatives. Man's dependence on restricted cultivar numbers is leading to a rapid loss of the genetic diversity in crop plants (Smartt 1984) and the use of pure breeding cultivars is creating insurmountable sterility barriers between modern crops and their wild relatives in secondary and tertiary gene pools.

In developing countries, the high cost of synthetic insectides combined with a lack of facilities and expertise for their safe use has led to a search among existing cultivars and landraces for crop legumes with greater insect resistance. Generally only low to moderate resistance has been found from these sources (see Table 22.3). Partial resistance, which acts by slowing down insect reproduction rates on the crop, can be useful as one of several components of integrated control schemes against field pests (Ponti 1982). However, higher levels of seed resistance are required to adequately protect the harvested crop from bruchids over several months of storage under rural conditions. Modern methods of plant breeding using somatic hybridisation and tissue culture are making possible and practicable the transfer of resistance factors from wild species outside the primary gene pool (Schieder *et al.* 1982). Developments in the fields of protoplast fusion and plant regeneration now permit not only the crossing of previously incompatible species but also the transference of only a few chromosomes or genes, creating many opportunities for crop improvement.

Known types of biochemical resistance to bruchids in legume seeds

In natural ecosystems most bruchids have extremely limited host ranges and are unable to develop in seed of other legumes (Janzen 1977, Johnson 1981) due, at least partially, to their inability to tolerate the seed chemistry of non-host species. From these and other studies it is clear that:

  (a) several different classes of seed chemical can prevent bruchid development (for examples see Table 22.4, p. 310).
  (b) it cannot be assumed that all bruchid species will be affected to the same extent by a given plant chemical, nor that all chemicals of that class will have identical effects on all bruchid species. The molecular heterogeneity and target specificity of broad chemical classes such as trypsin inhibitors and phytohaemagglutinins should be taken into account in screens for bruchid resistance factors.
  (c) synergistic interactions between different types of defensive compounds or between these chemicals and limiting insect nutrients can improve the seed's resistance to bruchid attack.
  (d) coevolutionary adaptations between bruchids and their host seed defenses have occurred, even among the less host-restricted pest species (for examples see Table 22.4).

Table 22.3 Examples of bruchid resistance levels previously found in cultivars (cvs) of grain legumes.

Crop	Callosobruchus	Type of resistance	References
Vigna aconitifolia	*C. maculatus*	Seed: low/no resistance in 20 cvs	Vir (1982)
Vigna mungo	*C. maculatus*	Seed: low resistance in 2/10 cvs	Dabi *et al.* (1978)
	C. maculatus	Seed: mod resistance in 14/17 cvs	Gibson & Raina (1972)
	C. maculatus)	Seed: mod resistance in 2 cvs	Singh & Sharma (1982)
	C. chinensis)		
	C. chinensis	Seed: high resistance in 1 cv	Yadav & Pant (1978)
	C. chinensis	Pod: decreased oviposition larval survival	Doria & Raros (1973)
Vigna radiata	*C. maculatus*)	Seed: v. low/no resistance in 3 cvs	Singh & Sharma (1982)
	C. chinensis)		
	C. chinensis	Seed: testa hardness/toxicity in 1 cv	Talekar & Lin (1981)
	C. chinensis	Pod: pubescence in 1 cv	Talekar & Lin (1981)
	C. chinensis	Pod: decreased oviposition in 3/46 cvs	Doria & Raros (1973)
Vigna umbellata	*C. chinensis*	Seed: high resistance within species	Chatterjee & Dana (1979)
Vigna unguiculata	*C. maculatus*	Seed: no resistance	Gokhale (1973)
	C. maculatus	Seed: low resistance in 2/6 local vars	Osuji (1976)
	C. maculatus	Seed: low resistance in 1/6 cvs	Nwanze & Horber (1975)
	C. maculatus	Seed: mod/high resistance in 1/5000 cvs	Gatehouse *et al.*(1979 & 1983)
	C. maculatus	Pod: low resistance; decreased emergence	Fatunla & Badaru (1983)
	C. maculatus	Pod: low resistance; decreased emergence 5/20 cvs	Akingbohunfbe (1976)

Table 22.3 – continued

Crop	Callosobruchus	Type of resistance	References
Phaseolus vulgaris	*C. maculatus*	Seed: high resistance, lectins	Janzen *et al.* (1976)
	C. chinensis	Seed: high resistance, pectosans	Ishii (1952)
	C. chinensis	Seed: high resistance, heteropolysaccharides	Applebaum *et al.* (1972)
Cicer arietinum	*C. maculatus*	Seed: rough testa reducing oviposition	Schalk (1973)
	C. maculatus)	Seed: rough testa reducing oviposition in 1/14 cvs.	Raina (1971)
	C. chinensis)		
	Callosobruchus spp.	Pod: no larval development	Reed *et al.* (1980)
	Callosobruchus spp.	Pod: malic acid exudate possibly implicated	Rembold (1981)
Glycine max	*C. maculatus*	Seed: no development	Gokhale (1973)
	C. chinensis	Seed: no development; saponins	Applebaum *et al.* (1965)

It is highly probable that many more types of anti-bruchid chemicals remain to be found. Those impossible to transfer in breeding programmes could have potential as exogenously applied insecticidal compounds.

The Kew bruchid project

This project was set up to evaluate wild and semi-cultivated legumes for bruchid resistance. It is multidisciplinary, involving entomologists, biochemists and plant taxonomists at Kew with seed banks, botanic gardens and agronomists around the world. The *Vigna/Phaseolus* complex, which contains five important semi-arid grain legumes susceptible to bruchid attack, was selected as a pilot project with the following aims:

(a) to screen a wide selection of wild relatives of crop species in *Vigna* and *Phaseolus,* taken from a range of taxonomic groupings and geographical locations, for seed resistance to infestation by two pest bruchids;

(b) to investigate the chemical basis of this resistance;

(c) to relate the distribution of bruchid resistance in these two genera to the taxonomy of the *Vigna/Phaseolus* complex at subgeneric and sectional levels in order to detect patterns of resistance which may reflect insect-plant coevolution;

(d) to select types of seed resistance which are compatible with mammalian consumption and to eventually collaborate with plant breeders in the introduction of multiple resistance factors for both pod and seed resistance into susceptible cultivars.

Methods

BIOASSAY OF SEED FOR BRUCHID RESISTANCE

The primary resistance screen was completed on a wide range of plant genotypes covering Old and New World taxa from the *Vigna/Phaseolus* complex, to assess interspecific variation in seed resistance. One laboratory-reared population of *C. maculatus* (origin IITA, Nigeria) and one population of *C. chinensis* (origin Japan) were used throughout the primary screen. (Intraspecific variability in plant and insect populations will be assessed in a secondary screening).

Since different parameters of insect population development are not always positively correlated with each other when screening wild plant material (Birch & Holt 1980) it is important not to rely on only one or two resistance indices in the primary screen. For example, oviposition by adult bruchids is sometimes experimental on seeds of species outside the normal host range but may not reflect the

Table 22.4 Examples of legume seed chemical toxicity to bruchids. A = *Acanthoscelides*, C = *Callosobruchus*, Z = *Zabrotes*.

Seed chemical type	Source	Effective against	Ineffective against	References
Protease inhibitors including trypsin inhibitors)	*Vigna unguiculata* (cv 2027)	*C. maculatus* (Brazilian pop)	*C. maculatus* (biotypes)	Gatehouse *et al.* (1979 & 1983, Dobie 1981, Birch in prep.)
	Glycine max	–	*C. maculatus*	Janzen *et al.* 1976, Gatehouse *et al.* 1983
	Phaseolus lunatus	–	*C. maculatus*	Gatehouse *et al.* 1983
	Phaseolus vulgaris	–	*C. maculatus*	Janzen *et al.* 1976
	Lathyrus sativus	–	*C. chinensis*	Roy & Bhat 1975
Phytohaemagglutins (lectins)	*Phaseolus vulgaris*	*C. maculatus*	*Z. subfasciatus**, *A. obtectus**	Janzen *et al.* 1976, Southgate 1978, Gatehouse *et al.* 1984
Saponins (integral)	*Cicer arietum*	–	*C. chinensis*)	
	Pisum sativum	–	–)	
	Lens culinaris		*C. chinensis*)	Applebaum *et al.* 1969
	Phaseolus vulgaris	*C. chinensis*	–)	
	Arachis hypogaea	*C. chinensis*	–)	
Non-protein amino acids (eg canavanine)	*Dioclea megacarpa*	*C. maculatus*	*Caryedes brasiliensis*	Janzen *et al.* 1977, Rosenthal *et al.* 1976
Alkaloids	*Erythrina flabelliformis*	*C. maculatus*	*Specularius* spp.	Janzen *et al.* 1977
Polysaccharides	*Phaseolus vulgaris*	*C. chinensis*	*A. obtectus*	Applebaum & Guez 1972

* Inferred from host range and seed chemistry records.

quality of the seed for larval survival and development (Janzen 1977, Wasserman & Futuyama 1981). In this bioassay an analysis of insect population development using lifetable studies (age-specific mortality of insect populations) was used to indicate which aspects of the insect-host-relationship prevented normal pest development on resistant seed accessions. This involved placing pairs of 0–24 hour old male and female adult bruchids on replicated batches of seeds previously equilibrated at 30°C, 70% R.H. for 28 days. Adults were removed after seven days' oviposition except when no oviposition occurred; in such cases failure to lay eggs was rechecked three times using new pairs of bruchids. Records were made of the numbers of eggs laid per seed, percentage and stage of incomplete egg development, percentage larval penetration of testa and cotyledon, individual adult emergence times and numbers emerging per seed and per gram of seed material. All bioassays were carried out at 30°C + 1°C and 65% + 5% R.H. Statistical analysis of this life-table data was used to reveal critical stages in the bruchids' life cycle which were used (mean emergence day and numbers emerging per gram of seed) as key resistance indices. Levels of seed resistance were designated very low, low, medium or high on this basis (see Table 22.5, p. 312). This screening methodology has been successfully used to screen wild and semi-cultivated relatives of the Faba bean (*Vicia faba*) for aphid resistance (Birch & Wratten 1984).

SCREENING FOR BRUCHID ANTIMETABOLITES IN SEEDS

Classes of legume chemicals claimed to confer resistance to bruchids in previous studies (see references in Table 22.4, p. 310) were assayed as follows:

TRYPSIN INHIBITOR

Trypsin inhibitor activity was assayed using the method of Schwert and Takenaka (1955). The degree of inhibition of bovine pancreatic trypsin by seed extracts was determined by monitoring the rate of hydrolysis of N-alpha-benzoyl-L-arginine ethyl ester hydrochloride (BAEE) spectrophotometrically. One BAEE unit is expressed as an increase at 253nm of 0.001 absorbance units per minute at 25°C in phosphate buffer pH 7.5. Trypsin inhibitor activity is defined in terms of BAEE units inhibited per mg seed extract.

PHYTOHAEMAGGLUTIN

Phytohaemagglutin (lectin) activity was measured by the agglutination of washed human group 0+ blood, based on the quantitative method of Bender (1983). Phytohaemagglutin activity is arbitrarily defined as the percentage of maximum agglutination (equivalent to zero absorption at 415nm of the lysed erythrocyte supernatant after agglutination) possible under the assay conditions.

Table 22.5 Arid-adapted *Vigna* and *Phaseolus* species: antimetabolites and bruchid resistance determined in the present study.

Species (Maréchal et al. 1978)	Uses (ref)	Trypsin[+] inhibitors	Phytohaemag-glutins[o]	Saponins	Resistance to C. maculatus	Resistance to C. chinensis
V. aconitifolia	seed, forage, anti-erosion (Kay 1979)	22	32	N/D	v. low	v. low
V. adenantha		51	N/D	high	high	high
V. ambacensis	forage (Kay 1979)	129	34	N/D	v. low	low-mod*
V. fischeri		222	32	N/D	v. low	mod.
V. frutescens	edible tuber (SEPASAT)	69	35	N/D	low	low
V. heterophylla		44	42	N/D	v. low	v. low
V. lobatifolia	edible tubers (NAS 1979)	136	49	N/D	high	high
V. marina		176	50	N/D	low-mod*	low-mod*
V. mungo	seed, vegetable, hay (Kay 1979)	25–70	25–40	N/D	low-mod*	low-mod*
V. radiata	seed, shoots, fodder (Kay 1976)	42	25	N/D	low-mod*	low-mod*
V. radiata var. *sublobata*	seed (Kay 1976)	52	53	N/D	mod	mod
V. reticulata	edible tuber (NAS 1979)	35	47	N/D	mod	low
V. subterranea	seed, fodder (SEPASAT)	113	50	N/D	low	low
V. trilobata	seed, fodder (SEPASAT)	31	53	N/D	v. low	v. low
V. unguiculata	seed, shoots, fodder (SEPASAT)	13	30	N/D	v. low	v. low
V. unguiculata subsp. *dekindtiana*	seed (Kay 1979)	25	41	N/D	v. low	v. low
V. umbellata	seed, shoots, disease resistance (Kay 1979)	22	39	N/D	high	high
V. vexillata	edible tubers, erosion control (NAS 1979)	97	34	N/D	high	high
P. acutifolius var. *latifolius*	seeds, hay, forage, drought & salt tolerance (SEPASAT)	22–42	83–93	N/D	high	high

Table 22.5 — continued

Species (Maréchal et al. 1978)	Uses (ref)	Trypsin[+] inhibitors	Phytohaemagglutins[o]	Saponins	Resistance to C. maculatus	Resistance to C. chinensis
P. angustissimus	ground cover, forage (SEPASAT)	31	50	N/D	low	low
P. grayanus	ground cover, forage (SEPASAT)	115	50	N/D	high	high
P. ritensis	ground cover, forage (SEPASAT)	220	28	N/D	high	high

N/D   not detectable

+   units = BAEE inhibited/mg seed; range for genus *Vigna* = 7–205, $\bar{x}$ = 65

    units. Range for genus *Phaseolus* 7–234, $\bar{x}$ = 110 units   see Methods

o   units = % max. agglutination using human type o[+] blood; range for genus

    Vigna = 8–92%, $\bar{x}$ = 42%. Range for genus *Phaseolus* = 26–95%, $\bar{x}$ = 58%

*   variable resistance amongst multiple accessions of species.

SAPONINS

Saponins were detected using the haemolytic method of Jones and Elliot (1969) and the qualitative TLC method, using anisaldehyde—acetic acid reagent of Fenwich and Oakenfall (1982).

Results and Discussion

Ninety species of *Vigna* and *Phaseolus* were screened in total. Of these, 18 *Vigna* and four *Phaseolus* species are or show potential as crop plants for semi-arid and arid areas, or as sources of useful characters for inclusion in arid crop improvement (Kay 1976, National Academy of Sciences 1979, SEPASAT). The results of bruchid bioassays and biochemical tests on those 22 species are given in Table 22.5, p. 312. (A further two *Vigna* species, *V. lanceolata* and *V. schimperi,* and two *Phaseolus* species, *P. parvulus* and *P. wrightii* are listed by SEPASAT as potentially useful arid plants but were not available for study).

SCREEN FOR SEED RESISTANCE TO BRUCHIDS

Among the species found most susceptible to *Callosobruchus maculatus* and *C. chinensis* are several currently grown as arid or semi-arid crops, where the need for new sources of resistance is particularly vital: *V. aconitifolia* (moth bean), *V. mungo* (black gram), *V. radiata* (green gram), *V. subterranea* (bambara groundnut) and *V. unguiculata* (cowpea). Resistance factors for some crops may be found by future intraspecific screenings of cultivars, landraces and wild forms (work in progress in this laboratory with *V. mungo* and *V. radiata* suggests this is so) but for others an interspecific transfer of seed resistance will be necessary. For example, in a survey at IITA of over 6000 accessions of cowpea, only one resistant cultivar was identified. Although high levels of seed trypsin inhibitor were thought to be responsible for the resistance of this cultivar (Gatehouse *et al.* 1979) later work has indicated that some *C. maculatus* populations can infest it (Dobie 1981) and that the expression of resistance is more complex than originally considered (Redden *et al.* 1983).

Those species listed in Table 22.5 as having high seed resistance to *C. maculatus* and *C. chinensis* (*V. adenantha, V. lobatifolia, V. umbellata, V. vexillata, P. acutifolius* var. *latifolius, P. grayanus* and *P. ritensis*) are likely donors of resistance genes. *Vigna umbellata* (rice bean), *V. vexillata* (wild mung) and *P. acutifolius* var. *latifolius* (tepary bean) may be of particular value to plant breeders since they are already cultivated to some extent and possess other agronomically useful characters. The tepary bean is of special interest since this traditional desert crop of the Sonoran desert Indians has potential both as a desert food source (Thorn *et al.* 1983) and as a source of drought tolerance and insect resistance for *Phaseolus* crops (Nabhan 1983).

The wide screen of 90 species, covering most subgenera and sections of *Vigna* and *Phaseolus* (as proposed by Maréchal *et al.* 1978) has revealed interesting correlations between patterns of seed resistance to the two *Callosobruchus* species and the taxonomic relationship of the plants. Most wild *Vigna* species in the two large Old World subgenera *Vigna* and *Ceratotropis* (containing crops *V. aconitifolia, V. angularis, V. mungo* and *V. radiata, V. subterranea* and *V. unguiculata*) were found to possess low resistance against these two Old World pest bruchids. Much higher levels of seed resistance were found in seeds of wild *Vigna* species from subgenus *Plectotropis* (linking Old and New World taxa) and in New World subgenera *Sigmoidotropis, Cochliasanthus, Leptospron,* and most *Phaseolus* species (all New World). (Full details to be published elsewhere).

CONTRIBUTION OF SEED TRYPSIN INHIBITORS, PHYTOHAEMAGGLUTINS AND SAPONINS TO BRUCHID RESISTANCE IN VIGNA AND PHASEOLUS

Levels of each chemical in arid-adapted *Vigna* and *Phaseolus* species are listed in Table 22.5, p. 312 (details of levels in all 90 species included in the wider screen will be published elsewhere). The difficulty in interpreting results of chemical screenings performing with only one assay method is that they may fail to reveal molecular heterogeneity within a given class of insecticidal chemical. However, since previous authors have claimed, largely on the basis of these restricted chemical assays, that the antimetabolites trypsin inhibitors, phytohaemagglutins and saponins do significantly contribute to bruchid resistance in various legume taxa (see examples in Table 22.4) it was felt worthwhile to assess to what extent such screens for these compounds could predict bruchid resistance within *Vigna/Phaseolus*.

Both trypsin inhibitor and phytohaemagglutin activity (PHA) were found to be widely distributed in the *Vigna/Phaseolus* complex. Most of the 70 wild *Vigna* species screened contained low PHA (with the exception of two New World species) and low to high trypsin inhibitor levels, while most of the 22 *Phaseolus* species contained higher levels of PHA and moderate to high levels of trypsin inhibitor activity. A correlation analysis of 56 *Vigna* and 18 *Phaseolus* species (see Table 22.6, p. 316) however failed to find any significant association between levels of either of these classes of compound and observed seed resistance to *C. maculatus* and *C. chinensis*. Furthermore, the wider screen revealed several wild *Vigna* and *Phaseolus* species containing high seed trypsin inhibitor activity which were susceptible to bruchid attack, and also one *Vigna* containing high seed PHA which was also susceptible. Therefore, although isolated trypsin inhibitor from *V. unguiculata* (Gatehouse *et al.* 1979 & 1983) and phytohaemagglutin from *P. vulgaris* (Janzen *et al.* 1976, Gatehouse *et al.* 1984) were found to be toxic to *C. maculatus* when incorporated into artifical diets, their contribution to resistance in the intact seeds of wild species requires further investigation.

Table 22.6 Statistical correlation analysis of the association between *Vigna* and *Phaseolus* spp. seed antimetabolites and bruchid resistance.

Chemical type	Correlation (r^2) with resistance to *C. maculatus*				Correlation (r^2) with resistance to *C. chinensis*			
	T.I.A. (*Vigna*)	P.H.A. (*Vigna*)	T.I.A. (*Phaseolus*)	P.H.A. (*Phaseolus*)	T.I.A. (*Vigna*)	P.H.A. (*Vigna*)	T.I.A. (*Phaseolus*)	P.H.A. (*Phaseolus*)
Mean emergence day	0.004 N.S.	0.04 N.S.	0.09 N.S.	0.13 N.S.	0.004 N.S.	0.00 N.S.	0.10 N.S.	0.10 N.S.
No. adults emerging per g. seed	0.021 N.S.	0.04 N.S.	0.03 N.S.	0.01 N.S.	0.03 N.S.	0.001 N.S.	0.23 N.S.	0.07 N.S.

Chemical type:

T.I.A. = Trypsin inhibitor activity
P.H.A. = Phytohaemagglutin activity

Vigna analyses : n = 56
Phaseolus analyses : n = 18
N.S. = not statistically significant; $p < 0.05$

High levels of seed saponins were only found in three closely related New World *Vigna* species, each highly resistant to both bruchids, including the marginally semi-arid species *V. adenantha* (see Table 22.5). Their restricted distribution within the *Vigna/Phaseolus* complex precluded any correlation analysis.

Several resistant *Vigna* and *Phaseolus* species did not contain high levels of any assayed antimetabolites and must contain other resistance factors. (Physical factors too cannot be excluded in some cases). Other chemicals with insecticidal activity have been reported in seeds of *Phaseolus,* including heteropolysaccharides (Applebaum *et al.* 1972) and alpha-amylase inhibitors (Powers & Culbertson 1983). Of interest too are recent reports of novel types of glycosidase inhibitors in legume seeds which are insecticidal at concentrations harmless to mammals (Saviano *et al.* 1977, Evans *et al.* in press).

Since seed resistance to bruchids in wild legumes almost certainly depends on the combined effects of several factors, chemical screens alone probably have little predictive value for the agriculturalist. A bioassay such as described in this project is likely to be a more reliable indicator of resistance potential in wide screens of crop relatives.

Future prospects

The present catastrophic rate of loss of wild plants from their natural habitats gives great urgency to programmes aimed at surveying and conserving the wild relatives of crops. The benefits of international collaboration are well illustrated by the Kew *Vigna/Phaseolus* screen which has revealed sources of pest bruchid resistance genes mainly in locations remote from the centre of distribution of the insects, as well as shedding light on the pattern of coevolutionary adaptations of bruchids and host plants in the Old and New World tropics.

Useful as laboratory studies are, however, field-based ecological surveys are still needed to assess the importance of wild plants to crop pests. For example, one of the few field studies on wild *Vigna* species (Alzouma 1983) suggests that some wild species can act as reservoirs for the pest bruchid *Callosobruchus maculatus.* Observations of wild legumes not attacked by pest bruchids might suggest sources of pod resistance factors which could be transferred to crop legumes to prevent initial infestation in the field.

The transfer of resistance from wild to crop species depends on the technological skills of plant breeders but progress in this field is encouraging (Machado *et al.* 1982). The aim must be the transfer of multiple component resistance which will be more durable than that based on a single factor only, even though its complexity might at present preclude a full genetic analysis.

Both the will and the technology for the study and exploitation of pest resistance in wild plants now exists. It is to be hoped that the means will also be found, since time is now on the side of the bulldozers.

Acknowledgements

This project was funded through the generosity of the Ward Blenkinsop Trust. We also wish to thank AVRDC, CSIRO, Gembloux, IITA and USDA for seed material, Mr J Harbour for biochemical and statistical analyses and Dr S Evans for helpful discussions.

The authors are especially indebted to Processor R. Maréchal, Gembloux, not only for his help in obtaining rare seed collections but also for taxonomic and ecological discussions on the *Vigna/Phaseolus* complex.

References

Akingbohunfbe, A.E. 1976. A note on the relative susceptibility of unshelled cowpeas to the cowpea weevil *(Callosobruchus maculatus). Trop. Grain Leg. Bull.* 5: 11–13.

Alzouma, I. 1983. Observations on the ecology of *Bruchidius atrolineatus* Pic. and *Callosobruchus maculatus* F. in Niger. In *The ecology of bruchids attacking legumes,* V. Labeyrie (ed.): 205– 215. The Hague: Junk.

Applebaum, S.W. and M. Guez 1972. Comparative resistance of *Phaseolus vulgaris* to *Callosobruchus maculatus* and *Acanthoscelides obtectus*: the differential digestion of soluble heteropolysaccharides. *Entomol. Expt. et Appl.* 15: 203–207.

Applebaum, S.W., B. Gestetner and Y. Birk 1965. Physiological aspects of host specificity in the Bruchidae. IV. Development incompatability of soybeans for *Callosobruchus. J. Insect Physiol.* 11: 611–616.

Applebaum, S.W., S. Marco and Y. Birk 1969. Saponins as possible factors of resistance of legume seeds to the attack of insects. *J. Agr. Food Chem.* 17: 618–622.

Bender, A.E. 1983. Haemagglutins (lectins) in beans. *Food Chem.* 11: 309–320.

Birch, N. and J. Holt 1980. *Aphid resistance in Vicia in relation to non-protein amino acids.* Proc. Eucarpia/IOBC meeting: IOBC/WPRS Bull. 1980/IV/I: 133–140.

Birch, N. and S.D. Wratten 1984. Patterns of aphid resistance in the genus *Vicia. Ann. Appl. Biol.* 104: 327–338.

Booker, R.H. 1967. Observations on three Bruchids associated with cowpea in northern Nigeria. *J. Stored Prod. Res.* 3: 1–5.

Caswell, G.H. 1970. *The storage of cowpea in the northern states of Nigeria.* Samaru Res. Bull. No. 120.

Chatterjee, B.N. and S. Dana 1979. Rice bean (*Vigna umbellata* Thunb.). *Trop. Grain Leg. Bull.* 10: 22–25.

Dabi, R.K., H.C. Gupta and S.K. Sharma 1978. Relative resistance of some black grain (*V. mungo* L.) varieties to the pulse beetle *Callosobruchus maculatus* F. *Bull. Grain Tech.* 16(2): 141–143.

Dobie, P. 1981. The use of resistant varieties of cowpeas (*Vigna unguiculata*) to reduce losses due to post-harvest attack by *Callosobruchus maculatus.* In *The ecology of bruchids attacking legumes,* V. Laberyie (ed.): 185–195. The Hague. Junk.

Doria, R.C. and R.S. Raros 1973. Varietal resistance of mungo to the bean weevil, *Callosobruchus chinensis* L. and some other characteristics of field infestation. *Philipp. Entom.* 2(6): 399–408.

Evans, S.V., A.M.R. Gatehouse and L.E. Fellows 1984. Toxicity of the secondary plant compound 2,5-dihydroxymethyl-3,4-dihydroxypyrrolidine to the seed-eating larvae of the bruchid beetle *Callosobruchus maculatus* and its *in vitro* effect on larval digestive carbohydrases. *Entomol. Expt et Appl.* (in press).

Fatunla, T. and K. Badaru 1983. Resistance of cowpea pods to *Callosobruchus maculatus* F. *J. Agric. Sci. Camb.* 100: 205–209.

Feeny, P. 1976. Plant apparency and chemical defence. *Rec. Adv. Phytochemistry* 10: 1–40.

Fenwick, D.E. and D. Oakenfall 1983. Saponin content of food plants and some prepared foods. *J. Sci. Food Agric.* 34: 186–191.

Gatehouse, A.M.R. and D. Boulter 1983. Assessment of the antimetabolic effects of trypsin inhibitors from cowpea (*Vigna unguiculata*) and other legumes on development of the bruchid beetle *Callosobruchus maculatus. J. Sci. Food Agric.* 34: 345–350.

Gatehouse, A.M.R., F.M. Dewey, J. Dove, K.A. Fenton and A. Puszta 1984. Effect of seed lectins from *Phaseolus vulgaris* on the development of larvae of *Callosobruchus maculatus*; mechanism of toxicity. *J. Sci. Food Agric.* 35: 373–380.

Gatehouse, A.M.R., J.A. Gatehouse, P. Dobie, A.M. Kilminster and D. Boulter 1979. Biochemical basis of insect resistance in *Vigna unguiculata. J. Sci. Food Agric.* 30: 948–958.

Gibson, K.E. and A.K. Raina 1972. A simple laboratory method of determining the seed host preference of Bruchidae. *J. Econ. Entomol.* 4(3): 1189–1190.

Gokhale, V.G. 1973. Development compatability of several pulses in the Bruchidae. I. Growth and development of *Callosobruchus maculatus* F. on host seeds. *Bull. Grain Tech.* 11(1): 28–31.

Gupta, S., S.K. Singhai and R.B. Doharey 1981. Studies on the chemical and nutritional changes in bengal gram (*Cicer arietum*) during storage caused by the attack of pulse beetle *Callosobruchus maculatus* F. *Bull. Grain Tech.* 19(3): 185–190.

Ishii, S. 1952. Studies on the host preference of cowpea weevil (*Callosobruchus chinensis* L.) *Bull. Nat. Inst. Agric. Sci. Jap.* 1: 185–256.

Janzen, D.H. 1977. How southern cowpea larvae (Bruchidae: *Callosobruchus maculatus*) die on nonhost seeds. *Ecology* 58: 921–927.

Janzen, D.H., H.B. Juster and E.A. Bell 1977. Toxicity of secondary compounds to seed-eating larvae of the bruchid beetle *Callosobruchus maculatus. Phytochemistry* 16: 223–227.

Janzen, D.H., H.B. Juster and I.E. Liener 1976. Insecticidal action of phytohaemagglutin in black beans on a bruchid beetle. *Science* 192: 795–6.

Johnson, C.D. 1981. Seed beetle host specificity and the systematics of the Leguminosae. In *Advances in legume systematics Part 2*, R.M. Polhill and P.H. Raven (eds): 995–1029. Kew: Royal Botanic Gardens.

Johnson, C.D. 1983. *Handbook on seed insects of Prosopis species*. Rome: FAO.

Jones, M. and F.C. Elliot 1969. Two rapid assays for saponins in individual alfalfa plants. *Crop Sci.* 9: 688–690.

Kay, D.E. 1979. *Food legumes*. TPI Crop Digest No. 3. London: Tropical Products Institute.

Machado, M., W. Tai and L.R. Baker (1982). Cytogenetic analysis of the interspecific hybrid *Vigna radiata* × *V. umbellata. J. Heredity* 73: 205–208.

Maréchal, R., J.M. Mascherpa and F. Stainier 1978. Etude taxonomique d'un groupe complexe d'especes des genres *Phaseolus* et *Vigna* sur la base de données morphologiques et polleniques, traitées par l'analyse informatique. *Boissiera* 28: 1–278.

Nabhan, G.P. 1983. The desert tepary as a food resource. *Desert Plants* 5(1): 1–10.

National Academy of Sciences 1979. *Tropical legumes: resources for the future*. Washington, D.C.: National Academy of Sciences.

Nwanze, K.F. and E. Horber 1975. Laboratory techniques for screening cowpeas for resistance to *Callosobruchus maculatus* F. *Environ. Entomol.* 4(3): 415–419.

Osuji, F.N. 1976. A comparison of the susceptibility of cowpea varieties to infestation by *Callosobruchus maculatus. Entomol. Exp. et Appl.* 20: 209–217.

Poleman, T.T. 1975. World food: a perspective. *Science* 188: 510–518.

Ponti, O.M.B. de 1982. Plant resistance to insects: a challenge to plant breeders and entomologists. In *Insect-Plant Relationships.* Proc 5th Int. Symp. J.H. Visser and A.K. Minks (eds): 337–349. Wageningen: Pudoc.

Powers, J.R. and J.D. Culbertson 1983. Interactions of purified bean (*Phaseolus vulgaris*) glycoprotein with an insect amylase. *Cereal Chem.* 60(6): 427–429.

Raina, A.K. 1971. Comparative resistance of three species of *Callosobruchus* in a strain of chickpea (*Cicer arietinum* L.). *J. Stored Prod. Res.* 7: 213–216.

Redden, R., P. Dobie and A. Gatehouse 1983. The inheritance of seed resistance to *Callosobruchus maculatus* F. in cowpea (*Vigna unguiculata* L. Walp.) I. Analyses of parental F_1, F_2, F_3 and backcross seed generations. *Aust. J. Agric. Res.* 34: 681–695.

Reed, W., S.S. Lateef and S. Sithanatham 1980. Are bruchids field pests of chickpea? *Int. Chickpea Newsletter* 3: 16.

Rembold, H. 1981. Malic acid in chickpea exudates – marker for *Heliothis* resistance. *Int. Chickpea Newsletter* 4: 18–19.

Rosenthal, G.A., D.L. Dahlman and D.H. Janzen 1976. A novel means for dealing with L-canavanine, a toxic metabolite. *Science* 192: 256–257.

Roy, D.N. and R.V. Bhat 1975. Variation in neurotoxin, trypsin inhibitors and susceptibility to insect attack in varieties of *Lathyrus sativus* seeds. *Environ. Physiol. Biochem.* 5: 172–177.

Saviano, D.A., J.R. Powers, M.J. Costello, J.R. Whittaker and A.J. Clifford 1977. The effect of an alpha-amylase inhibitor on the growth rate of weanling rats. *Nut. Rpts Int.* 15(4): 443–449.

Schieder, O., P.P. Gupta, G. Krumbiegel and T. Hein 1982. Protoplast fusion and transformation. In *Better crops for food.* Ciba Foundation Symp. No. 97: 213–224. London: Pitman.

Schalk, J.M. 1973. Chickpea resistance to *Callosobruchus maculatus* in Iran. *J. Econ. Entomol.* 66(2): 578–579.

Singh, D.P. and S.S. Sharma 1982. Studies on grain damage and germination loss caused by *Callosobruchus maculatus* F. in different varieties to moong and mash in storage. *Bull. Grain Tech.* 16(2): 141–143.

Smartt, J. 1984. Gene pools in grain legumes. *Econ Bot.* 38(1): 24–35.

Southgate, B.J. 1978. The importance of Bruchidae as pests of grain legumes, their distribution and control. In *Pests of grain legumes: ecology and control,* S.R. Singh and H.F. van Emden (eds): 219–229. London: Academic Press.

Schwert, G.W. and Y. Takenaka 1955. A spectrophotometric determination of trypsin and chymotrypsin. *Biochem. Phys.* 16: 570–575.

Talekar, N.S. and Y.H. Lin 1981. Two sources with differing modes of resistance to *Callosobruchus chinensis* in mungbean. *J. Econ. Entomol.* 74(5): 639–642.

Thorn, K.A., A.M. Tinsley, C.W. Weber and J.W. Berry 1983. Antinutritional factors in legumes of the Sonarian desert. *Ecol. Food Nutr.* 13: 251–256.

Vir, S. 1982. Varietal preference in moth (*Vigna aconitifolia* Jacq.) for the pulse beetle *Callosobruchus maculatus* F. *Bull. Grain Tech.* 20(1): 3–7.

Wasserman, S.S. and D.J. Futuyama 1981. Evolution of host plant utilisation in laboratory populations of the southern cowpea weevil, *Callosobruchus maculatus* F. *Evolution* 35(4): 605–617.

Yadav, T.D. and N.C. Pant 1978. Development response of *Callosobruchus maculatus* F. and *C. chinensis* L. on different pulses. *Ind. J. Entomol.* 40(1): 7–15.

23 Seed banks: a useful tool in conservative plant evaluation and exploitation

R. D. Smith

Jodrell Laboratory, Royal Botanic Gardens, Wakehurst Place, Ardingly, Haywards Heath, Sussex RH17 6TN, UK

Introduction

Throughout the world, much time and money is spent assembling seed collections for evaluation in field trials aimed at identifying either novel or improved crop genotypes. All too frequently such collections are quickly lost through falling seed viability. Further time and effort will then be needed to secure fresh supplies. Yet by adopting simple and cheap precautions, seed viability can be maintained and the usefulness of such collections prolonged. Collections conserved in this way are now known as Seed Banks. This paper is devoted to a review of the factors controlling seed longevity and their practical manipulation to ensure acceptable periods of seed survival. It is hoped that those responsible for making and maintaining seed collections will be stimulated to review their current practices and introduce any necessary improvements.

Classification of seed storage physiology

On the basis of their storage behaviour, seeds are classified as either 'orthodox' or 'recalcitrant'. The great majority of species so far studied have shown the orthodox behaviour.

Orthodox seed can be dried to low moisture contents (*c.* 5% on a wet weight basis) without loss of viability, provided that the drying technique used is non-damaging. Drying seed to moisture contents below 20% decreases the rate of loss of viability logarithmically, thus the more the seed is dried the greater is the improvement in seed longevity. Orthodox seeds are suitable for long-term storage in seed banks.

Drying recalcitrant seeds to moisture contents of 20% or below kills them immediately. Maintaining the viability of recalcitrant seeds involves keeping them hydrated and oxygenated yet preventing them from either germinating or becoming fungally infected. Recalcitrant seeds cannot be stored for long periods without loss of viability by any known method even in seed banks.

Fortunately a correlation exists between storage physiology and both seed size and structure which makes the separation of the two seed types reasonably certain without detailed physiological tests. Recalcitrant seeds are large and fleshy with individual seed weights measured in grams. Comparable large orthodox seeds are, by contrast, rarely fleshy, while orthodox fleshy seeds are rarely large.

RECALCITRANT SEED CONSERVATION

In view of their failure to survive drying, the suspected occurrence of recalcitrant seed storage characteristics amongst species originating from the arid and semi-arid regions, such as *Cordyla somalensis,* is perhaps surprising. That they do occur poses the question of what can be done to conserve the seeds if such species are to be included in evaluation trials.

It is currently recommended that the following procedures be carried out with recalcitrant seeds:—

(a) minimise the time between collection and sowing;
(b) maintain the seeds in a light gauge inflated polythene bag which will reduce moisture loss and allow for some gas exchange and
(c) reduce the risk of an anaerobic environment occurring within the bag by only half filling it with seed and opening and closing the bag once a day to change the air.

Any long-term conservation of recalcitrant seeds is likely to be through the establishment of seed orchards which exploit the fact that all known species, with seeds of this kind, are also long-lived woody perennials.

ORTHODOX SEED CONSERVATION

As has been mentioned, the overwhelming majority of species possess orthodox seeds, *Acacia* and *Prosopis* being just two of the many genera with this type of seeds.

Orthodox seed are those whose longevity is predictively controlled by two factors, seed moisture content and storage temperature. These two factors act independently so that at all temperatures the relative effects of moisture content are the same, and at all moisture contents the relative effects of temperature are the same.

In all of the species studied in detail, the relationship between seed longevity and these two factors can be quantified in the equation

$$\log \sigma = K_E - C_W \log m - C_H t - C_Q t^2$$

Where σ is the time taken for viability of the seed population to fall one Probit unit. The relationship between Probits and the more usual percentage values is shown in Figure 23.3, p. 329. . If the true initial viability, expressed in Probit units is known, then the final viability after any period of storage under known conditions can be calculated by simple proportions. K_E is a constant reflecting the inherent viability of that particular species. C_W is a constant reflecting the relative effects of seed moisture content. m is the seed moisture content. C_H and C_Q are constants reflecting the relative effects of seed temperature. These latter two constants are substantially similar for all species so far studied. This is not surprising given the universal applicability of thermodynamics. t is the storage temperature.

Table 23.1 shows the viability constants which have been determined for some of the species so far. They are the result of the work of Ellis and Roberts at Reading University, U.K.

Table 23.1 Values of the seed viability constants (Cromarty *et al.* 1982).

Crop	K_E	C_W	C_H	C_Q
barley	9.983	5.896	0.040	0.000428
chickpea	9.070	4.829	0.045	0.000324
cowpea	8.690	4.715	0.026	0.000498
onion	6.975	3.470	0.040	0.000428
soya bean	7.748	3.979	0.053	0.000228

The logarithmic nature of the moisture content term indicates that the benefits of decreasing moisture content by a single percentage point will increase as the absolute value of the moisture content falls. This relationship will not hold all the way to 0%. The lower limit of applicability is generally in the region of 1–3% moisture content. A general standard of 5% moisture content is currently recommended for long-term storage, although there would appear to be one or two tropical gymnosperm trees where this limit is in the region of 7%.

The quadratic nature of the temperature term determines that the more the temperature is lowered, the less is the benefit of that lowering. Lowering the temperature by 10°C from 70°C will increase longevity more than nine-fold whilst a similar 10° fall from -10° to -20°C only increases longevity by less than two-fold.

Comparison of the moisture content term with that of the temperature term reveals drying to be the more efficient way of preserving seed viability: a fall of 1% moisture content very approximately increases viability to the same level as lowering the temperature by 10°C.

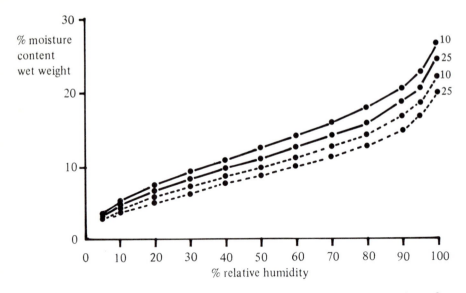

Figure 23.1 Equilibrium moisture content/relative humidity relationships for barley — and onion --- at 10° and 25 °C.

Maintaining seed viability in arid and semi-arid climates

If low seed moisture content is the dominant factor in preserving seed longevity, the arid and semi-arid zones could be expected to provide good ambient conditions for seed storage.

Seeds are hygroscopic in that they absorb or desorb the water until their moisture content is in equilibrium with the ambient relative humidity (Fig. 23.1). This property can be combined with the meteorological data for stations in the arid and semi-arid regions to allow this *a priori* supposition to be investigated by calculation for those species with known viability constants. The validity of the results of such calculations in modelling actual seed behaviour has been offered elsewhere (Smith 1984).

Figure 23.2 shows the meteorological data and the expected seed moisture contents and viability losses which have been calculated for three locations in Niger, namely Bilma, N'Guigmi and Tahoua. Bilma with an annual precipitation of 18 mm was taken as representative of a truly arid climate, whilst N'Guigmi with 247 mm precipitation and Tahoua with 444 mm precipitation were taken as representative of drier and more humid semi-arid climates.

Barley seed, with low oil content and high inherent viability, has been used as the test species at all locations because of the wealth of information available. The behaviour of onion seed with high oil content and low inherent viability has also

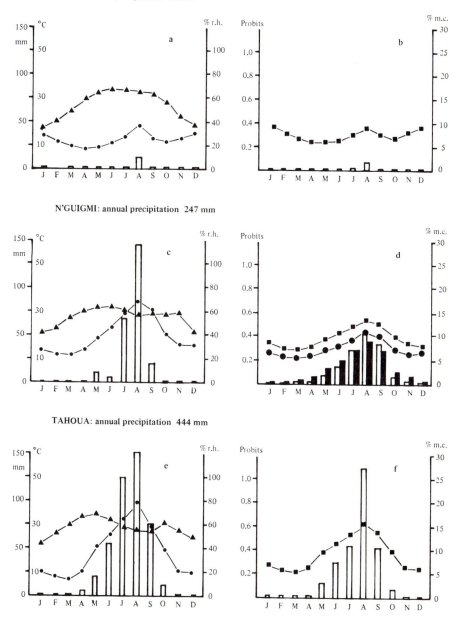

Figure 23.2 a,c,e Meteorological data for three arid–semi-arid locations in Niger: ▲ mean monthly temperatures, ● mean monthly humidities, ∏ mean monthly rainfall. **b,d,f** Calculated moisture content and viability losses of barley (and onion in N'Guigmi) at these locations: ■ barley moisture content, ● onion moisture content, ∏ loss of viability for barley, ▮ loss of viability for onion.

been calculated for N'Guigmi to allow comparison with the results for barley. The pattern of viability loss is similar for these widely differing species and therefore can be considered to model the likely behaviour of those species which are more usually found in the arid and semi-arid areas.

From the results presented, two predictable observations are confirmed: (a) the higher the annual rainfall the greater the annual loss of viability and (b) the highest monthly losses of viability occur during the months of the rainy season.

Further consideration of the annual total losses suggest that if seeds are to be kept at reasonable viability levels for, e.g. a five year period, then only storage under ambient conditions at Bilma will be suitable. Seeds of high initial viability would show only slight losses in the region of 2 to 5%. However, similar seeds held at N'Guigmi and Tahoua would have lost all viability.

In semi-arid regions, storage of seeds under ambient conditions therefore seems unlikely. Attempts are frequently made to solve this problem by placing seeds in cold storage in order to gain advantage from the effects of temperature on viability. This ignores the fact that cooling warm, dry air to cold room temperatures increases its relative humidity. Such behaviour can be easily predicted from psychrometric charts. For example, ambient air at Bilma in April when viability losses are at their lowest, cooled to 5°C will raise the relative humidity from 16.5% to 78%. In August, the worst month for viability losses, ambient air at 37% relative humidity cooled to 5°C will reach 100% relative humidity. Again, considering barley seed, re-equilibration would take place in the cold room and an equilibrium moisture content of 18% would occur in April and of 40% in August. Even at refrigerated temperatures, 18% moisture content is so high that the rates of loss of viability will be 10–20 times greater at 5°C than would have been achieved under ambient conditions. At 40% moisture content, many practical problems suddenly become involved if seed viability is to be maintained. Storage at such high moisture contents should be avoided. If these moisture content increases and their attendant viability losses are to be avoided, storage in sealed containers to prevent the uptake of moisture will be necessary.

Adopting sealed storage raises two further questions. What moisture contents are necessary to achieve the required viability period? Can they be achieved by passive drying under ambient conditions or will active drying, involving either sorption driers or warm air drying, be necessary? Consideration of the calculated viability losses and the time of seed maturity at N'Guigmi and Tahoua will help to frame solutions.

Any species whose seeds mature early in the rains will need active drying as the total loss of viability between June and October, before the seed falls to its lowest moisture content, is unacceptably high and will severely reduce longevity. For those species whose seed ripens later in or at the end of the rains, viability losses between November and April are so low that their effects on subsequent longevity can be discounted. It is suggested, therefore, that passive drying could be adopted with the seed being placed into sealed storage at the time of lowest moisture content and before the onset of the next rains.

If such an approach is adopted, then the annual losses of viability for barley at Bilma, N'Guigmi and Tahoua could be reduced to 15%, 9% and 1% respectively of those which would occur under open storage at the same ambient temperatures. For onion at N'Guigmi a reduction to 25% of open storage losses would be achieved. In all cases, the viability losses would be acceptable in practice, with seeds of high initial viability only losing some 15–20% over five years.

Viability losses will be further reduced if the sealed seeds were then placed in a refrigerator at 5°C. Under such a regime the losses would be *c.* 2% of those recorded under ambient conditions at all the locations considered (Table 23.2). For all practical purposes, no loss of viability would occur over a five year period.

Table 23.2 Annual viability losses expressed in Probit units.

	Bilma	N'Guigmi	Tahoua
Open storage under ambient conditions	0.195	1.468	2.272
Sealed at lowest ambient moisture content: ambient temperature	0.029	0.128	0.025
Sealed at lowest ambient moisture content followed by refrigerated storage	0.002	0.008	0.002

SEED DRYING

Seed drying is at its most rapid when seed is held in a thin layer through which the air can easily circulate. Thus spreading seed thinly on a simple elevated frame is sufficient to ensure acceptable drying rates. However, holding seeds even in cotton bags will greatly increase the time taken for seeds to dry and consequently greater losses of viability will occur before the seed reaches equilibrium moisture content. This should be avoided wherever possible. Efforts should also be made to shade the seeds to prevent any unnecessary heating by the sun.

THE EFFECTS OF IMPERMEABLE SEEDCOATS ON VIABILITY LOSSES

Most legume seeds have hard impermeable seedcoats which allow the seeds to desorb but prevent the absorption of moisture if the seed is placed at higher relative humidities. Therefore, they will behave as though under open storage until they reach the lowest equilibrium moisture content. Any subsequent rise in ambient relative humidity will have no effect on the seed moisture content. The seeds will behave, therefore, as though in sealed storage at the lowest moisture content. It is the ability of the impermeable seedcoat to prevent adsorption of water, rather than any unusual viability constants, which has given rise to the reputation of legume seeds for extreme longevity.

THE EFFECT OF INITIAL SEED QUALITY ON LONGEVITY

In addition to the relative effects of seed moisture and storage temperature, the absolute longevity is the most valuable criterion to those involved in maintaining seed collections. Absolute longevity of stored seed is controlled by the true initial viability of each seed batch when placed in storage. Figure 23.3 demonstrates this relationship: those with truly higher initial viabilities on the Probit scale taking longer to reach the same final viability level. However, if the equivalent percentage values are considered, then the practical difficulties in managing a seed store becomes clear. All seed batches with a true initial viability of 2.5 or more Probit units cannot give significantly different germination test results if the usual sample of 400 seeds is used. All will give results of 99–100%. Yet the differences in absolute longevity can be more than two-fold.

As initial viability levels control the absolute longevity for seeds held under identical conditions, the losses of viability which occur between seed maturity, harvest and the completion of seed drying to acceptably low moisture contents will determine the storage life of the seed. Limiting this damage will result in much enhanced seed longevities.

Collecting high quality seeds

Figure 23.4, p. 330, diagrammatically represents the effect of seed moisture content at all levels from that at which the seed reaches physiological maturity until it is sufficiently dry to have an acceptable storage life. Seeds are considered to have reached physiological maturity when maximum dry weight is achieved following fertilisation. The moisture content of seeds at physiological maturity is high, between 40–80% depending on species. Two points should be noted:–

(a) At moisture contents above the 'critical' (15–25% moisture content depending on the species) the oxygen available to the seed exerts a considerable influence on seed viability. Under anaerobic conditions very rapid losses of viability occur, in the manner predicted by the viability equation. However, under well-ventilated conditions, which allow a proper oxygen supply to the seed, damage appears to be repaired and longevities are much greater than those which would be predicted by the equations.

(b) Under properly ventilated conditions seed drying must also occur at all relative humidities other than 100%. This reduction in seed moisture content increases the rate of loss of viability. Thus, the time spent drying,

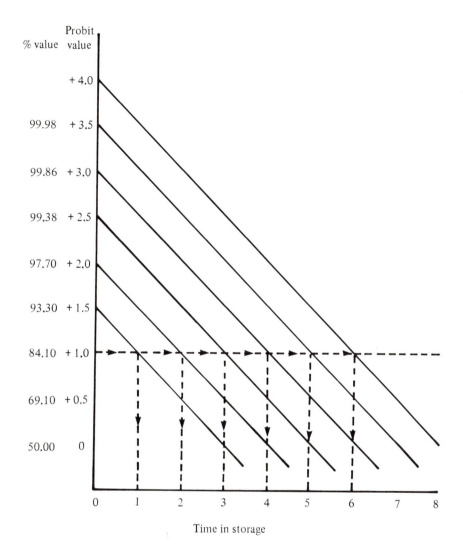

Figure 23.3 Relationship between initial quality and absolute longevity of seed lots of a species held under identical conditions.

before reaching acceptably low moisture contents are achieved, will be critical.

Fortunately, within the arid and semi-arid regions, drying will be sufficiently rapid to be benign provided that the seed is dried in ventilated thin layers rather than kept in a tight mass in a plastic bag where anaerobic conditions can soon result.

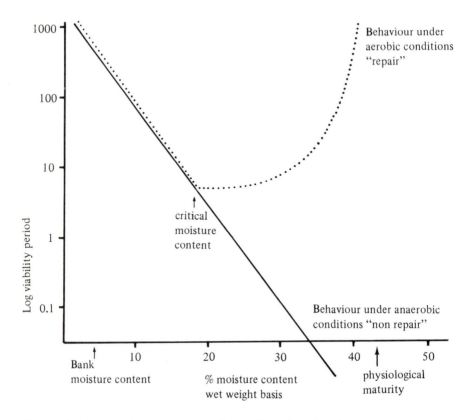

Figure 23.4 Schematic presentation of the effect of seed moisture content between 3–50% wet weight basis on the rate of loss of seed viability under anaerobic conditions.

The twin risks to initial seed quality of poor drying and/or anaerobic conditions will be at their greatest when seed is being collected in remote locations where providing adequate conditions will be difficult. Unless active efforts are made to overcome these risks, the value of making such collections must be in doubt.

Loss of seed viability during subsequent processing

Over vigorous cleaning and treatment of seeds with fungicides and pesticides have been shown to reduce initial seed viability and thus reduce absolute longevity. Such treatment is often necessary in order to comply with seed health regulations. Nonetheless, whenever and wherever such treatments can be avoided without contravening the phytosanitary regulations, this should be done in order to avoid reducing absolute longevity.

Over vigorous or careless seed cleaning which results in physical damage to the seed should be avoided. Instances are known of legume seeds being removed from their pods by repeatedly driving over them in a Land Rover.

Conclusions

The following five main conclusions can be drawn from this study:

(a) Holding seed collections under ambient open storage conditions will result in unacceptably high losses of viability in all but the most arid of locations.

(b) For species which fruit during the early part of the rainy season, active drying will be necessary to avoid the high viability losses which occur in these months.

(c) For species whose seeds mature late in or after the rainy season, passive drying in thin layers under ambient conditions can be practiced provided the seeds are placed under sealed storage conditions before the onset of the next rainy season. Once placed in field storage, viability losses at ambient temperatures are, however, low enough to allow a five year storage period.

(d) Holding such seeds under refrigerated and sealed storage conditions will further increase seed longevity so that losses of viability will be undetectable over five years.

(e) The time and effort spent in ensuring that seed collections arrive for storage at high initial viabilities will be repaid with greatly increased storage lives.

References

Two comprehensive references are offered:

Cromarty, A.S., R.H. Ellis and E.H. Roberts 1982. *The design of seed storage facilites for genetic conservation.* Rome: IBPGR, FAO.

Dickie, J.B., S. Linington and J.T. Williams 1984. *Seed management techniques for genebanks.* Rome: IBPGR, FAO.

Smith, R. 1984. The influence of collecting, harvesting and processing on viability of seed. In *Seed management techniques for genebanks,* J.B. Dickie, S. Linington and J.T. Williams (eds): 42–87. Rome: IBPGR, FAO.

24 The potential for the *in vitro* propagation of a number of economically important plants for arid areas

Anne Woods

*formerly Micropropagation Unit, Royal Botanic Gardens, Kew, Richmond, Surrey TW9 3AB, UK
now 38 Sunbury Road, Feltham, Middlesex TW12 4PG, UK*

Introduction

The Micropropagation Unit at The Royal Botanic Gardens, Kew, is primarily concerned with the *in vitro* propagation of plant species which have been found difficult to propagate using conventional methods. The function of the Unit is therefore not one of pure research but rather to act as an extension to the propagation facilities of the nursery, dealing with a wide range of botanic garden species and producing a sufficient number of plants for Kew's living collections and distributing surplus material to other centres.

Preferably, plants are raised from seed to maintain genetic variability in the botanic garden population. Embryo culture has proved particularly useful where specific dormancy problems exist, enabling such plants as *Melia volkensii,* a tropical timber tree, *Carica candamarcensis* and various members of the Proteaceae to be successfully added to the collections.

All orchid seed is also raised *in vitro* and currently approximately 200 different species exist in culture. The problems of tiny seed size and dormancy of many orchid species make them ideal subjects for *in vitro* culture where competition from microfauna and flora can be eliminated.

In vitro culture of vegetative material mainly by using nodal cuttings has been employed successfully when no seed has been available or where a particularly desirable clone exists and attempts to conventionally propagate the species has failed. Plants in this category include a wide range of succulents and a number of woody species e.g. *Cotinus obovatus* and *Emmenopterys henryi.*

Many rare and endangered species have likewise been cultured from both seed and vegetative pieces with the aim of rapidly multiplying the material for distribution to other centres. A number of rare Canary Island species have recently been cultured and distributed, thus providing a greater chance for their survival in cultivation.

In December 1982 and January 1983 seeds of ten different species of plants, known to have economic uses in arid areas, were received by the Unit with the brief to investigate their potential as subjects for micropropagation. The species were

Centrosema brasilianum, C. pascuorum, Geoffraea decorticans, Macroptilium atropurpureum, Prosopis alba, Stylosanthes hamata, S. sympodialis and *Tylosema esculentum.* All these species are members of the family Leguminosae. The other two species were *Moringa oleifera* (Moringaceae) and *Argania spinosa* (Sapotaceae). In the majority of cases, the samples of seed were small, some not exceeding three seeds. This, therefore, limited the numbers of replicate cultures which could be set up and also restricted the choice of nutrient medium and growth regulators. In each case the potential for rapid multiplication, either by bud formation through callus induced on seedlings and hypocotyl and root segments or by shoot multiplication through induced axillary bud break, were investigated. Vegetative material of *Ceratonia oreothauma* (Leguminosae) was obtained from a young potted plant from the glasshouse in the nursery with the aim of inducing axillary bud break and multiple shoot formation on single node explants followed by root induction on the single shoots.

Members of the Leguminosae have been found difficult to culture using *in vitro* techniques. It has been reported that the perennial forage legumes perform better than the seed legumes (Thomas & Wernicke 1978). Most research is concentrated in the areas of legume tissue, cell and protoplast culture, where not only is there the potential for producing large numbers of plants in a relatively short time, but also providing the opportunity to genetically manipulate that material for economic improvement.

In vitro plantlet formation has been reported for only a limited number of legume species and of those under investigation in the Micropropagation Unit, reference was found only for *Stylosanthes hamata* (Scowcroft & Adamson 1976).

Materials and method

All seeds were sterilised by dipping for five minutes in industrial methylated spirit (96% O.P.) followed by 20–30 minutes in a 0.5% sodium hypochlorite solution with 0.1% Tween 80 added as a wetting agent. After sterilisation, the seeds were rinsed three times in sterilised deionised water. Further treatments necessary for germination are outlined under the species concerned. Single nodal cuttings with the leaves removed of *Ceratonia oreothauma* were sterilised for ten minutes in 0.5% sodium hypochlorite solution plus 0.1% Tween 80 and rinsed twice in sterilised deionised water.

All seeds were placed in 24 X 150 mm Pyrex test tubes containing approximately 20 mls of Murashige and Skoog (1962) medium (Flow Laboratories, Irvine, Scotland) with the addition of 3% sucrose. The growth regulators investigated were the cytokinin 6-benzyl-aminopurine (BAP) at 0.25, 0.5, 1 and 2 mg/litre either alone or in combination with the auxin napthaleneacetic acid (NAA) at 0.1 and 1 mg/litre. Explants of *Ceratonia oreothauma* were not only set on the above but also Kinetin (K) at 0.5, 1 and 2 mg/litre and 3-indoleacetic acid (IAA) at 0.1 and

1 mg/litre were also investigated in half-strength Murashige and Skoog medium containing 1.5% sucrose (MS/2). A number of species were transferred to MS/2 and MS + 0.25 and 5 mg/litre NAA for rooting.

After the addition of growth regulators all media were adjusted to approximately pH 5.7 and solidified with 0.6% agar (BDH) and autoclaved at 121°C at 15p.s.i. for 15 minutes. All cultures were maintained at 25–30°C and illuminated on a 16 hour photoperiod by three 65–80W Thorn 'Warm White' fluorescent tubes suspended 45 cm above them.

Resulting plantlets were transferred to pots containing 50 per cent sterilised peat-based compost and 50 per cent perlite, placed in a glasshouse under mist propagators and maintained at 100% humidity. Once the plants were established they were removed from the propagators and grown on in a glasshouse environment maintained at *c.* 25°C.

Results

A summary of the performance of each plant in culture now follows; the main uses, together with the distribution of each species are also mentioned.

Centrosema brasilianum

Native to Brazil, Venezuela, Peru, Argentina and introduced to Australia, the principal use of *Centrosema brasilianum* is as a fodder plant.

Because of the limited number of seeds the range of MS media had to be restricted to: MS; MS + 0.5, 1 and 2 mg/litre BAP; MS + 0.5 and 1 mg/litre BAP + 0.1 mg/litre NAA and MS + 1 and 2 mg/litre BAP + 1 mg/litre NAA.

Fifty per cent of the seeds germinated within seven days. Hypocotyl and root sections, 3 mm long, were cut from two of the seedlings and placed on the range of media stated above.

Maximum axillary bud break and multiple shoot formation on the seedling was obtained with 0.5–2 mg/litre BAP. BAP with the addition of the auxin NAA at both concentrations stimulated callus formation. The individual shoots rooted after one subculture to MS without growth regulators. Difficulty was experienced in transferring the plantlets to compost and only a few survived.

Callus only was initiated on all hypocotyl and root sections on media containing growth regulators. Regeneration of plantlets through adventitious bud formation was not observed.

Centrosema pascuorum

Native to Brazil, Venezuela, Costa Rica and introduced to Australia; as with *C. brazilianum* its use is as a fodder crop.

The range of media was limited to MS, MS + 0.5, 1 and 2 mg/litre BAP; MS + 1 mg/litre BAP + 0.1 mg/litre NAA; MS + 1 and 2 mg/litre BAP + 1 mg/litre NAA. Forty three per cent of the seeds germinated within six days. Seven days after germination, one seedling on MS was cut into approximately 3 mm long hypocotyl and root sections and set on the full range of media excluding those containing 0.25 mg/litre BAP.

Callus was initiated at the base of the hypocotyl on all seedlings on media containing growth regulators. Axillary bud break and multiple shoot formation was best at the lower concentrations of BAP (0.5 and 1 mg/litre). At the higher concentrations of BAP and in all combinations with NAA, stunting of the shoot was observed. Individual shoots rooted after one subculture to MS and the plantlets transferred well to compost. Callus only was initiated on all hypocotyl and root sections with no signs of regeneration of shoots.

Ceratonia oreothauma subsp. *oreothauma*

Native to Oman, the species is used as browse and is potentially important as a source of genetic material for crossing with the carob, *Ceratonia siliqua*.

Sterilised nodal cuttings from a young nursery plant were set not only on the full range of media but also on MS/2 with 0.5, 1 and 2 mg/litre K alone or in combination with 0.1 and 1 mg/litre IAA. Callus was initiated at the base of all cultures on the MS + BAP + NAA media although axillary bud break and shoot proliferation were observed in the cultures containing BAP alone. However, it must be stressed that this development was slow and rarely more than six shoots were produced per culture. Figure 24.1d.

After eight weeks the cultures were transferred to MS/2 containing no growth regulators, when a small number of shoots were continually produced over the next four months. Fifty per cent of these shoots rooted on MS/2 with 0.5 or 1 mg/litre IAA. Although callus formation was reduced using MS/2 supplemented with K and IAA, little stimulation of axillary bud proliferation was observed. Approximately 50 per cent of the plantlets survived transfer to sterilised compost.

Geoffraea decorticans

Native to Argentina, Chile, Peru, Bolivia and Paraguay, *Geoffraea decorticans* is used as browse for cattle, fuelwood and as an ornamental tree. The leaves are used in local medicine as an anthelmintic.

Germination of this species was eventually achieved by removing the embryo which was cultured on basic MS. At the time of writing this report, no further results have been achieved.

Fig. 24.1 a. *Stylosanthes hamata,* formation of callus and buds from hypocotyl explants; **b.** *S. hamata,* entire plantlet formation from callus; **c.** *Macroptilium atropurpureum,* shoot formation from callus stimulated on entire seedling; **d.** *Ceratonia oreothauma,* axillary bud break and shoot elongation on nodal explants.

Macroptilium atropurpureum

This species is native to Mexico, southern Texas, New Mexico and Arizona and is widely distributed in South America. It has also been introduced into South Africa and Australia. The plant produces a palatable, high protein forage which is highly drought-resistant even when heavily grazed.

Because only seven seeds were available the range of media was restricted to MS and MS + 0.25, 0.5, 1 and 2 mg/litre BAP. The variation between the seedlings was very marked, making it impossible to confirm the results with any degree of certainty. Obviously it would be necessary to repeat the work with a larger sample of seeds. However, it appeared that shoots were initiated from callus stimulated at the base of the hypocotyl using 0.5 mg/litre BAP. The shoots rooted after one subculture to MS. It also appeared that multiple shoots formed with 2 mg/litre BAP; these shoots also rooted well after one subculture to MS. All plantlets thus produced transferred well to compost. Figure 24.1c.

Prosopis alba

An important native tree of Argentina, Uruguay, Paraguay, southern Bolivia, northern Chile and Peru and introduced to Ecuador and Arizona. It is cultivated for windbreaks, shade and shelter, for fodder and as a timber tree to supply building poles and fuel-wood. The fruit can be milled to a flour and cooked.

The five seeds germinated approximately seven days after sowing; 14 days later, the seedlings were cut into 3 mm long hypocotyl and root sections which were set on the entire range of media. The five terminal shoots were placed one on each of the following media:– MS; MS + 0.25, 0.5, 1 and 2 mg/litre BAP. Hard, granular, green proliferating callus developed on the hypocotyls on the media containing NAA. The callus was particularly prolific at the higher NAA concentration but rapidly turned brown and ceased to grow after four weeks. No adventitious bud formation was observed.

Only a slight amount of callusing occurred on the root sections and again, callus growth ceased after four weeks. Growth of the shoot tips was extremely poor with no multiple shoot formation. Two shoots rooted after one subculture to MS/2 but died after being transferred to compost.

Stylosanthes species

Shoot formation from cultures derived from hypocotyl, root and leaf tissues has been reported for three cultivars of *Stylosanthes guianensis* (Meijer & Broughton 1981; Meijer 1982a) and also in *Stylosanthes humilis* (Meijer 1982b). The long-term regeneration of plantlets from a callus of *Stylosanthes hamata* has previously been reported (Scowcroft & Adamson 1976). Callus was initiated on a basal medium supplemented with 2,4-dichlorophenoxyacetic acid. Shoots and roots were induced readily from the callus; entire plantlets were produced on a medium supplemented with kinetin. In this report, similar results to those of Scowcroft and Adamson were achieved with *S. hamata,* although the media and growth regulators differed.

Stylosanthes hamata

This species is native to the West Indies, Florida, Guatemala, Colombia, Venezuela and Panama and has been introduced to Australia and India, where it has been developed and used extensively as a pasture legume.

Ninety seven per cent of the seeds germinated within four days of sowing on the full range of media. Eight days after germination, four seedlings on MS were cut into hypocotyl and root sections approximately 1 cm long and set on the entire range of media, excluding those containing 0.25 mg/litre BAP.

Seedlings
Callus was initiated at the base of the hypocotyls on all cultures. Buds developed from the callus within six weeks of the sowing date on all MS + BAP concentrations only, the greatest number of buds being produced with 2 mg/litre BPA. Only callus was initiated on all cultures containing 1 mg/litre NAA.

Axillary bud break and multiple shoot formation was minimal and best in those cultures containing 0.25, 0.5 and 1 mg/litre BAP. All buds and shoots developed well and rooted on one subculture to either MS or MS/2.

Hypocotyl sections
Callus was initiated on all the cut ends of hypocotyl sections. Buds were observed in the callus four weeks after sowing, the greatest number being initiated with 1 and 2 mg/litre BAP. Callus only was observed in all cultures containing 0.1 mg/litre NAA and 0.5 mg/litre BAP + 1 mg/litre NAA. All buds developed and rooted well on subculture to MS/2. Figure 24.1a, b.

Weaning
All plantlets thus produced transferred well to pots of sterilised compost.

Stylosanthes sympodialis

As with the previous species, *Stylosanthes sympodialis* is a pasture legume and is native to Ecuador, Peru and the Galapagos.

Germination did not readily occur until the testa of each seed was scarified with a sterilised scalpel blade after which 50 per cent germinated. The seeds were initially sown onto a full range of MS media. The hypocotyls of three seedlings germinated on MS were cut into 1 cm sections and placed on the full range of MS media with the exception of the media containing 0.25 mg/litre BAP. There was insufficient material to investigate the effects of the growth regulators on root sections.

Seedlings
As few seeds germinated the effects of the entire range of growth regulators could not be investigated. However, it would appear that results are comparable with those of the previous species in that buds arose from the callus stimulated at the base of the hypocotyls in cultures containing BAP only and BAP in combination with 0.1 mg/litre NAA. The shoots developed slowly and rooted after one subculture to MS/2.

Hypocotyl sections
Callus was initiated on the cut ends of all hypocotyl sections and buds developed in the calli in all cultures containing BAP only, BAP + 0.1 mg/litre NAA; 1 mg/litre BAP + 1 mg/litre NAA. 1 mg/litre BAP induced the greatest number of buds. The shoots developed roots slowly on one subculture to MS/2.

Weaning
All plantlets transferred well to pots of sterilised compost.

Tylosema esculentum

Native to the Kalahari and neighbouring sandy regions of South Africa and introduced to Texas and Israel, it is rare in cultivation. The plant produces a tuber which is not only cooked and eaten but is a source of water. The seeds have a protein content similar to that of soybean and can also be cooked and eaten.

The seed coats of the four seeds were scarified (after sterilisation) with a sterilised hack-saw blade. Three seeds germinated within seven days. Two seedlings were transferred to compost but did not survive.

The remaining seedling was cut into nodal, stem and root sections, as well as approximately 1 cm square cotyledon and leaf pieces. The full range of media excluding those containing 0.25 mg/litre BAP was tested. Severe browning of the culture medium by exudation of phenolic compounds from the explants occurred in all cultures, causing death of some of the material. The addition of activated charcoal at 0.1% to the culture medium was not effective in absorbing these compounds. Callus formation only was observed on all stem, leaf and cotyledon explants and also to a lesser extent on the root sections. Axillary bud break occurred on the nodal sections with 0.5 and 2 mg/litre BAP but may not have been initiated by the cytokinin. A large amount of callus formed on the basal cut ends of all the nodal sections. Roots were initiated in 25 per cent of the cultures at 0.25 and 5 mg/litre NAA.

Argania spinosa

Native to Morocco and introduced to Egypt, Sudan, Israel, Canary Islands and U.S.A., *Argania spinosa* is cultivated for fodder and the timber is used for fuelwood and as a source of charcoal. An oil similar to olive oil can be extracted from the seeds and used for cooking; other uses include seed cake for cattle, soap-making, as an illumant and in folk medicine.

The hard testa of each of the three seeds was chipped with a sterilised scalpel blade. Splitting of the testa occurred four weeks after sowing and was then easily removed. Further removal of the endosperm exposed the large embryo which was then cultured on MS only. To date, one large, well developed seedling exists *in vitro*. No further work has yet been carried out on this species.

Moringa oleifera

Native to India this species has a variety of uses. The young pods, seeds, flowers and leaves are eaten. The seeds are used to clarify turbid water and yield ben oil which is not only edible but is also used as an illuminant, a lubricant in watchmaking and in cosmetics. The trees are used as live fence posts and the bark fibre for cordage, for mats and in paper making.

The testa was removed before sterilisation of the seeds after which approximately 90 per cent germinated. No further work has yet been undertaken on this species.

Discussion

In vitro germination of a number of species may well prove beneficial when a species is known to be difficult to germinate conventionally, such as *Argania spinosa*. However there are problems of establishment of seedling plants from *in vitro* conditions to compost and this has proved particularly difficult with a number of the arid-land species. A detailed investigation of the establishment of the correct environmental conditions necessary to ensure survival of the plantlets is needed.

The marked reluctance for leguminous species to undergo organo-genesis from callus has been repeatedly highlighted by many workers. The only species readily producing shoots from callus in this study were *Stylosanthes hamata* and *Stylosanthes sympodialis* and some evidence was observed with *Macroptilium atropurpureum*. It is doubtful whether results would be much improved using different basal media and growth regulators.

In vitro propagation of vegetative material using single nodal explants from existing plants could prove useful to multiply and distribute material of a particularly valuable clone where only limited vegetative material exists. Results with *Ceratonia oreothauma* were encouraging and repetition of the work using MS/2 supplemented with BAP alone or in combination with IAA may well prove worthwhile.

Acknowledgements

The author would like to thank Dr G.E. Wickens of the Herbarium, Royal Botanic Gardens, Kew, for providing the information on the distribution and uses of the plants mentioned and Mr J.B. Simmons, Curator of Living Collections, for his helpful advice.

References

Meijer, E.G.M. 1982a. High frequency plant regeneration from hypocotyl-derived and leaf-derived tissue cultures of the tropical pasture legume *Stylosanthes humilis*. *Physiol. Plant.* 56: 381–385.

Meijer, E.G.M. 1982b. Shoot formation in tissue cultures of three cultivars of the tropical pasture legume *Stylosanthes guyanensis*. *Z. Pflanzenzüchtg.* 89: 169–172.

Meijer, E.G.M. and W.J. Broughton 1981. Regeneration of whole plants from hypocotyl, root and leaf-derived tissue cultures of the pasture legume *Stylosanthes guyanensis* cultivar. Cook. *Physiol. Plant.* 52: 280–284.

Murashige, T. and F. Skoog 1962. A revised medium for rapid growth and bioassays with tobacco tissue cultures. *Phsyiol. Plant.* 15: 473–497.

Scowcroft, W.R. and J.A. Adamson 1976. Organogenesis from callus cultures of the legume *Stylosanthes hamata. Plant Sci. Lett.* 7: 39–42.

Thomas, E. and W. Wernicke 1978. Morphogenesis in herbaceous crop plants in *Frontiers of plant tissue culture*, T.A. Thorpe (ed.): 403–410. Calgary: International Association Plant Tissue Culture.

25 Gums and resins, and factors influencing their economic development

D. M. W. Anderson

Chemistry Department, The University, Edinburgh EH9 3JJ, UK

Introduction

Gums and resins from a wide variety of botanical sources have been important items of international trade for centuries. In common with all other natural products, the extent of general scientific knowledge and understanding of these complex chemical substances has increased greatly during the past 30 years. Nevertheless, there is still a great deal that is not yet known; studies of various aspects of gums and resins currently form an active area of multidisciplinary research. For the purposes of this Conference, and because of the restrictions on time and space that apply, it is hoped that the main objective of this contribution can best be achieved by providing an essentially non-chemical but up-to-date guide to the specialised literature, together with a discussion of several important developments that have influenced the international supply and demand for gums and resins in recent years.

Factors affecting international supply and demand

Following World War II, trade in natural water-soluble gum polysaccharides and in water-insoluble terpenoid resins grew steadily until the end of the sixties or early seventies, when the first signs of a general world-wide trade recession were detected. Unfortunately for gum markets, a number of other factors also started to become apparent at that time. The combined effects of these factors, to be discussed briefly below, was to affect supply and demand in ways which, in general, have led to decreasing demand for natural gums and resins throughout the seventies and early eighties.

ECOLOGICAL, SOCIOLOGICAL, ECONOMIC AND NATIONAL FACTORS

Not so long ago, the sale of the gum collected by nomads and villagers in Africa constituted their major single source of annual income; gum trees were treasured possessions and fights to the death to retain ownership were not infrequent. Nowadays, the nationalised policies of fixed farm-gate prices result in the gum farmers and villagers claiming that they can derive a greater income and reward for their labours by producing other crops, e.g. sugar, oil-seeds. In contrast, the imposition of relatively large (40%) export taxes results in the gum being offered for export at prices that are not cost-effective, in relation to the alternative commodities available, for at least a significant number of the potential industrial uses for gums.

The devastating droughts in the Sahel in the early seventies also undoubtedly affected the market for gum arabic adversely. For several years, there was insufficient gum available to satisfy commercial demands. When supplies did become more readily available again, the selling price asked made the gum a non-viable economic proposition; many former major industrial gum users changed formulations, or devised new products, and started to use alternative, more cost-effective natural products. The price of gum karaya also became non-competitive recently when Indian authorities decided, at short notice, to restrict gum production by giving a proportion of the trees a rest from tapping every year in order to try to increase their long-term survival. The price for locust-bean gum rose suddenly and sharply last year as a result of crop failure through unusual seasonal weather conditions. Locust-bean gum, like gum karaya, is derived from well-matured trees and gum production cannot be increased quickly in the short term. In contrast, genetic breeding programmes for the annual crop, *Cyamopsis tetragonoloba,* which produces guar gum, have produced faster-growing strains that can be used to secure two crops per season in suitable locations. For several successive seasons at the end of the seventies the demand for guar gum for oil-drilling purposes, etc. exceeded supply and the selling price rose. The number of producing countries increased, but the oil exploration industry went into recession and consequently led to a world glut in the past year with guar prices falling drastically. Natural gums are subject to the economic fluctuations well known in all natural commodity trading.

For ecological reasons, it is essential that increased numbers of trees are planted, particularly in arid zones. The factors contributing to desertification are well-known, and the need to increase the number of trees in the Sahel is particularly strong. It is deeply disappointing that the joint action of many nations over the past six years is now seen to have made no real impact in redressing the prevailing balance, despite the expenditure of much effort and money (Cross 1984). Excellent surveys and suggestions for the planting of rapidly developing trees as firewood crops (National Academy Sciences 1983) and trees with under-exploited economic potential (National Academy Sciences 1979) have been published. When trees must be introduced or regenerated in the hottest, most arid zones, the most suitable are frequently gum-bearing leguminous species, e.g. *Acacia, Prosopis, Astragalus,* etc. Within the Sahel there are many international development and aid programmes; in the Sudan alone 1 million new *Acacia* trees have been planted annually in the past three years; these schemes are expected to continue in future years. These programmes are essential to restore ecological balance. Similar programmes are in existence in other Sahelian zone countries. Thus regeneration of the gum belt is in progress. Unfortunately, for the reasons outlined, and for others to be described below, it must not be expected to follow automatically that all the gum available from increased numbers of trees will be a marketable proposition.

INDUSTRIAL, TOXICOLOGICAL AND REGULATORY FACTORS

Manufacturing and trading practices, world-wide, change steadily. Where crops are liable to be unreliable in supply for variable climatological and other factors, industry in the consuming countries tends to seek ways of establishing more reliable, alternative sources of supply or more reliable alternative raw materials. Over the years, chemical and industrial research and development programmes have led to the marketing of new, cost-effective, novel products; as a result, the marketability of many traditional natural products of previous economic importance has decreased. Such patterns of change must be expected to continue and it is the rate of change that is now increasing very rapidly.

The requirements for modern industrial companies to comply with 'good manufacturing practice' and 'good laboratory practice' regulations has also led to rapid change. The trend is for raw materials that are variable in quality or in availability, or for raw materials that are of variable bacteriological quality, to be replaced by safer commercial commodities.

There is now in existence extensive international legislation involving toxicological evaluations of all foodstuffs additives (e.g. gums, resins) and also the more general safety-at-work requirements that substances, with which plant operatives come into contact, must not present hazards to their health or safety. Such requirements are formulated by national safety committees and by powerful international organisations vested with legal powers. Restrictions, already imposed by national committees and by joint FAO/WHO expert committees, and effective in the EEC, USA, Japan, etc. will inevitably be imposed within the developing gum-producing countries. There are particularly strong moves to ensure that food-borne infections and other microbiological hazards should be reduced. For the natural gums of major commercial importance microbiological standards were imposed in 1983; gums must now be free from *Eschericha coli* and from *Salmonella* type organisms. It appears likely that materials exported from tropical countries will be required to be treated or processed, to achieve compliance with the standards set, prior to their exportation. Within the EEC, importers are deemed to be the manufacturers and hence liable in law for the safety-in-use of the products they subsequently retail.

Whilst the rules applicable to foodstuffs and to foodstuffs additives are particularly but justifiably severe, the European Inventory of Existing Chemical Substances (EINECS), which came into existence at the end of 1982 after postponements to the provisional European Core Inventory (ECOIN), lists all chemical substances used commercially prior to September 1981. Any substance not listed is deemed to be a new product, and cannot be offered for sale before it has been subjected to defined and expensive toxicological testing to establish its safety-in-use. As such tests cost a minimum of £50 000 per substance — with complete foodstuffs clearance costing some £250 000 minimum — it has been apparent for several years that trade in the future can be expected to be based very largely on those natural products that have been used commercially for many years, have been proved to be safe in use, and are therefore included in approved international permitted lists. As long as the aetiology of cancers in humans remains unknown, and hitherto unsuspected or undetected chemicals or their metabolites in food chains and/or in the gaseous and aqueous environment are suspected to be a contributing cause, all chemical raw materials used in the production of food and drinks will continue to be scrutinised. Toxicological evaluations of all substances are kept under review; checks of earlier proofs of safety (e.g. for guar gum) and re-evaluations of status (e.g. of gum arabic) have recurred quite frequently. Of the natural gums, only gum arabic (the exudate from *Acacia senegal*), gum guar, and locust bean gum have been subjected successfully to the full range of tests.

RESEARCH FACTORS

Further factors affecting the supply and demand for natural gums and resins involve recent research programmes and technological advances. There has been very rapid growth of the fermentation technology industries, based on the use of cheap waste products or by-products that would otherwise present pollution problems but which can be used as industrial feed stocks for suitable micro-organisms. Such fermentation processes have led to the marketing of a wide range of cost-effective modern products of reliable and reproducible quality, attributes that are favoured by modern industrialists. These products have included new, novel polysaccharides that act as efficient emulsifiers, viscosifiers and stabilisers. Their introduction has very considerably reduced the demand for natural gums, particularly gums karaya and tragacanth. Other modern industrial products, e.g. modified starches and chemical derivatives of cellulose, have also seized a considerable share of the market. Other products that may seriously affect the market for natural gum arabic and other gums are under development. It is not necessary to synthesise such a complex molecule as natural gum arabic in order to reproduce, reasonably well, the properties shown by gum arabic or its solutions. It must be emphasised that all such modern products are 'new' products that have had to be submitted to the full range of toxicological evaluations before they could be marketed. Their attraction to the modern manufacturer is that they are usually of modern technological origin, cost-effective, standardised in quality from batch to batch, homogeneous, free from foreign matter and sterile as a result of the manufacturing processes that led to their production. They are therefore acceptable for storage along with other ingredients in modern high-technology processing plants, as opposed to natural products imported direct in original sacks, etc. from tropical countries, frequently associated with unacceptably high levels of microbiological organisms. In addition, the political uncertainties associated with some producing countries, transportation delays, and risks of seasonal crop failures are minimised. The failures through the droughts of the early seventies will not easily be forgotten.

The effects of research and development programmes, customarily commencing with academic or purely scientific research programmes, have therefore been positive and far-reaching so far as the effect on the demand for natural gums is concerned. Equally significant, and in marked contrast, has been the lack of any research activity devoted to increasing the marketability of the natural gums. Lack of activity within the producing countries has been virtually complete; there has been little financial interest on the part of gum producers or merchants to partici-pate in research programmes designed to lead to the discovery of new uses of natural gums and hence increased consumption. The fundamental research progress made over the past 30 years has been extensive, but has very largely involved the application of pure rather than applied science and has been carried out within universities supported by government grants and financial support from certain industrial gum users. It is well-established industrial practice to make the research and development costs of future products a first charge against current, taxable income. In the face of these modern trends the gum-producing countries have for too long persisted in producing, collecting, and marketing natural gums in the traditional primitive ways. These factors, in addition to othe other factors that lie unfortunately outside the control of the gum producers, have undoubtedly partly contributed to the decreased demand for natural gums. When trees must be planted for essential ecological reasons, and when the unique (but replaceable) secondary products, e.g. gums and resins, from these trees can earn essential overseas currency for the developing gum-producing countries, it is poor business practice if these overseas earnings are not maximised.

The water-soluble gum polysaccharides of international commercial importance

Attention will be devoted to five polysaccharides, gum arabic, gum karaya, gum tragacanth, guar gum and locust bean gum. A distinction in chemical terms must be made: gums arabic, karaya and tragacanth are acidic, complex heteropolysaccha-rides of high molecular weight obtained as exudates from trees and shrubs in response to tapping. Gum guar and locust bean gum are essentially simpler, neutral galactomannans extracted from the seed endosperm. A third galactomannan, tara gum, is not of such great commercial importance; it will be mentioned later, together with some other tree exudates, in a section devoted to less common gums of local but not international interest, p. 353.

GUM ARABIC

In terms of tonnages, gum arabic is the major natural gum exudate of commerce. The current price ex Port Sudan is 1500 US dollars per tonne. The Sudan produces about 85% of the world's supply. Details of the tonnages exported by the various producing countries, and of the main outlets for gum arabic, are given in a recent report (International Trade Centre 1983).

In the past, 'gum arabic' was virtually synonymous with 'gum acacia' and the exudate from any species of *Acacia* was acceptable, although Sudanese production, and hence the bulk of the world supply, has always been based on two *Acacia* species only, *A. senegal* (gum hashab) and *A. seyal* (gum talha). For many years the Sudan has offered these two *Acacia* gums as separate commodities, gum talha being regarded as inferior and cheaper in price. There is a wide range of industrial uses for poorer grades of gum arabic, e.g. in foundry sands, printing inks, lithography, etc., but some 70% of all gum arabic goes into confectionery, foodstuffs or pharmaceutical applications, supplies for which command the top price.

Within the EEC however, foodstuffs grade gum arabic (E414) is now rigorously defined as the exudate from *A. senegal* and from related species of *Acacia*. On taxonomic grounds, this definition includes *A. laeta, A. polyacantha, A. mellifera* and the members of the recognised *A. senegal* complex (Ross 1979, International Trade Centre 1983). The amounts of gum collected from *A. laeta, A. mellifera* and *A. polyacantha* are very small as these species are not cultivated, they are also difficult to tap.

In the comprehensive toxicological and other tests carried out on gum arabic within the period 1978–1982, including a dietary study in humans (Ross *et al.* 1983), which led to the acceptance of gum arabic as a foodstuffs additive with the ADI category 'not specified' (i.e. there is no specified limit to the amount that can be used, subject to the understanding that 'foodstuffs additives' are components present to the extent of less than 2% in a foodstuff), the Test Article was clearly specified as the exudate from *A. senegal* (Anderson, Bridgeman, Farquhar & McNab 1983). Other *Acacia* exudates have not been subjected to any form of toxicological evaluation; their use, deliberately or inadvertently, in foodstuffs is not permitted. Now that there is a rigorous specification of identity and purity for 'gum arabic E414', and now that a labelling declaration is mandatory, foodstuffs processors would be wise to seek written declarations from merchants to the effect that the gum arabic supplied is derived from *A. senegal* and is not a composite blend containing gum talha or, indeed, gums from other plant genera that are not on permitted foodstuffs lists.

The chemical composition and properties of *A. senegal* gum samples from a wide range of producing regions has been published (Anderson *et al.* 1983). The structure of the gum molecule is very complex, and there have been recent re-evaluations (Street & Anderson 1983, Churms *et al.* 1983) of the data obtained some 20 years earlier (Anderson, Hirst & Stoddart 1966). These re-evaluations have, however, been concerned only with the polysaccharide portion of the gum molecule. It has long been established (Anderson & Herbich 1963) that proteinaceous matter is an integral part of the complex gum molecule, as emphasised recently in a Japanese report (Akiyama *et al.* 1984).

At present, gum arabic production exceeds supply; despite the pegging of the controlled export price from Port Sudan for the past few years, sales have remained static at around 35 000–40 000 tonnes. For tonnages to increase, the price of gum arabic would have to fall by at least 20% to make it comparable in cost-effectiveness with starch and modified starches for confectionery use. But all of the manufacturing, economic and research factors outlined above must be taken into account in assessing the situation. It is unfortunate that demand for the secondary product of an essential arid zone tree such as *A. senegal* cannot be stimulated further, but, as with all natural products, international trade factors and the cost-effectiveness of the product predominate. This should not prohibit the extensive planting of *A. senegal* in arid zones; gum collected will always be marketable and sought by the world's gum merchants provided the price asked is cost-effective and competitive with that fixed by the Sudan. Of course, the gum from other *Acacia* spp., such as *A. seyal, A. tortilis, A. drepanolobium, A. hockii,* etc. is now of very little commercial interest; such species, which are not related to *A. senegal* and give gums having positive rotations, could only be sold for technological purposes, in competition with the darker and dirtier grades of *A. senegal* gum, at lower prices. Where re-afforestation programmes are planned, with the intention and hope that gum can be tapped, collected, and sold, only *A. senegal* should be considered. A recent trade report (International Trade Centre 1983) indicates clearly that *A. senegal* is the only *Acacia* species involved in re-afforestation programmes in the recognised Sahelian gum belt.

At present the precise mode of biosynthesis of gum, and the gum precursors within the tree, are not known, although this is a line of research that has been pursued for several years. It would be of academic interest to gain more insight into the enzymatic processes involved but the information may not be exploitable commercially, particularly if the composition of any resulting modified gum were to differ in any respect from that of the gum exuded naturally or in response to traditional tapping methods.

GUM KARAYA

Gum karaya is obtained, almost exclusively, from Indian plantations of *Sterculia urens* and smaller plantations of *S. villosa*. In the Sudan, and elsewhere in Africa, gum karaya can be obtained from *S. setigera*. The exudates from these three species have been completely characterised chemically (Anderson *et al.* 1982) and shown to be very similar in terms of chemical composition and physico-chemical characteristics. In terms of the current legal definitions of identity and trade specifications, gum karaya is defined as the exudate obtained from species of *Sterculia*; thus the exudate from any *Sterculia,* or admixtures, can be offered for sale as gum karaya. Gum karaya is obtained by traditional tapping processes; it has been mentioned that Indian government agencies are now controlling, on conservation grounds, the proportion of available trees tapped in any one season. The effect has been to create a moderate shortage of gum karaya and a resulting tendency towards increased prices. This has merely led to trade switching to available cost-effective alternative products that can be used to replace gum karaya; thus such moves are not, in the end, beneficial in terms of overseas earnings for the producing country.

Gum karaya is hand-cleaned, picked, sorted and graded by local labour and sold in five different grades, based solely on the criteria of pale colour and freedom from external bark, sand, etc. The different grades meet separate demands for various end-uses; prices range from £1000–£2000 sterling per tonne. Exports are controlled very largely by a large number of long-established gum merchants based in Bombay. The principal uses of gum karaya are in pharmaceutical applications as an emulsifier, an adhesive for colostomy appliances, a fixative for synthetic dentures, and as a bulk laxative preparation. Gum karaya, itself a strongly acidic polysaccharide, shows good stability in acidic preparations and is used as an additive in foodstuffs.

At present, gum karaya is classified as 'generally recognised as safe' (GRAS) within the U.S.A. and its use is permitted, as additive E416, within the EEC on a provisional and temporary basis, with a recommended ADI of 12.5 mg/kg body weight/day. Despite this general EEC classification, the use of gum karaya has never been permitted within Germany. Although karaya has performed convincingly in the limited toxicological evaluations conducted within the past 3 years, including a study of its effects when ingested by humans (Eastwood *et al.* 1983), the Scientific Committee for Food of the EEC and the Joint FAO/WHO Expert Committee on Foodstuffs Additives made demands in 1983 for further toxicological tests to be conducted in a non-rodent species of laboratory animal. Such tests are very expensive. It is reasonable for the cost to be met very largely by the gum-producing countries, which retain by far the major share of the income derived from the sale of gum. At present the International Natural Gums Association for Research (INGAR), for which the writer acts as Secretary and general scientific adviser, is trying to ensure that the funds necessary will be raised so that the studies demanded can be carried out.

GUM TRAGACANTH

Gum tragacanth is the dried exudate obtained by slashing the stems of shrubby plants of the genus *Astragalus*. Production is labour-intensive, carried out in remote, hostile areas of Iran and Turkey. After collection the gum is selected by hand into five grades of 'ribbons' (superior quality) and five grades of 'flakes' (inferior quality). The wide range of grades is reflected in the price, which ranges from some £4000 to £40 000 sterling per tonne, with demand for only a few tonnes at the top price. Turkish tragacanth is, in general, inferior to Iranian; this doubtless reflects the fact that different *Astragalus* spp. yield the gum in the different locations. In trade description terms, gum tragacanth is derived from *A. gummifer* or other Asiatic species; but it is now believed that *A. gummifer* is not itself the main source. A botanical survey in Turkey is planned to try to gain more detailed information concerning the principal gum-yielding species.

Gum tragacanth is an extremely complex polysaccharide. It consists of at least two components which differ in their solubility. Gum tragacanth is the most stable of the natural gums in acidic solution and is the preferred natural emulsifier/stabiliser for salad dressings and preparations containing major amounts of acetic acid. It also has applications in pharmaceuticals, but, because of its high cost, gum tracanth has been replaced in many formulations by modern fermentation-type products.

Gum tragacanth is currently classed as GRAS in the USA and is included in a temporary, provisional list of foodstuffs additives in the EEC. A study of the effects of dietary gum tragacanth has been carried out in humans (Eastwood *et al.* 1984) and a major toxicological and reproduction study is known to have been carried out in laboratory rats in the USA in 1981. A report is now expected during 1985. Meantime the international foodstuffs regulatory authorities have requested that the results from further toxicological studies in a non-rodent laboratory animal be made available in 1985. It would be unwise to finalise a protocol for this study before the results of the study in rats becomes available.

A further major problem exists with regard to gum tragacanth. For foodstuffs and pharmaceutical purposes there is no mandatory total plate count per gram for micro-organisms, but the material must be free from *E. coli* and *Salmonella* spp. For some samples this can only be achieved through sterilisation procedures involving treatment with ethylene oxide, a process which is currently the subject of official evaluation from a safety point of view as ethylene oxide is clearly established as a potent mutagen.

GUAR GUM

This is obtained from the endosperm of seeds of the annual forage crop *Cyamopsis tetragonoloba*. The endosperm (*c.* 40%) has to be separated from the germ (45%) and the seed coat (15%) by a careful milling process during which heat must not be generated if degradation of the gum is to be avoided. World consumption for technological/foodstuffs purposes is probably in excess of 100 000 tonnes (i.e. very much greater than any of the exudate gums) and this figure does not include the massive tonnages used for domestic and animal feeding purposes in India, Pakistan, and other countries of origin.

Guar gum is now also produced in many other countries e.g. Zaire, and this has led to over-production, particularly as demand has simultaneously decreased. In 1980—81 good quality guar fetched *c.* £1000 per tonne: in 1983—4 the price fell to around £500.

As a result of plant genetic advances and breeding programmes, many newer cultivars of this annual forage crop are available, giving increased yields, disease resistance, etc. details being given in a recent book (Whistler & Hymowitz 1979). Figures for the production and details of the industrial uses of guar gum are given in a recent report (International Trade Centre 1982) which is, unfortunately, classified as restricted to the Indian Government and Trade Associations. Efforts are being made to have this information made available generally. The main applications of guar are in foodstuffs, pet foods, paper manufacture, textile printing, tobacco products, and in drilling/mining muds and flotation preparations.

The chemistry of the guar galactomannan polysaccharide, which is proteinaceous, has been studied extensively. It is a linear β-linked chain of mannose units, every second of which, statistically, carries one galactose residue. Gum guar can form gels, and give synergistic interactions with other polysaccharides, e.g. agarose, carrageenan. Guar gum is also made into chemical derivatives. Although the market is in recession at present, there are prospects for larger tonnages of guar to be marketed in the future.

LOCUST-BEAN (CAROB) GUM

This is also a seed galactomannan, but with a different ratio of mannose to galactose (2:1 in guar; 4:1 in locust bean gum). This leads to a range of specialised technological applications; synergistic effects are stronger than those given by guar.

The source of this gum, *Ceratonia siliqua,* only starts to bear economic quantities of pods after 15 years. Thus trees cannot be planted to meet rapid fluctuations in demand. The price of carob gum has been increasing steadily, partly through seasonal crop failures and through a tendency for demand to increase. Production appears to be around 20 000 tonnes, mainly from hotter areas bounding the Mediterranean. For good yields of pods, hot, dry weather is necessary. Large trees can yield 0.5 tonne of pods which reach full size in July but are left until October to ripen. The gum is extracted from the endosperm of the seed kernels; the yield of gum does not exceed *c.* 3% of the original weight of beans collected. Careful technological processing, separation, and milling of the kernels is necessary to obtain gum of good colour and viscosity.

Resins of international commercial importance

Because so many plant genera yield water-insoluble exudates, their classification is almost impossible, although attempts have been made to distinguish resins and oleoresins (*Pinus* spp.); varnish resins (dammars, copals, kauri fossils, mastics, sandaracs (*Callitris* spp.)); oriental lacquers (*Rhus* spp.) etc.; medical resins and balsams, e.g. from *Myroxylon* spp., *Styrax* spp., *Guiacum* spp. Other resins are used in perfumery (*Opopanax* spp.) or in religious ceremonies, such as *Boswellia* spp. (frankincense, olibanum) and *Commiphora* spp. (myrrh). Extensive details concerning rosins (Enos *et al.* 1976) and resins (Wagner 1976) have been published.

As with the water-soluble gums, the marketing and utilisation of resins is a specialised business. The market for copals and resins is still considerable, but as the producing countries, e.g. Portugal, USA, Scandinavia, etc. are not strictly 'arid zones', further consideration is not appropriate here. The market for resins formerly used in paints, varnishes, lacquers and pharmaceutical formulations has decreased over the past 20 years because of the increased use of petroleum-based products (emulsion paints, alkyd and acrylic resins, etc.) and synthesised chemicals (flavours, perfumery, cosmetic and pharmaceutical products). For those resins that still meet a limited demand, e.g. olibanum, frankincense, myrrh, shipments are often only a few tonnes, rather than hundreds (e.g. for gum karaya, gum tragacanth, locust bean gum) or thousands of tonnes (e.g. gum arabic, gum guar). Because trees giving a particular resin usually occur extensively over wide geographical areas, there is a great deal of small local production; transport costs to the ports for export may be a critical factor. Those resins that are used for religious festivals, or in folk-medicine, tend to be used largely within the countries of origin. They cannot be assessed as having any significant economic potential. Companies that formerly specialised in resins have been forced in recent years to diversify. Furthermore there seems little prospect of market demand increasing, particularly for resins exuded by the Burseraceae, which occur widely throughout arid zones; supply exceeds demand.

Frankincense and olibanum (*Boswellia* spp.) and myrrh (*Commiphora* spp.) are interesting chemically; they are mixtures containing a water-soluble polysaccharide and a water-insoluble resin (Anderson *et al.* 1965). The exudates from *Araucaria* spp. (Anderson & Munro 1969) are similar in this respect. Unfortunately, there is insufficient industrial interest in these materials to stimulate further studies of their composition, structure, or uses.

In recent years, colourless or pale yellow grades of gum dammar, of East Indian and Malaysian origin, have been used to import cloudification effects, through their almost complete insolubility, in citrus-based soft drinks. Such gum dammar is apparently obtained from the Dipterocarpaceae, e.g. *Shorea* and *Anisoptera* spp., from *Vatica rassak,* and also, according to trade literature, from *Agathis* spp. There is no toxicological data available for such resins. Because of the very small tonnages involved, there is insufficient revenue generated to support the cost of toxicological testing. Another adverse consideration is the present imprecise information regarding the true botanical source(s) of these resins and their chemical composition; substances included in permitted lists for foodstuffs and pharmaceutical purposes must be capable of precise description in terms of their origin, specifications of identity, and criteria of purity. Thus modern toxicological and other factors have had a depressing influence on some of the markets for resins.

Water-soluble gum polysaccharides of local interest

For the reasons outlined above, very few natural gums are of modern international commercial interest and economic potential. Some exudates, formerly of some interest internationally but no longer in demand, have decreased in value and may now be more readily utilised in the producing countries themselves, particularly for technological purposes; the main reason for their decreased demand is because they have not been evaluated toxicologically.

Tara gum (*Caesalpinia* spp.) is perhaps the sole exception; it is a seed galacto-mannan, mainly of Peruvian origin, intermediate in structure between guar gum and carob gum. Tara gum has been subjected to some toxicological evaluations (Food and Drug Administration 1981), and is being offered commercially, particularly in deals involving exchange for other goods.

Mesquite gum (*Prosopis* spp.) is the subject of a separate paper at this Confer-ence; it will therefore be referred to very briefly here. The gum, used extensively for a range of purposes, is of good solubility and has similarities in composition and properties to some *Acacia* gums. Gums from several *Prosopis* spp. have been com-pletely characterised chemically (Anderson & Farquhar 1982).

The gum from *Azadirachta indica* (neem, margosa) is highly proteinaceous (Anderson & Hendrie 1971). Its amino acid composition is known (Anderson *et al.* 1972); this gum may be a useful foodstuffs supplement in regions where foodstuffs are scarce.

Gums from *Combretum* spp., which are abundant in West Africa markets, have been completely characterised chemically (Anderson & Bell 1977; Anderson 1978a). They are much more viscous than *Acacia* gums.

The gum from *Albizia amara* subsp. *sericocephala* (Anderson, Cree, Marshall & Rahman 1966) is similar in solubility and function to some *Acacia* gums, but other *Albizia* spp. yield gums that are of dark colour, poor viscosity and solubility. They are of little potential.

The gums from several species of *Lannea*, e.g. *L. humilis, L. coromandelica,* have been completely characterised chemically (Anderson & Hendrie 1973). *Lannea* gum is used in Bangladesh in paper manufacture; this may be exploited elsewhere.

Gum ghatti (*Anogeissus* spp.), of considerable importance in India, can be of excellent quality, giving solutions of good colour and viscosity if not adulterated with poorer types of gum sent from other countries for blending purposes. Because of lack of any toxicological evidence, gum ghatti has been deleted from European permitted lists; considerable quantities of gum ghatti, formerly purchased for blend-ing with *Acacia* gums, will now have to be absorbed in the producing countries.

Several exudates from *Grevillea* spp. have been characterised chemically (Anderson & Pinto 1982). Specimens of Australian origin are amongst the most viscous gums studied so far and are worthy of consideration for local industrial use. Regular availability could make them of wider international interest.

Anacardium occidentale yields a poor-grade, water-soluble gum (Anderson *et al.* 1974) used in textile processing. Analytical data for some gums from *Parkia* spp. (Anderson & Pinto 1984) and from some *Julbernardia* and *Brachystegia* spp. (Anderson *et al.* 1984) have been reported recently.

Despite all these interesting genera, *Acacia* has provided the major interest for studies in gum chemistry. Progress was summarised (Anderson 1978b) when the gum from some 100 *Acacia* spp. had been studied. Data for a further 20 species, mostly of Australian (Anderson *et al.* 1984) or South African (Churms *et al.* 1980) origin have been published more recently. Nevertheless only *Acacia senegal* gum remains likely to have major international economic potential for the future; the gums from relatively few other *Acacia* spp. are likely to be useful technologically within the producing countries. Many *Acacias,* e.g. *A. tortilis, A. nilotica* and *A. drepanolobium,* yield gums of such poor solubility, colour, or viscosity as to be of little economic value. From the chemotaxonomic point of view, however, there undoubtedly remains a great deal of interest in the Juliflorae (Anderson, Farquhar & McNab 1983), particularly species from Northern Territory and Western Australia. Such gums contain up to 45% of protein and are of interest for studies of the tertiary chemical structure of gums, which were regarded over-simplistically, in earlier studies, as non-proteinaceous polysaccharides.

Acknowledgements

It is a pleasure to acknowledge the contributions of my research students; the assistance of all those who have collected gum samples; and the advice and comments provided over many years by Professor J.P.M. Brenan, Dr G.E. Wickens, Dr Mary Tindale, Les Pedley, Bruce Maslin and the late J.R. Maconochie.

References

Akiyama, Y., S. Eda and K. Kato 1984. Gum arabic is an arabinogalactan-protein. *Agric. Biol. Chem.* 48: 235–237.

Anderson, D.M.W. 1978a. Water-soluble plant gum exudates – Part 2: The *Combretum* gums. *Process Biochem.* 13: 4–18.

Anderson, D.M.W. 1978b. Chemotaxonomic aspects of *Acacia* exudates. *Kew Bull.* 32: 529–536.

Anderson, D.M.W. and P.C. Bell 1977. Gum exudates from some *Combretum* species; the botanical nomenclature and systematics of the Combretaceae. *Carbohyd. Res.* 57: 215–221.

Anderson, D.M.W. and J.G.K. Farquhar 1982. Gum exudates from the genus *Prosopis. Int. Tree Crops J.* 2: 15–24.

Anderson, D.M.W. and A. Hendrie 1971. The proteinaceous gum from *Azadirachta indica* A. Juss. *Carbohyd. Res.* 20: 259–268.

Anderson, D.M.W. and A. Hendrie 1973. *Lannea coromandelica* gum. *Carbohyd. Res.* 26: 105–115.

Anderson, D.M.W. and M.A. Herbich 1963. *Acacia seyal* gum. *J. Chem. Soc.* 1963: 1–6.

Anderson, D.M.W. and A.C. Munro 1969. Gum exudates from the genus *Araucaria. Carbohyd. Res.* 11: 43–51.

Anderson, D.M.W. and G. Pinto 1982. Gums from the genus *Grevillea. Carbohyd. Polymers* 2: 19–24.

Anderson, D.M.W. and G. Pinto 1984. Gum polysaccharides from some *Parkia* spp. *Phytochem.* (in press).

Anderson, D.M.W., P.C. Bell and J.R.A. Millar 1974. The gum from *Anacardium occidentale. Phytochem.* 13: 2189–2192.

Anderson, D.M.W., J.G.K. Farquhar and C.G.A. McNab 1983. Some highly proteinaceous *Acacia* exudates from the subseries *Juliflorae. Phytochem.* 22: 2481–2484.

Anderson, D.M.W., J.G.K. Farquhar and C.G.A. McNab 1984. Gums from some *Acacia* spp. of the subseries *Botryocephalae. Phytochem.* 23: 579–580.

Anderson, D.M.W., A. Hendrie and A.C. Munro 1972. The amino acid composition of some plant gums. *Phytochem.* 11: 733–736.

Anderson, D.M.W., E.L. Hirst and J.F. Stoddart 1966. The structure of gum arabic (*Acacia senegal*). *J. Chem. Soc.* C: 1959–1966.

Anderson, D.M.W., P.C. Bell, N.C.L. Gill and C.W. Yeconemi 1984. Gum exudates from *Brachystegia* and *Julbernardia* species. *Phytochem.* 23: in press.

Anderson, D.M.W., M.M.E. Bridgeman, J.G.K. Farquhar and C.G.A. McNab 1983. Chemical characterisation of the test article used in toxicological studies of gum arabic (*Acacia senegal* (L.) Willd.) *Int. Tree Crops J.* 2: 245–254.

Anderson, D.M.W., G.M. Cree, J.J. Marshall and S. Rahman 1965. The gum resin from *Boswellia papyrifera* (Del.) Hochst. *Carbohyd. Res.* 1: 320–323.

Anderson, D.M.W., G.M. Cree, J.J. Marshall and S. Rahman 1966. *Albizia* gum exudates. *Carbohyd. Res.* 2: 63–69.

Anderson, D.M.W., C.G.A. McNab, C.G. Anderson, P.M. Brown and M.A. Pringuer 1982. Studies of uronic acid materials, Part 58: Gum exudates from the genus *Sterculia* (gum karaya). *Int. Tree Crops J.* 2: 147–154.

Churms, S.C., E.H. Merrifield and A.M. Stephen 1980. *Acacia longifolia* gum. *S. Afr. J. Chem.* 34: 8–12.

Churms, S.C., E.H. Merrifield and A.M. Stephen 1983. Some new aspects of the molecular structure of *A. senegal* gum. *Carbohyd. Res.* 123: 267–279.

Cross, M. 1984. U.N. admits failure to halt deserts. *New Scientist* 1409: 3.

Eastwood, M.A., W.G. Brydon and D.M.W. Anderson 1983. The effects of dietary gum karaya in man. *Toxicology Lett.* 17: 159–166.

Eastwood, M.A., W.G. Brydon and D.M.W. Anderson 1984. The effects of dietary gum tragacanth in man. *Toxicology Lett.* 21: 73–81.

Enos, H.I., G.C. Harris and G.W. Hedrick 1976. Rosin and rosin derivatives. *Kirk-Othmer encyclopaedia chem. technology,* 2nd edn. Vol. 17: 475–508. New York: Interscience.

Food and Drug Administration 1981. *Carcinogenesis Bioassay of Tara Gum.* Washington D.C.: Nat. Instn. Health Publ. No. 81–1780.

International Trade Centre 1982. *Major markets for guar gum.* Geneva: Restricted Report ITC/DTC/436.

International Trade Centre 1983. *The gum arabic market and development of production.* Geneva: ITC.

National Academy of Sciences 1979. *Tropical legumes: resources for the future.* Washington, D.C.: BOSTID Report No. 25.

National Academy of Sciences 1983. *Firewood crops.* Vol. 2. Washington D.C.: BOSTID Report No. 40.

Ross, J.H. 1979. A conspectus of the African *Acacia* species. *Memoirs Bot. Survey S. Africa,* No. 44.

Ross, A.H.M., M.A. Eastwood, W.G. Brydon, D.M.W. Anderson and J.R. Anderson 1983. The effects of dietary gum arabic in man. *Amer. J. Clin. Nutrition* 37: 368–375.

Street, C.A. and D.M.W. Anderson 1983. Refinement of structures previously proposed for gum arabic and other *Acacia* exudates. *Talanta* 30: 887–893.

Wagner, F.A. 1076. Resins, natural. *Kirk-Othmer encyclopaedia chem. technology,* 2nd edn. Vol. 17: 379–410. New York: Interscience.

Whistler, R.L. and T. Hymowitz 1979. *Guar: agronomy, production, industrial use and nutrition.* Lafayette: Purdue Univ. Press.

26 Resins from *Grindelia*: a model for renewable resources in arid environments

Barbara N. Timmermann and Joseph J. Hoffmann

Office of Arid Lands Studies, College of Agriculture, University of Arizona,
Bioresources Research Facility, 250 E. Valencia Road, Tucson, AZ 85706, USA

Introduction

As the availability of abundant, inexpensive petroleum fuels and industrial feed-stocks lessens, it is important that consideration be given to renewable sources of organic compounds and fuels as alternatives to non-renewable petroleum-based substances. The actual use of bioenergy has been increasing steadily in recent years indicating that biomass may be destined to become a major source of energy in the world.

Calvin (1979) and others have suggested the establishment of 'energy farms' as places for cultivation of plants that produce highly reduced hydrocarbon-like compounds, which could serve as substitutes for crude oil. Since many arid regions of the world offer the advantages of high intensity solar radiation and longer growing seasons, many researchers now propose that energy crops be developed specifically for arid and semi-arid lands (Bassham 1977, Calvin 1979, Lipinsky & Kresovich 1979).

The recent revival of interest in renewable resources for solid and liquid fuels as well as chemicals has directed attention to natural plant products (Wang & Huffman 1981, McLaughlin & Hoffmann 1982, Princen 1982, Adams & McChesney 1983, Adams *et al.* 1984). These plant constituents, which are readily extracted with organic solvents, have been called hydrocarbons, resins, whole plant oils or bio-crude. They contain a complex mixture of terpenoids (mono-, sesqui-, di-, tri-terpenes and isoprenoid polymers), long-chain aliphatics (waxes, tri-glycerides, fatty acids) and phenolics (phenols, flavonoids, polyphenols) which can be up-graded to liquid fuels or used as chemical feedstocks (Weisz *et al.* 1979, Haag *et al.* 1980, Davis *et al.* 1984). The chemical composition of the mixture varies among different plant species.

The majority of attention in the field of biocrude production has been devoted to the mesophytic *Euphorbia lathyris,* which was proposed by Calvin (1979) as a bioenergy crop for arid lands. Three years of extensive agronomic research con-ducted at the University of Arizona indicated that *E. lathyris* is not a suitable plant for bioenergy production for xeric areas (Kingsolver 1982). Our recent phyto-chemical studies suggest that the plants with greatest potential as energy and chemical crops will likely be found among the latex and resinous species native to the arid and semi-arid environments that receive little rainfall. Our survey of native species of southwestern USA and northwestern Mexico, which included more than four hundred plant collections representing 195 species and varieties in 107 genera from 35 families, identified several xerophytic species that product significant amounts of whole plant oils (McLaughlin & Hoffmann 1982). The plants that appear to be economically promising are resinous members of the family Asteraceae, tribe Astereae, subtribe Soladiginae, including *Grindelia, Xanthocephalum* and *Chrysothamnus.*

The purpose of this paper is to draw attention to the chemical investigations of the arid-adapted *Grindelia camporum*, its potential as an alternative source of the commercial resin used in the naval stores industry, ecological significance of the resins, and the need for development of native crops yielding extractable materials that could be used as feedstocks for specialty chemicals or other commodities in the arid and semi-arid lands of the world.

The genus Grindelia

The New World *Grindelia*, with approximately 60 species, belongs to the tribe Astereae, subtribe Soladiginae in the large family Asteraceae. This genus occurs in western North America from southern Mexico north to Canada and Alaska, and from the Great Plains to the Pacific coast. In South America, it occurs at high elevations in the southern part of the continent. Although the greatest concentration of species (i.e. 22) is in the xerophytic sites of the south-western USA, only three species are widespread in this area: *G. camporum, G. aphanactis*, and *G. squarrosa*. (Fig. 26.1).

The basic chromosome number in *Grindelia* is six. Chromosome counts have shown the existence of diploid (2n = 12) and tetraploid (2n = 24) chromosomes in both North American and South American populations, including those of *G. camporum*. There are close genetic relationships between the two ploidy levels, and geographical, ecological and seasonal barriers to gene exchange exist in these obligate outcrossers (Dunford 1964).

Grindelia camporum is a herbaceous resinous perennial that typically stands between 0.5 and 1.5 m tall. Its diploid populations occur in the Sierra Nevada foothills and the North Coast ranges; the tetraploids grow in xerophytic habitats in the Central Valley region of California, where it is often found in saline flats and in disturbed fields and roadsides. The salt-tolerant tetraploid has a woody base that produces several erect, open-branched, and herbaceous stems with yellow flower heads. The diploid can be distinguished by its more sparsely branched stems, fewer flower heads, and leaves that are fewer and further apart.

Grindelia camporum produces large amounts of characteristically aromatic resins that exude naturally to thickly cover the surface of the plant. This exudate makes the plant sticky to the touch and gives it the common name, 'gumweed'.

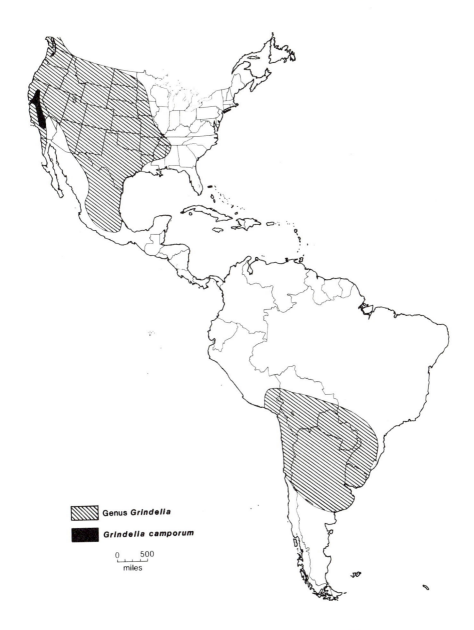

Figure 26.1   Geographic distribution of *Grindelia*.

The term 'resin' is defined here on the basis of simple physico-chemical properties different enough to distinguish them from other types of plant products
such as latex, essential oils, gums, mucilages and waxes. *Grindelia* resins are clear,

non-volatile terpenoid mixtures that are insoluble in water but soluble in organic solvents. Stable, inert and amorphous, they become very sticky when heated at low temperatures and are fusible with no sharp melting point.

Resin chemistry

Although reports on the natural products chemistry have been growing rather rapidly in the literature during the last few years for members of the tribe Astereae, only 14 species from the large genus *Grindelia* have been chemically investigated. The occurrence of C_{10} acetylenes (Bohlmann *et al.* 1973 & 1982), flavonoids (Wagner *et al.* 1972, Pinkas *et al.* 1978, Ruiz *et al.* 1981), simple phenolic acids, monoterpenoids (Pinkas *et al.* 1978) as well as sterols (Forliard 1969) and rare tropones and modhephenepoxides (Bohlmann *et al.* 1982) have been reported.

Grindelic acid derivatives were isolated from several species of *Grindelia* and are typical of the labdane-type diterpenoids found in few other genera of the subtribe Soladiginae (Mangoni & Belardini 1962, Bruun *et al.* 1962, Pinkas *et al.* 1978, Rose *et al.* 1981, Guerreiro *et al.* 1981, Bohlmann *et al.* 1982, Hoffmann *et al.* 1982 Timmermann *et al.* 1982, 1983).

In connection with our phytochemical investigations of the genus *Grindelia* we have separated the crude resin of *G. camporum* into its acidic and neutral components. Twelve labdane-type diterpene acids were isolated from the acidic portion including grindelic acid and its derivatives 6-oxo grindelic acid, 6-hydroxy grindelic acid, five C_{17} substituted homologs (methoxy, acetoxy, propionyloxy, isobutyryloxy and isovaleroyloxy), $7\alpha,8\alpha$-epoxy grindelic acid; 7-hydroxy 8(17)-dehydro grindelic acid, 6,8(17)diene grindelic acid and strictanonoic acid (Timmermann *et al.* 1983). Additional grindelane terpenoids in the acidic resin are presently under investigation.

The neutral resin fraction afforded naturally occurring methyl esters of grindelic acid and its 6,8(17)-diene, $7\alpha,8\alpha$-epoxy; 7-hydroxy 8(17)-dehydro; 6-hydroxy; 6-oxo; 17-acetoxy and 17-propionyloxy derivatives in addition to kaempferol, quercetin, apigenin and their methyl ethers and a mixture of C_{25} to C_{35} paraffin waxes and steroids (Hoffmann *et al.* 1984).

In our studies we have also observed a large amount of variability in the quality and quantity of resin produced between and within *Grindelia* species. The amount of crude resin extracted with dichloromethane has ranged from 5−18% of the dried biomass. When the crude resin is partitioned with sodium carbonate, its resin acid fraction may vary from 80−55% and the neutral component, correspondingly, from 20−45%. Moreover, the relative amounts of the various diterpene acids also vary within the species; grindelic acid, for example, has been found to constitute any amount from 20−60% of the resin acid fraction.

Gas chromatography has been found to be very effective in determining the constituents and variation in the resin acid composition of *Grindelia* species, varieties, populations and their different phenological stages. A typical 'fingerprint' of methyl ester derivatives of grindelic acid is shown in Figure 26.2.

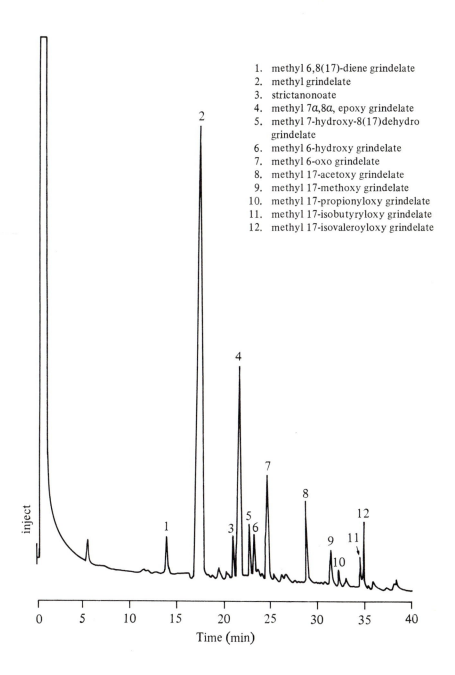

1. methyl 6,8(17)-diene grindelate
2. methyl grindelate
3. strictanonoate
4. methyl 7a,8a, epoxy grindelate
5. methyl 7-hydroxy-8(17)dehydro grindelate
6. methyl 6-hydroxy grindelate
7. methyl 6-oxo grindelate
8. methyl 17-acetoxy grindelate
9. methyl 17-methoxy grindelate
10. methyl 17-propionyloxy grindelate
11. methyl 17-isobutyryloxy grindelate
12. methyl 17-isovaleroyloxy grindelate

Figure 26.2 Gas chromatography of acid methyl esters from *Grindelia camporum*.

Preliminary genetic and breeding studies have shown that much of the variation in the crude resin yield and composition is genetic, although phenology, population structure, climate and soil types may also contribute to this variation (McLaughlin pers. comm.).

Potential uses of Grindelia resins

The discovery of numerous labdane-derived acids in *Grindelia* could be of considerable economic importance. These bicyclic compounds have chemical and physical properties similar to the tricyclic terpenoids in the wood and gum rosins used in the naval stores industry, and are related to abietic-derived acids (Fig. 26.3) that are the major constituents of pine (*Pinus sylvestris*) rosin (Zinkel 1975, Hannus 1976). In recent years much research has been directed towards finding alternative sources of resins, since many industrial uses place a high demand on these materials. The two principal types of high-quality resins, wood rosin and gum rosin, are unlikely to supply future demands since the former is extracted from a non-renewable resource (virgin pine stumps) and the latter is obtained by tapping living trees, a procedure that is extremely labor-intensive. Chemical substitution in this area may become necessary in the near future because of increasing costs and declining supplies of conventional sources of naval stores (Zinkel 1981). Although the shortage may be alleviated somewhat by imports, a domestic, renewable source of terpenoids would have a great possibility to be used as substitutes for wood rosin (Hoffmann 1983, Hoffmann *et al.* 1984; Hoffmann & McLaughlin in press).

Natural resin acid products are commercially used as adhesives, varnishes, paper sizings, printing inks, soaps, resinates and other numerous industrial applications. Extracts from *Grindelia squarrosa* have been processed and patented in the past for applications in the food, rubber, coatings, textile and polymer industries (McNay 1964, McNay & Peterson 1964 & 1965).

Sites of resin production

Our preliminary microscopic observations of stems, leaves and capitula (flower heads) of *G. camporum* have shown the presence of two types of resin-producing structures: multicellular resin glands and resin ducts. The resin glands occur in shallow pits on the surface of all above-ground organs; they are sparse on stems, more abundant on leaves and very dense on the phyllaries (bracts surrounding the capitula), and lacking on the flowers themselves. The resin ducts, on the other hand, were found only in the leaf mesophyll and in the stem cortex (Hoffmann *et al.* 1984).

Such glands and ducts have been noted in several Asteraceae genera (Metcalfe & Chalk 1979) and were reported in 13 *Grindelia* species (Hohmann 1967, Forliard 1969) although their function and biological significance have not been discussed.

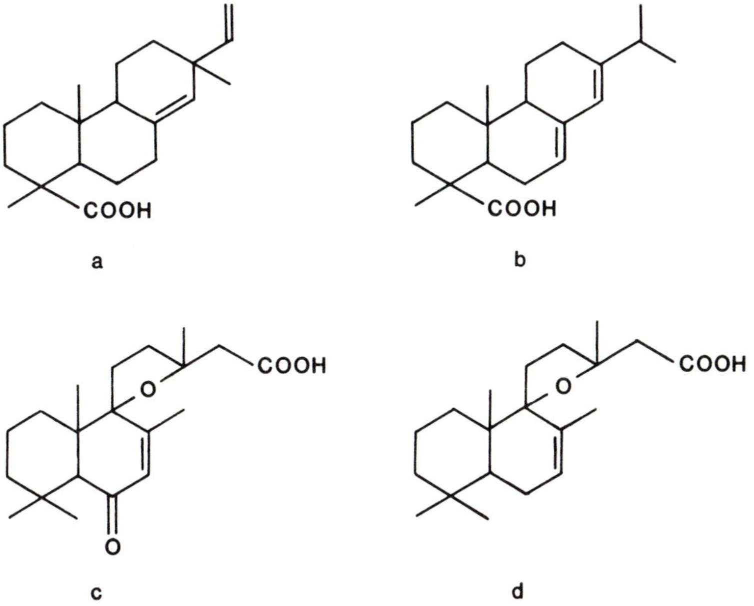

Figure 26.3 Diterpene resin acids in *Pinus* (a,b) and *Grindelia* (c,d).

An external wash of resins in *G. camporum* individual plant parts using dichloro-methane, showed a positive correlation between resin yields and secretory struc-tures; the capitula were found to contain 20–30% of resins while the leaves and stems accounted for less than 10%. The surface washes removed most of the crude resin and nearly all of the grindelic acid.

Although the work is preliminary, we now have evidence that most of the resin produced by *G. camporum* occurs on the plant surface as part of the cuticular layer and that the resin glands are directly and quantitatively responsible for resin pro-duction (Hoffmann *et al*. 1984).

A similar finding that the occurrence of surface resins is closely correlated with the occurrence of glands that produce most, if not all, of the resin was demon-strated in a recent survey of 32 resinous genera belonging to 22 arid-adapted families of Australian plants (Dell 1977); only five were found which lacked obvious glands.

Resins functions

While resins may be found on the surface of plants growing in diverse habitats, resin coatings are very prominent in xeric plants, supporting the hypothesis that resins in herbaceous plants represent a phytochemical adaptation to arid and semi-arid environments. Resins might reasonably be included among the xeromorphic features of xerophytes.

In agreement with this evidence, the survey of North American species having resinous coatings showed that this feature is associated with arid-adapted plants (McLaughlin & Hoffman 1982). Moreover, diterpene acids also appear to be fairly common in other resinous members of the tribe Astereae, including *Xanthocephalum, Gutierrezia, Chrysothamnus* and *Erigeron* (Rose 1980, Bohlmann *et al.* 1981, Hoffmann *et al.* 1982, Waddell *et al.* 1983).

Several Western Australian genera in the families Boraginaceae, Dicrastylidaceae, Euphorbiaceae, Goodeniaceae, Lamiaceae, Mimosaceae, Myoporaceae, Sapindaceae and Scrophulariaceae show that a considerable portion of the dry weight of the plant biomass is also made up of resins of terpenoid-type (Dell & McComb 1974, 1975, Dell 1975).

Resinous exudates, although of different chemical types, occur in members of other arid-adapted plants, including members of the Zygophyllaceae, where the lignan nordihydroguaiaretic acid (NDGA) makes up to 50% of the resin in *Larrea tridentata*, a dominant shrub in the North American deserts (Sakakibara *et al.* 1976).

In arid regions, the resins are characteristic of perennial, evergreen shrubs, not the drought-evading perennials or annuals. This characteristic suggests that resin function may be concerned with the extremes of climate experienced in arid regions.

Several theories exist regarding the biological and ecological significance of resins. First, what appears to be important is that the plant is able to secrete resinous coatings, the general properties of which are that they are hydrophobic and non-volatile at the temperatures encountered in the field. The other possibility is that the resin is secreted onto the leaf surface making the cuticle less permeable to water and, as a consequence, reduces water loss through its hydrophobic properties.

It has long been recognized that leaf temperature might be reduced by the presence of surface features such as scales or hairs (Pearman 1966, Ehleringer *et al.* 1976). It has also been noted that a shiny leaf surface can reduce the leaf temperature by 10–15°C under stress conditions (Slatyer 1964, Waggoner 1966). Resin surfaces are often shiny, and the removal of resins from leaves in *Eremophila fraseri* resulted in a marked reduction of total reflectance (Dell 1977).

The other possibility is the ecological significance of resins as important plant defense compounds. Terpenoid compounds have been implicated as feeding deterrents in a number of instances (Burnett *et al.* 1974, Elliger *et al.* 1976, Mabry & Gill 1979, Bell 1984). This evidence indicates that diterpenoids are probable sources of pest resistance observed in plants in which they occur. Two grindelic acid derivatives (18-hydroxy grindelic acid and 18-succinyloxy grindelic acid) isolated from *Chrysothamnus nauseosus* were found to exhibit significant antifeeding properties against third instar larvae of the Colorado potatoe beetle (Rose 1980). In addition, a series of diterpene acids including 6-hydroxy grindelic acid, isolated from *Grindelia humilis*, showed deterrent activity towards the aphid *Schizaphis graminum* (Rose *et al*. 1981).

It is possible that the different functions of the resins are not mutually exclusive, on the contrary, all the types of adaptations are likely to occur in the same plant.

The concentration of resins in the leaves and particularly in the capitula of *Grindelia* and other resinous species in the Astereae, presumably reflects the importance of protecting these plant parts from excessive dessication and insect damage. This property may enable the plant to produce at least some seeds; thus, maintaining some reproductive potential even during periods of extreme and prolonged drought and stress when only few other plants in the hot areas are physiologically active.

Conclusions

In practical terms, the arid and semi-arid lands of the world are vast and underutilized, and the development of native crops as renewable sources of specialty chemicals or other commodities that could be grown in these regions would be an accomplishment of considerable economic importance. Moreover, crops of this type would not compete with conventional food and fiber crops that require more productive land.

Although the production of liquid fuels from desert plants was originally viewed as a near-term economic possibility due to the potential shortage of crude oil, the bioenergy systems that would produce the liquid fuels now appear to be uneconomical due to the decreased prices of petroleum. A recent approach that seems more feasible for the immediate future, at least in the United States, is the development of new crops that could provide chemical commodities that are considerably more valuable than crude oil (Palsson *et al*. 1981, Lipinsky 1981). The resin-producing *Grindelia* cultivar which is currently being investigated at the University of Arizona is a promising candidate for near-term commercial development. In addition, the few surveys that have been conducted in order to identify potential chemical crops have identified a surprisingly large number of candidate species (Buchanan *et al*. 1978, Buchanan & Otey 1979, Adams *et al*. 1984).

The economic advantage of localized specialty chemicals and energy sources, such as those supplied by biomass, will become important in the arid and semi-arid lands of the world.

The ultimate success of developing a renewable, biomass-based industry will depend on a detailed knowledge of native plants, water use, plant productivity, chemical composition, and their potential to produce useful natural products. Moreover, in order to enhance the economic development in energy-importing areas, multi-product crops could be established for use as diesel fuel, upgraded liquid fuels, feedstocks for fermentation, along with specialized commodities for a lower volume, higher priced market. The discovery of novel and safer chemical products with antifeedant properties, which could become a new source of pesticides and antibiotics, would also be beneficial for the arid lands economic scenario.

After the long-term development to recover valuable chemicals, the bagasse or plant residue should be considered in order to have a positive impact on the local economic balance of the system. Several options exist, including densification for direct combustion to produce steam and electricity, or use as low-cost fermentation feedstocks, animal feed supplements, soil amendments or synthetic fuels.

Acknowledgements

This work was funded in part by research agreements with the Diamond Shamrock Corporation (Dallas, Texas) and Hercules Inc. (Wilmington, Delaware).

References

Adams, R.P. and J.D. McChesney 1983. Phytochemicals for liquid fuels and petrochemical substitution: extraction procedures and screening results. *Econ. Bot.* 37: 207–215.

Adams, R.P., M.F. Balandrin and J.R. Martineau 1984. The showy milkweed, *Asclepias speciosa*: a potential new semi-arid land crop for energy and chemicals. *Biomass* 4: 81–104.

Bassham, J.A. 1977. Increasing crop production through more controlled photosynthesis. *Science* 197: 630–638.

Bell, A.A. 1984. Morphology, chemistry, and genetics of *Gossypium* adaptations to pests. In *Recent advances in phytochemistry*, Vol. 18, B.N. Timmermann, C. Steelink and F. Loewus (eds): 197–230. New York: Plenum Press.

Bohlmann, F., T. Burkhardt and C. Zdero 1973. *Naturally occurring acetylenes*. London: Academic Press.

Bohlmann, F., M. Grenz, A. Dahr and M. Goodman 1981. Labdane derivatives and flavones from *Gutierrezia dracunculoides*. *Phytochem.* 20: 105–107.

Bohlmann, F., M. Ahmed, N. Borthakur, M. Wallmeyer, J. Jakupovic, R.M. King and H. Robinson 1982. Diterpenes related to grindelic acid and further constituents from *Grindelia* species. *Phytochem.* 21: 167–172.

Bruun, T., L. Jackman and E. Stenhagen 1962. Grindelic and oxy-grindelic acids. *Acta Chem. Scand.* 16: 1675–1681.

Buchanan, R.A. and R.H. Otey 1979. Multi-use oil and hydrocarbon producing crops in adaptive systems for food, material and energy production. *Bioresources Digest* 1: 176–200.

Buchanan, R.A., I.M. Cull, F.H. Otey and C.R. Russell 1978. Hydrocarbon and rubber producing crops: evaluation of 100 U.S. plant species. *Econ. Bot.* 32: 146–153.

Burnett, W.C., S.B. Jones, T.J. Mabry and W.G. Padolina 1974. Sesquiterpene lactones: insect feeding deterrence in *Vernonia. Biochem. Syst. Ecol.* 2: 25–29.

Calvin, M. 1979. Petroleum plantations for fuel and materials. *Bioscience* 29: 533–538.

Davis, E.A., J.L. Kuester and M.O. Bagby 1984. Biomass conversion to liquid fuels – potential of some Arizona chaparral brush and tree species. *Nature* 307: 726–728.

Dell, B. 1975. Geographical differences in leaf resin components of *Eremophila fraseri* F. Muell. (Myoporaceae) *Aust. J. Bot.* 23: 889–897.

Dell, B. 1977. Distribution and function of resins and glandular hairs in Western Australian plants. *J. Roy. Soc. W. Austr.* 59: 119–123.

Dell, B. and A.J. McComb 1974. Resin production and glandular hairs in *Beyeria viscosa* (Labill.) Miq. (Euphorbiaceae). *Aust. J. Bot.* 22: 195–210.

Dell, B. and A.J. McComb 1975. Glandular hairs, resin production, and habitat of *Newcastelia viscida* E. Pritzel (Dicrastylidaceae). *Aust. J. Bot.* 23: 373–390.

Dunford, M.P. 1964. A cytogenetic analysis of certain polyploids in *Grindelia* (Compositae). *Am. J. Bot.* 51: 41–56.

Ehleringer, J., O. Bjorkman and H.A. Money 1976. Leaf pubescence: effects on absorptance and photosynthesis in a desert shrub. *Science* 192: 376–377.

Elliger, C.A., D.F. Zinkel, G.B. Chan and A.C. Waiss 1976. Diterpene acids as larval growth inhibitors. *Experientia* 32: 1364–1366.

Forliard, N. 1969. Contribution a l'etude de la composition chimique de l'herbe de *Grindelia. J. Pharm. Belg.* 24: 397–414.

Guerreiro, E., J. Kavka, J. Saad, M. Oriental and O. Giordana 1981. Acidos diterpenicos en *Grindelia pulchella* y *G. chiloensis. Revista Latinoamericana Quimica* 12: 77–81.

Haag, W.O., P.G. Rodewald and P.B. Weisz 1980. *Catalytic production of aromatics and olefins from plant materials.* Symposium on alternative feedstocks for petrochemicals. American Chemical Society Meeting, Las Vegas, Nevada, August 24–25 (mimeo.).

Hannus, K. 1976. Lipophilic extractives in technical foliage of pine (*Pinus sylvestris*). In *Applied Polymer Symposia* No. 28, T.E. Timell (ed.): 485–501. New York: J. Wiley.

Hoffmann, J.J. 1983. Arid lands plants as feedstocks for fuels and chemicals. *Critical reviews in plant science* 1: 95–116.

Hoffmann, J.J. and S.P. McLaughlin (in press). *Grindelia camporum*: a potential cash crop for the arid Southwest. *Desert Plants.*

Hoffmann, J.J., B.K. Kingsolver, S.P. McLaughlin and B.N. Timmermann 1984. Production of resins by arid-adapted Astereae. In *Recent advances in phytochemistry,* Vol. 18, B.N. Timmermann, C. Steelink and F. Loewus (eds): 251–272. New York: Plenum Press.

Hoffmann, J.J., S.P. McLaughlin, S.D. Jolad, K.H. Schram, M.S. Tempesta and R.B. Bates 1982. Constituents of *Chrysothamnus paniculatus* (Compositae) 1: Chrysothame, a new diterpene, and 6-oxygrindelic acid. *J. Org. Chem.* 47: 1725–1727.

Hohmann, V.B. 1967. Botanisch-varenkundliche untersuchungen inner-half der gattung *Grindelia. Planta medica* 15: 255–263.

Kingsolver, B.K. 1982. *Euphorbia lathyris* reconsidered: its potential as an energy crop for arid lands. *Biomass* 2: 281–298.

Lipinsky, E.S. 1981. Chemicals from biomass: petrochemical substitution options. *Science* 212: 1465–1471.

Lipinsky, E.S. and S. Kresovich 1979. Fuels from biomass systems for arid lands environments. In *Arid land plant resources,* J.R. Goodin and D.K. Northington (eds): 294–306. Lubbock: Texas Tech University Press.

Mabry, T.J. and J.E. Gill 1979. Sesquiterpene lactones and other terpenoids. In *Herbivores: their interaction with secondary plant metabolites,* G.A. Rosenthal and D.H. Janzen (eds): 501–537. New York: Academic Press.

Mangoni, L. and M. Belardini 1962. Constituents of *Grindelia robusta. Gaz. Chimica Ital.* 92: 983–994.

McLaughlin, S.P. and J.J. Hoffmann 1982. Survey of biocrude-producing plants from the southwest. *Econ. Bot.* 36: 323–339.

McNay, R.E. 1964. *Emulsion polymerization.* US Patent 3 157 608.

McNay, R.E. and W.R. Peterson 1964. *Treatment of synthetic rubber.* US Patent 3 157 609.

McNay, R.E. and W.R. Peterson 1965. *Method of sizing cellulose fibers with resinous material from the plant Grindelia and products thereof.* US Patent 3 186 901.

Metcalfe, C.R. and L. Chalk 1979. *Anatomy of the Dicotyledons.* Vol. 2. Oxford: Clarendon Press.

Palsson, B.O., S. Fathi-Afshar, D.F. Rudd and E.N. Lightfoot 1981. Biomass as source of chemical feedstocks: an economic evaluation. *Science* 213: 513–517.

Pearman, G.I. 1966. The reflection of visible radiation from leaves of some western Australian species. *Austr. J. Biol. Sci.* 19: 97–103.

Pinkas, M., N. Didry, M. Torck, L. Bezanger and J.C. Cazin 1978. Recherches sur les polyphenols de quelques especes de *Grindelia. Ann. Pharm. Françaises* 36: 97–104.

Princen, L.A. 1982. Alternate industrial feedstocks from agriculture. *Econ. Bot.* 36: 302–312.

Rose, A. 1980. Grindelane terpenoids from *Chrysothamnus nauseosus. Phytochem.* 19: 2689–2693.

Rose, A., K. Jones, W. Haddon and D. Dreyer 1981. Grindelane diterpenoid acids from *Grindelia humilis*: feeding deterrency of diterpene acids toward aphids. *Phytochem.* 20: 2249–2253.

Ruiz, S.O., E. Guerreiro and O.S. Giordano 1981. Flavonoides en tres especies del genero *Grindelia. Anales Asoc. Quimica Argentina* 69: 293–295.

Sakakibara, M., D. DiFeo, N. Nakatani, B.N. Timmermann and T.J. Mabry 1976. Flavonoid methyl ethers on the external leaf surface of *Larrea tridentata* and *L. divaricata. Phytochem.* 15: 727–731.

Slatyer, R.O. 1964. Efficiency of water utilization by arid zone vegetation. *Annals Arid Zone* 3: 1–12.

Timmermann, B.N., J.J. Hoffmann, S.D. Jolad, K.H. Schram, R.E. Klenck and R.B. Bates 1982. Constituents of *Chrysothamnus paniculatus* (Compositae) 2: Chrysolic acid, a new labdane-derived diterpene with an aromatic B-ring. *J. Org. Chem.* 47: 4114–4116.

Timmermann, B.N., D.J. Luzbetak, J.J. Hoffmann, S.D. Jolad, K.H. Schram, R.B. Bates and R.E. Klenck 1983.Grindelane diterpenoids from *Grindelia camporum* and *Chrysothamnus paniculatus. Phytochem.* 22: 523–525.

Waddell, T.G., C.B. Osborne, R. Collison, M.J. Levine, M.C. Cross, J.V. Silverton, H.M. Fales and E.A. Sokoloski 1983. Erigerol, a new labdane diterpene from *Erigeron philadelphicus. J. Org. Chem.* 48: 4450–4453.

Waggonner, P.E. 1966. Decreasing transpiration and the effect upon growth. In *Plant environment and efficient water use*, W.H. Pierre, D. Kirkham, J. Pesek and R. Shaw (eds): 49–72. Madison, Wisconsin: American Society of Agronomy and Soil Science Society of America.

Wagner, H., M. Iyengar, O. Seligmann, L. Horhammer and W. Herz 1972. Chrysoeriol-7-glucuronoid in *Grindelia squarrosa. Phytochem.* 11: 2350.

Wang, S. and J.B. Huffman 1981. Botanochemicals: supplements to petrochemicals. *Econ. Bot.* 35: 369–382.

Weisz, P.B., W.O. Haag and P.G. Rodewald, 1979. *Catalytic production of aromatics and olefins from plant materials*, presented at 2nd Chemical Congress of American Chemical Society, Las Vegas, Nevade, August 26, 1980 (mimeo.).

Zinkel, D.F. 1975. Naval stores: silvichemicals from pine. In *Applied polymer symposia* No. 28, T.E. Timell (ed.): 309–327. New York: J. Wiley.

Zinkel, D.F. 1981. Turpentine, rosin and fatty acids from conifers. In *Organic chemicals from biomass*, I.S. Goldstein (ed.): 163–187. Boca Raton, Florida: CRC Press.

27 Plant hydrocarbon resources in arid and semi-arid lands

D. O. Hall

Department of Plant Sciences, King's College, London SE24 9JF, UK

Introduction

The overuse and undersupply of biomass is currently a serious problem and potentially a greater long term danger than ready lack of food. Today 14 per cent of the world's primary energy is derived from biomass — equivalent to 20 million barrels oil/day. Predominant use is in the rural areas of developing countries where half the world's population lives, e.g. Ethiopia, Tanzania and Nepal derive nearly all, Sudan four-fifths, Kenya three-quarters, India half, China one-third, Brazil one-quarter and Egypt and Morocco one-fifth of their total energy from biomass. A number of developed countries also derive a considerable amount of energy from biomass, e.g. Sweden 9 per cent, Canada 5 per cent, and USA and Australia 3 per cent each. An especially valuable contribution could be in the form of liquid fuels which have become so prone to fluctuating price and supply. The resources available, the effect of large agricultural surpluses (especially in North America and Europe) and the factors which will influence biomass energy schemes around the world are currently hotly debated.

Worldwide government expenditure on biomass energy systems is over 2 billion US dollars a year while the costs of surplus food production is over 60 billion US dollars a year. However, biomass energy is not necessarily the panacea to any country's energy problems even though at present the process of photosynthesis produces an amount of stored energy in the form of biomass which is almost ten times the world's annual use of energy. Additionally, the productivity of biomass-for-energy species can be dramatically increased as has been already shown in a number of demonstrations around the world; such improvements have been accomplished with a number of well known agricultural species such as maize, wheat, soya and rice.

The world produces 10—20 per cent more food than is required to feed its 4.5 billion people an adequate diet. In North America and Europe the main problem with food is its easy overproduction and general over-consumption. However, there are an estimated 450 million undernourished people, mostly in Asia and Africa. Simplistically, if available food production was increased by 1.5 per cent (equivalent to *c.* 25 million tonnes of grains) and if this food was distributed equitably to who need it, there would be no undernourished people in the world. The same argument applies if only 10 per cent of the developed countries' grain production was diverted away from animals to humans. Health authorities have recommended lower meat and sugar consumption in the USA and UK and such changes are already occurring in some developed countries. These diet and other biotechnological changes will have long term socio-economic consequences, both in developed and developing countries.

'Food versus Fuel?' This is the wrong question. It should rather be 'How can we distribute equitably the existing ample supply of food?' The approach should be on how to achieve both food and biomass fuel production locally on a sustainable basis. Both are required – thus planning and provision of the appropriate infrastructure and incentives must be provided. Increased support of research and development training and firm establishment of top priority to agriculture and forestry are essential in many countries of the world – if necessary, with significant help from abroad. These ideas are certainly not novel, but unless real recognition of the importance of agriculture, forestry and revegetation in many countries of the world is forthcoming very soon, they will continue to suffer national shortages of food and/or fuel which may have a debilitating effect on their development.

In considering the use of hydrocarbon plants in any specific region or country these biomass fuel and food supply problems are also supplemented by the problem of the current surplus of oil. Since 1979 when the world oil demand peaked and since 1981 when the price of oil peaked, there has been a significant decline in demand and prices which undoubtedly affects decisions in implementing schemes for new sources of hydrocarbons – whether they be shale oils, tar sands, coal or growing plants. Recent statements from oil companies do however predict price instability and perceived shortages (Raisman 1984). Some groups also predict long-term surpluses of oil and other forms of energy on a global basis with possibly some regional problems (Odell & Rosing 1983). However, to a country, such as Brazil or Kenya, with no indigenous fossil fuels and which is spending half to three-quarters of its total foreign income on importing oil, these surpluses of oil are theoretical and internally the costs entail great hardship on the population. This has certainly been an incentive to providing local sources of liquid fuels (and also possibly chemicals) as has been so successfully done for alcohol in Zimbabwe and Brazil.

Biomass as a source of energy has problems and it has advantages. Like every other energy source one must realise that it is not the universal panacea. Some advantages and disadvantages are listed in Table 27.1. I wish to emphasize one that is very interesting to me, namely the large biological and engineering development potential which is available for biomass. Presently we are using knowledge and experience which has been static for very many years; the efficiency and production and use of biomass as a source of energy has not progressed in the way agricultural yields for food have increased. Research in agriculture has paid off very well (Table 27.2) and this could also be the case for biomass research and development as has been shown from recent fuelwood schemes. Thus there is an undoubted

Table 27.1 Some advantages and problems foreseen in Biomass for Energy schemes (Hall 1984).

Advantages	Problems
1. Stores energy	1. Land and water use competition
2. Renewable	2. Land areas required
3. Versatile conversion and products with high energy content	3. Supply uncertainty in initial phases
	4. Costs often uncertain
4. Dependent on technology already available with minimum capital input; available to all income levels	5. Fertilizer, soil and water requirements
	6. Existing agricultural, forestry and social practices
5. Can be developed with present man-power and material resources	7. Bulky resource; transport and storage can be a problem
6. Large biological and engineering development potential	8. Subject to climatic variability
	9. Low conversion efficiencies
7. Creates employment and develops skills	10. Seasonal (sometimes)
8. Reasonably priced in many instances	
9. Ecologically inoffensive and safe	
10. Does not increase atmospheric CO_2	

Table 27.2 Studies of agricultural research productivity — direct cost-benefit type studies (World Bank 1981a).

Commodity	Country	Time period	Annual internal rate of return (%)
Hybrid corn	USA	1940–55	35–40
Hybrid sorghum	USA	1940–57	20
Poultry	USA	1915–60	21–25
Sugarcane	South Africa	1945–62	40
Wheat	Mexico	1943–63	90
Maize	Mexico	1943–63	35
Cotton	Brazil	1924–67	77+
Tomato harvester	USA	1958–69	37–46
Maize	Peru	1954–67	35–40
Rice	Japan	1915–50	25–27
Rice	Japan	1930–61	73–75
Rice	Colombia	1957–72	60–82
Soybeans	Colombia	1960–71	79–96
Wheat	Colombia	1953–73	11–12
Cotton	Colombia	1953–72	None
Aggregate	USA	1937–42	50
		1947–52	51
		1957–62	49

potential to increase biomass energy yields. Another advantage is the versatility of the biomass production and conversion technologies such that any one project can select the routes most suited to the prevailing conditions and requirements of that area. The most obvious problems that immediately come to mind are land use in competition with food production. Existing agricultural, forestry and social practices are also certainly a hindrance to promoting biomass as a source of energy, whether in a developing or a developed country.

In North America and Europe the problem in agriculture and nutrition is over-production, excessive consumption of animal products via feeding of grains, and surpluses affecting world trade, especially in relation to commodity prices for products from many developing countries. For example, the international traded price of sugar today (August 1984) is the lowest in real terms since 1945 and thus a tonne of sugar has very low purchasing power compared to buying a tonne of oil. Obviously commodity incomes and surpluses fluctuate but the medium-term trends are important as they are likely to be aggravated by increased productivity, the influence of new biotechnological processes and changes in diet.

Today two thirds of the world's people depend on plants for nearly all their food and for the majority of their energy. The question should be how can we increase the productivity of plant-based agriculture (and forestry) in order to provide the required levels of both food and fuel at the national or regional level. Since only less than 10 per cent of the world's food and hardly any of the world's biomass fuel enters transnational trade it is local production of plant products which is all important. Fortunately for the world as a whole per capita food production has been increasing at a compound rate of 0.5 per cent in developing countries and 0.8 per cent worldwide since 1950; even though the world's population has increased by two thirds in the last 30 years (increased from 2.5–4.5 billion) the production of food has doubled. Thus there is a precedence for antici-pating the selection and development of high-yielding hydrocarbon plants if the necessary incentives exist.

Major references which should be consulted are Pimentel and Pimentel (1979), Flaim and Hertzmark (1981), Foley *et al.* (1981), Holdgate *et al.* (1981), Hall *et al.* (1982), World Bank (1982), Smil (1983), Borwn (1984), Hall (1984), and Simon and Kahn (1984).

Surveys of Hydrocarbon Plants

Many species of plants produce hydrocarbons which can be used as fuels and chemicals but a well known quotation from Martin (1944) should be noted: 'Far more plants contain rubber than is generally realized, but few contain enough to make extraction worth while'. The best known is natural rubber from the *Hevea* rubber tree. Such hydrocarbons are chemically more reduced than carbohydrates, i.e. they contain less oxygen per carbon and thus can be used more directly, rather than in the case of carbohydrates where a microbiological or thermochemical conversion is required before use as a fuel. Currently *Hevea* natural rubber provides about one-third of the world's rubber requirement. With increasing costs of synthetic rubber (derived from petroleum) and increasing yields from the rubber trees the proportion is rising. Even 40 years ago the 'outstanding advantages of *Hevea* as a source of rubber . . . [were recognized as] . . . the large yield per acre, the low cost of collection, the low maintenance costs and the ease with which rubber of the best quality is separated from the latex' (Martin 1944).

Natural rubber can also be produced from the desert shrub *Parthenium argentatum* (guayule) and large quantities have been so produced in the past in Mexico — about 9 million kg in 1910 and again in 1944 in response to prevailing strategic and economic conditions (Solo 1980). Possibilities for indigenous natural rubber production is being seriously re-examined both in Mexico and the USA, and in other countries (see p. 375).

Concurrently, there are efforts to select and establish trial plantations of crops which produce hydrocarbons of lower molecular weight than rubber. The idea is to extract liquids from such plants which will have properties very close to those of petroleum. Various solvent extraction methods have been used for assaying and removing 'hydrocarbons', 'resins', 'polyphenols', 'biocrude' from plants and plant parts — these are well discussed by Stewart *et al.* (1982) and Williams and Home (1984). The best known work is that of Calvin in California using *Euphorbia* species with the aim to produce the equivalent of about 20 barrels of 'oil' per hectare per year in a semiarid (desert-type) environment. Additionally, trees have been identified in Brazil (*Cobafeira* sp. and *Croton* sp.) which produce 'oils' which can either be used directly or require some processing before being used in engines.

Systematic searches for species of plants with high hydrocarbon contents have been made sporadically in the past, mostly in the Americas. Such efforts have been for latex-producing plants since latex is a milky emulsion of about 30% hydrocarbons, and the rest water. Natural rubber is the best known latex product but many species of plants, e.g. Euphorbiaceae, produce hydrocarbons ('oils') in their latex with much lower molecular weights than rubber (10–20 000 rather than 1–2 million).

Little is totally new in this area (see Hall (1980) for a historical review): (a) In about 1874, workers at Kew Gardens screened many *Euphorbia* spp. for their oil content, and samples still exist 100 years later; (b) Thomas Edison in 1928–32 examined over 2000 plants for their rubber and resin contents in his efforts to find an American rubber plant. He found many hydrocarbon-containing plants, but only one or two in which the molecular weights of the hydrocarbon were large enough to be considered as a possible candidate for substitution for natural rubber; (c) Hall and Long (1921) published a monograph on the rubber content of North American plants; (d) Martin (1944) cites surveys in India, various parts of Africa, Australia, Sri Lanka and the West Indies.

In the 1970's with the advent of the oil crisis there was a resurgence of studies on the possibilities of deriving economic yields of oils and rubber from so-called hydrocarbon plants. The progenitor was Calvin who reasoned that if plants produce hydrocarbon-like materials and if the yields could be improved, such as with *Hevea,* other latex-producing plants might be found which could be grown under less humid and tropical conditions, perhaps on land which would actually be arid or semi-arid and otherwise unproductive. *Hevea,* which grows only in tropical climates, is a member of the family Euphorbiaceae and in the genus *Euphorbia* almost every species is a latex-producing plant. Latex from the rubber tree was used as a standard in analytical studies of extraction, analysis, chromatography, etc. About a dozen species were examined, most of which contain hydrocarbons of a much lower molecular weight than rubber.

Calvin later (1979) concluded that this type of exploration gives rise to two practical approaches to renewable resources:— firstly to use the hydrocarbon as it comes from the plant itself as a crude oil, refine it, remove the sterols which it contains, crack the rest of the compounds to ethylene, propylene, etc. and then reconstruct other chemicals from those products; this approach can be developed immediately; and secondly to learn how the molecular weight is controlled and to manipulate the plant to construct materials of the desired molecular weight, an approach which will be longer and more complex, using the plant as the collecting and constructing vehicle.

As the USDA research laboratories in Peoria, Illinois, Buchanan *et al.* (1983) evaluated some 300 plant species, and developed a scenario for the introduction of new oil and hydrocarbon-producing crops into the USA agricultural scene. They concluded that, for practical agricultural production of hydrocarbons and natural rubber in the USA, highly productive, adaptable plant species must be identified and selected for genetic improvement. The selection should be made from the viewpoint of entire-plant utilization for fibre, protein, and carbohydrate. Thus, procedures and criteria have been established for the preliminary evaluation of plant species as potential multi-use hydrocarbon-producing crops; and it was envisaged that, for oil- and hydrocarbon-producing crops, primary processing factories ('botanochemical factories') might be situated to handle crops grown within a 25 km radius. Each factory would process 280 000—550 000 tonnes/yr of whole-plant produce grown on about 24 500 ha. The product mix could be adapted to changing economic or social needs; possible primary products might include whole-plant oils, soluble polyphenols, rubber, gutta rubber, cattle feed, high-protein feeds and food supplements, paper-making fibre, glucose, xylose, fuel alcohols, methane and soil-amendment chemicals.

These authors anticipate that a 50 per cent improvement in dry-matter yield and a two- to three-fold increase in oil and hydrocarbon content should be possible during domestication. Such projected increases in yields seem justified in view of past accomplishments by plant breeders and agronomists working on conventional crops such as *Hevea,* but they must produce at least 2 tonnes/ha per year of oil plus hydrocarbon to be comparable with *Hevea*. A few of the large number of species examined were selected for more detailed studies as potential rubber crops (13 spp.), potential oil plus rubber crops (11 spp.), oil crops (9 spp.) and gutta rubber crops (3 spp.).

Two further extensive study surveys have been conducted by groups in Arizona (McLaughlin & Hoffmann 1982) and Australia (Stewart *et al.* 1982). In the Arizona study desert plants from south-west USA and north-west Mexico were studied to determine their possible economic potential as producers of 'biocrude' (the combined extracts from cyclohexane and ethanol treatment). This study came to somewhat different conclusions than that of Buchanan *et al.* (1983) especially with regard to individual species. The summary of the Arizona study very adequately draws the main conclusion from the paper. 'One hundred ninety-five species of plants native to the southwestern United States and northwestern Mexico were surveyed for potential feedstocks for biocrude production in arid lands. Biocrude is the hydrocarbon and hydrocarbon-like chemical fraction of plants which may be extracted by organic solvents and upgraded to liquid fuels and chemical feedstocks (Table 27.3). Plants were evaluated using a set of models which provide estimates of oil and energy production costs. Plants producing either latex or resinous exudates had the highest percentage of high energy extracts. Total extracts were highest in smaller, potentially less productive plants. The optimum combination of percentage biocrude and potential yield occurred in plants of intermediate size having higher than average extractables. High biomass yields do not appear necessary for the economic production of biocrude in irrigated, arid regions. Eleven desert plants might produce biocrude for between $10–15 per million BTU without by-product credits' (McLaughlin & Hoffmann 1982).

Table 27.3 Summary of percentage extractables in 195 species of southwestern plants (McLaughlin & Hoffmann 1982).

Plant group	No. spp.	Cyclohexane	Ethanol	Total
Latex-bearing plants	69	4.9a[a]	17.4a	22.3a
Euphorbia spp.	26	4.5	16.5	21.0
Asclepias spp.	16	5.4	15.0	20.4
Amsonia spp.	9	5.5	26.1	31.6
Resinous plants	23	8.1b	14.2b	22.3a
All other plants	103	2.5c	14.9b	17.4b
Legumes	10	1.8	15.5	17.3

[a]Means not followed by same letter are significantly different (P < .05).

In this study it is important to note that in no case do the projected yields exceed those commonly reported for irrigated crops in the south-west USA. Bioenergy projects dependent on extremely high yields with consequent high water use have little chance of succeeding in the arid southwest. Biocrude yields vary from 12–30 barrels/ha/yr among the plants listed. Imported crude oil at $42 per barrel costs approximately $7.80 per million BTU. An economical use of the bagasse after solvent extraction would be required before biocrude could compete as a substitute for imported crude oil. Several options exist, including direct combustion to produce steam and electricity, the manufacture of animal feeds or soil amendments, or further conversion to other energy products.

In the USA additional studies are those done by Sonalysts (1981) for the Department of Energy on economic critera, by Lemaine (1982) for the Bioenergy Council, by Foster and Karpiscak (1983) for the Electric Power Research Institute and by Adams *et al.* (1984) of the Plant Resources Institute where they have concentrated on *Asclepias* (milkweed).

The Australian study is an analysis of the potential for production of liquid fuels from four resin-containing crops. The data are summarized in Table 27.4. The immediate conclusion is that, except for guayule, the cost of extracted resins from the other three species would be three to ten times the cost of oil under local conditions. The byproduct lignocellulosic residues have a value dependent on

Table 27.4 Candidate crop species for various regions in Australia, and their estimated yields and costs (Stewart *et al.* 1982).

Candidate crop species	Region rainfall season	Estimated yield t(DW)/ha per year	Estimated foodstock cost $/t(DW)	Proportion recoverable resin ‡ in DW	Cost of resin $/barrel o.e. §
Euphorbia lathyris	southern, winter	3	57	.07	147
Euphorbia lathyris	southern, irrigated	10	51	.07	134
Asclepias arborea (= *A. rotundifolia*)	eastern year round	7.5	58	.09	116
Calotropis procera	northern, summer	10	46	.045	190
Calotropis procera	northern, irrigated	20	48	.045	196
Parthenium argentatum (guayule)	Tara Shire	5	60*	.10	-79¶
			184†	.10	104

* Cost estimates of Stewart *et al.* 1982
† Cost estimates for Texas, USA
‡ Recoverable resin includes resin I and resin II for *Euphorbia lathyris* and *Parthenium argentatum*, and resin I plus hydrocarbons for other species.
§ Barrel o.e. = barrel oil equivalent = 5.8GJ. Resins assumed 40GJ/t = 6.9 barrels oil equivalent per tonne.
¶ Revenue from other products exceed total costs by $79/barrel o.e.

conversion to methanol, but the scale of resin extraction would be too small to be economic for methanol production. If the recoverable resin content of the plants could be increased from a range of 0.045–0.10, to values of 0.15–0.20 of dry matter they could 'become competitive with seed oils'. The authors are careful to emphasize that the lack of commercial experience and field experimental data makes the estimation of crop yields, cost of crop production, and cost of processing more tentative than for previous assessments of alcohols (methanol and ethanol) and seed oils.

Numerous other studies on extractable hydrocarbons have been published over the last decade. Many are reported in articles in 'Biomass', 'Economic Botany', 'Biomass Abstracts' (IEA), 'International Bio-Energy Directory and Handbook' (Bio-Energy Council), 'Pudoc Press' (Seegeler 1983), etc. These studies are from many different countries such as Brazil, Ethiopia, South Africa, Greece, Australia, USA, Chile, etc.

For further information see Manassah and Briskey (1981) and Campos-Lopez and Anderson (1983).

Potential hydrocarbon crops

There are undoubtedly many plants which might be included where research and some commercial development had been carried out. The problems that guayule, euphorbias and algae have encountered over many years and the upheavals in trying to establish a solid economic base are a salutory experience for those trying to further develop this field. Unfortunately, only in the case of guayule has there been any progress in selecting and manipulating the yields of the crop. Until substantial development has been made on increasing the yields of hydrocarbon crop plants, they will always suffer by comparison with other species which have had concerted efforts made to increase their yeilds.

GUAYULE

Guayule (*Parthenium argentatum*) is native to the semi-arid regions of north-central Mexico and southern Texas where it is scattered over a 338 000 km² plateau. Like many arid region plants it needs a great deal of sunlight and low night temperatures; it is, however, one arid plant that has been cultivated as much in as out of its normal habitat.

A two-year-plant normally produces some 10% rubber by dry weight; some varieties yield as much as 25% and, with chemical stimulants, rubber production can be increased at early stages of growth to 30%. Guayule rubber is found not in a specialized lactifer system, but in the parenchyma of stems and roots as latex particles similar in size to those obtained from *Hevea*. For this reason it cannot be 'tapped' but must be extracted from the tissues, and because it contains no natural antioxidant it degrades rapidly in contact with air, so the plant must be processed within a few days of harvesting.

In the early part of the century many prominent Mexican and American industrialists were associated with the development of guayule, but the amount of rubber produced, mostly by Mexican mills, varied widely over the years. Starting with 385 000 kg in 1905, it reached a maximum of more than 9.5 million kg in 1910 from 14 factories, providing 10 per cent of the world's rubber and 50 per cent of USA consumption. It dropped to a low of 29 500 kg in 1921, but again increased to a peak of about 5.4 million kg in 1927, when the British government restricted the sale of rubber from Malayan plantations. After a drop in the early 1930s because of low rubber prices, production increased to a peak of nearly 9 million kg in 1944, practically all of which was shipped to the United States.

In 1942 following the loss of some 90 per cent of the rubber supply to the Japanese forces in SE Asia, the Emergency Rubber Project started in the USA with a reconnaissance survey of some 13 million ha of land and the classification of over 2 million ha for guayule culture. Land acquisition was intensified and some 24 000 ha were leased. Because of later curtailment of the project, the planting goal was never reached and only 13 000 ha were planted, but a great deal of expertise was acquired. Temperatures below $16°C$ were considered unsuitable (below $4°C$ the plant becomes dormant). A mean annual rainfall of 280–640 mm was considered suitable; below 280 mm supplementary water was necessary for a worthwhile yield, while above 640 mm excessive vegetative growth occurred.

But in 1946, on the return to peace-time conditions which included a by-now highly competitive synthetic elastomer industry, all guayule plantations were ordered to be destroyed and the land turned back to the owners, 85 per cent of the shrub that had been grown on the land was unharvested, resulting in the destruction of 9.5 million kg of rubber. Also during the liquidation period, two mills for the extraction of rubber, various laboratories and other buildings were disposed of.

Though many countries have experimented with guayule plantations, world rubber economics militated against success, and there are now no commercial plantations in existence; only in Mexico was there a continuum of development which has persisted to the present day. However, interest has recently been rekindled in the USA, Australia, South Africa, Argentina, Italy and probably other countries. Currently there is a 1 tonne/day pilot processing plant nearing completion in Texas and 600 ha of guayule are being planted in Arizona at Gila River (harvesting in 1987) both under Department of Defense sponsorship. The Australian study (Stewart *et al.* 1982) concluded that 'The production of resin from guayle appears very favourable, primarily because guayule produces twice as much rubber as resin and the rubber is five times as valuable as crude oil'. As yet, however, there are no extensive trials planned in Australia.

Guayule rubber produced before 1950 contained a high percentage of impurities (resin: 20–25%) and insoluble material (5–10%). Systems for handling elastomers in solution, and modern technology for the extraction of oils using solvents have only been developed since 1950. Such processes can produce guayule rubber with the necessary economic and technical specifications to compete with synthetic alternatives to natural rubber.

Whether or not any country can establish guayule as an economically viable crop, however, depends on many factors. Firstly, price increases for petroleum and therefore greater competitiveness with synthetic rubbers. Guayule can be a locally produced source of polyisoprene rubbers. It seems likely that in coming decades there will be markets for all the natural rubber than can be produced, whether *Hevea,* guayule, or other plants. There is a continuing rise in world consumption of rubber, and natural rubber is still preferred in many applications. Also, *Hevea* can be cultivated only in a limited tropical zone, which makes it vulnerable to political, economic, or biological problems. Secondly, the need to stabilize desert margins, to find crops adapted to desert environments, and to provide jobs and incomes for desert dwellers where farming conventional crops is risky or impossible, may be justifications outside the realm of conventional economics for the cultivation of guayule.

EUPHORBIA SPECIES

There are more than 2000 species of *Euphorbia* but only two have been 'developed' in the last decade; there are probably other species which should also be closely examined. Recent test plantings are known to have been made in California and Arizona of *Euphorbia lathyris,* a biennial about 1–2 m high and in Kenya and Japan of *E. tirucalli,* a perennial 3–10 m high taking years to come to harvest. *E. lathyris* is a more temperate climate plant and is not as well adapted to arid environments as is *E. tirucalli.*

E. lathyris has been proposed by Calvin (1979) as a hydrocarbon and sugar crop, that may yield 25 tonne/ha/yr dry matter, containing approximately 2 tonnes of oil and 5 tonnes of sugar with a total energy value of 4 million kcal. Other authors are more pessimistic concerning the potential of *E. lathyris* (see discussion by Kohan & Wilhelm 1980, Ayerbe *et al.* 1984); for example, Kingsolver (1982) has suggested that it is not suitable to be grown in arid zones, whereas Sachs *et al.* (1981) have estimated that the price would reach between $150–200 per barrel. In Ward's opinion economic resources could better be dedicated to looking at new species rather than to studying *E. lathyris.* Cost estimates by various researchers have varied from $18 to $90–117 per barrel of oil. Calvin's projected price of about $40 per barrel for the finished product, based on 25 barrels/ha was, in 1978, roughly twice the price of crude oil. However, the price of crude oil has escalated rapidly since then and the hydrocarbons from *Euphorbia* contain a high proportion of C_{15} compounds such as terpene trimers with a molecular weight of around 20 000, which can be made to yield products similar to those obtained from naptha, one of the principal raw materials that the chemical industry derives from petroleum. At the 1978 price of $50 a barrel for napththa, hydrocarbons from

Euphorbia were then considered to be competitive but more recent cost estimates seem lacking.

Besides differences of opinion on yields and cost estimates there have been varying claims about the irrigation and fertilizer requirements, the net energy benefits, the yields of valuable byproducts, susceptibilities to disease, etc. At present it appears that demonstration trials with *E. lathyris* have been discontinued, at least in the USA. However, an extensive and thorough trial is being conducted in Spain examining planting densities, harvest date, water and fertilizer requirements and yields of dry matter, sugars and oils. It has been concluded that low plant density, lack of irrigation and prolonged growth result in optimum hydrocarbon yields (Ayerbe *et al.* 1984).

In Kenya a 100 ha project with *E. tirucallii* started in 1980. Over 300 species from six genera of latex bearing plants were screened before *E. tirucalli* was selected. This species is widely grown in the arid and semi-arid regions of Africa for live fencing for stock and around houses. The trials are to obtain a techno-economic evaluation of the potential to produce liquid and solid fuels without irrigation and fertilizers. After 18–24 months growth yields of up to 20 tonne oven dry matter/ha/yr have been reported. Establishment of plantations appears relatively easy and regrowth after the first cutting seems possible. However, long-term trials are essential to establish fertilizer requirements especially with the removal of so much plant matter by harvesting. Intercropping with nitrogen-fixing legumes may be one answer. The yields of liquid fuels alone may be a problem and an alternative might be chemicals and solid fuel. Whatever the outcome of these trials they will be an important milestone in establishing the validity of hydrocarbon plants in such dry environments (J. Smets & J. Roman, pers. comm.)

ALGAE

The algae are often overlooked as a source of food, feed and chemicals in arid and semi-arid regions. For example, an alga, *Botryococcus braunii* has been shown to yield 70% of its extract as a hydrocarbon liquid closely resembling crude oil. This has led to the work on immobilizing these algae in solid matrices such as alginates and polyurethane and using a flow-through system to produce hydrocarbons. A species of the green alga *Dunaliella* discovered in the Dead Sea produces glycerol, β-carotene, and also protein. This alga does not have a cell wall and grows in these very high salt concentrations; thus to compensate for the high salt externally it produces glycerol internally. The alga *Phaeodactylum tricornutum* has been shown to have 25% lipid and 50% protein, while *Neochloris oleoabundans* contains 35–45% oily lipids. The blue-green alga *Spirulina* contains 75% protein, has good yields and grows well at high pH in hot climates such as Chad and Mexico where it has been eaten for centuries. There are economic problems with large scale culture of algae but costs are being lowered by the use of simply-constructed ponds and the use of brackish or sea water (Raymond 1982, Richmond 1984).

For further information see Hall (1980), Johnson and Hinman (1980), Sonalysts (1981), Stewart *et al.* (1982), Campos-Lopez and Anderson (1983), Davis *et al.* (1983) and Calvin (1983).

Implementation and conclusions

It must be realised that dependence on biomass as a source of energy has both advantages and problems as shown in Table 27.3. These advantages and disadvantages must be recognised in any regional development programme which endeavours to improve and implement renewable energy resources. The fact that most of the world's population relies on biomass does not mean that it is always a good thing!

One of the main problems with biomass energy is that it involves such a diverse 'technology', that actually implementing or improving biomass-for-energy schemes is often extremely difficult. In fact this might be one of the reasons why so many planners have shied away from actual recognition of the problem and stopped at half-hearted attempts to implement significant schemes.

In my estimation the main requirements for successful implementation are: (a) involvement of local people, local expertise and artisans, entrepreneurs, creditors and leaders; (b) political, public, law, bureaucratic and media involvement and co-operation; (c) a continuing commitment by leaders at national, international, regional and local levels; and (d) incentives like credits, guaranteed purchase, and salaries in the initial stages especially. I fully realise that it is not easy to have all these factors coming together at the same time, and the same can be said for developed as well as developing countries.

One of the first things that must be done is a local energy analysis. One needs to establish what the current (and possibly future) needs (energy and other) of the population are, what biomass resources are used now and what could be made available, what infrastructure is possible to implement new schemes, what are the land and tree tenure situations and their protection mechanisms — and so on. These local energy analyses must be done by multidisciplinary teams of people including engineers, scientists, economists and others. One needs to involve local leaders, social workers, women's groups, extension officers, etc., in order to have a diverse group of people trying to establish what is really needed by the community and what is available to implement any new schemes. The traditional role of women in 'energy' acquiring and use needs to be understood. Again, just formulating such ideas is very easy, but how one actually gets teams of people with some degree of expertise in local areas is not that easy. Such people need to have some degree of training and direction, and they need to know what is occurring in other similar localities of their country. The survey also needs to establish why people are making existing decisions. The manner in which the surveys (and interviews) are conducted must conform with local social habits and understanding and must be unbiased in the questioning and sampling; again this is not always easy to accomplish. The community surveyed should also be where the field tests are conducted and the follow-through is important.

How can regional development in semi-arid/arid zones be assisted by implementing renewable energy schemes? There must be adequate initial financing from the central or regional governments. There is also a prerequisite for financing on a continuing basis with the costs and beneficiaries clearly understood; this is an essential requirement before proceeding. There is no doubt that demonstration and pilot schemes in a country itself are important incentives in raising adequate financing. Successful examples of similar schemes in other regions and countries should also be carefully analysed.

In my opinion, the following six factors are very important in the successful implementation of renewable energy on a regional basis. A top-level, highly visible commitment must be obtained, and facilities and (a) prestige created to persuade people of adequate calibre to work towards biomass fuels implementation. Appropriate (b) incentives such as credits, guaranteed purchase and salaries should be made especially in the initial stages. Regional development staff should be given certain perquisites to prevent their work becoming 'hardship posts'. (c) Training is an absolute necessity to provide extension workers, engineers, agriculturists, foresters and other people who can implement pilot and developmental research programmes. (d) Infrastructure, in the form of an adequate number of full-time posts, with adequate auxillary fundings, is necessary for the maintenance, extension and administration of the programme. (e) Monitoring during the implementation phase to assess progress and the attitudes of local people and leaders. Finally, (f) demonstration schemes and realistic timing should be built into the scheme from the beginning. People should have employment and other forms of income during the initial stages and expectations should not be excessive, either by the local population or central planners, otherwise disillusionment can set in. Schemes which are well integrated into local work patterns and social habits will be most acceptable and successful.

In conclsuion, if hydrocarbon plants are to be successfully used in arid and semi-arid regions a host of implementation decisions have to be integrated into the overall problems of (a) desertification involving overcultivation, overgrazing, salinisation, deforestation, etc., (b) the low priority which agriculture and forestry so often have in a country's development plans, (c) the reluctance to take on long-term projects which require sustained implementation, (d) the lack of trained persons at various levels and expertise, (e) the reluctance to involve non-governmental organisations who often have such relevant experience, and (f) the difficulty of devolving centralized control to local authorities who have the appropriate incentives.

Thus hydrocarbon fuel and chemical projects are certainly not new and will have severe problems of implementation. However, if there is a genuine commitment to using arid and semi-arid regions for such purposes there is a large bank of data and commercial experience which can be drawn upon. But we must beware of extrapolating previous experiences to present day circumstances without understanding all the factors which influenced prior schemes and those which will contribute to the success of any new projects. Botanical, ecological, economic, social and political factors can wreak havoc on projects which take at least 5–10 years to come to fruition.

For further information see World Bank (1981b), Hall *et al.* (1982), Breman and de Wit (1983), Clayton (1983), Maunder and Ohkawa (1983), Baker (1984), Malhotra (1984), Neu and Bain (1984), Karrar (1984) and Walsh (1984).

References

Adams, R.P., M.F. Balandrin and J.R. Martineau 1984. The showy milkweed *Asclepias speciosa*; a potential new semi-arid land crop for energy and chemicals. *Biomass* 4: 81–104.

Ayerbe, L., J.L. Tenorio, P. Ventas, E. Fures and L. Mellado 1984. *Euphorbia lathyris* as an energy crop. I & II. *Biomass* 4: 283–293, 5: 37–42.

Baker, R. 1984. Protecting the environment against the poor. *Ecologist* 14: 53–60.

Bio-Energy Council 1981 and 1984. *International bio-energy directory and handbook.* Washington, DC.: Bio-Energy Council.

Breman, H. and C.T. de Wit (1983). Rangeland productivity and exploitation in the Sahel. *Science* 221: 1341–1347.

Brown, L.R. (ed.) 1984. *State of the World 1984.* Washington, DC.: Worldwatch Institute.

Buchanan, R.A., E.G. Know, A.A. Theisen and G.L. Laidig 1983. Botanochemicals from arid lands; towards development of production, marketing and consumption systems. In *Natural resources and development in arid regions,* E. Campos-López and R.J. Anderson (eds): 195–230. Boulder, Colorado: Westview Press.

Calvin, M. 1979. Petroleum plantations and synthetic chloroplasts. *Energy* 4: 851–870.

Calvin, M. 1983. New sources of fuels and materials. *Science* 219: 24–26.

Campos-López, E. and R.J. Anderson (eds) 1983. *Natural resources and development in arid regions.* Boulder, Colorado: Westview Press.

Clayton, E. 1983. *Agriculture, poverty and freedom in developing countries.* London: Macmillan.

Davis, J.B., D.E. Kay and V. Clark 1983. *Plants tolerant of arid, or semi-arid, conditions with non-food constituents of potential use.* London: Tropical Development and Research Institute.

Flaim, S. and D. Hertzmark 1981. Agricultural policies and biomass fuels. *An. Rev. Energy* 6: 89–121.

Foley, G., G. Barnard and E. Eckholm 1981. *Fuelwood: which way out?* London: Earthscan.

Foster, K.E. and M.M. Karpiscak 1983. Arid land plants for fuel. *Biomass* 3: 269–285.

Hall, D.O. 1980. Renewable resources (hydrocarbons). *Outlook Agric.* 10: 246–254.

Hall, D.O. 1984. Food versus fuel: a world problem? In *Economics of ecosystem management,* D.O. Hall, N. Myers and N. Margaris (eds). The Hague: Junk.

Hall, D.O., G.W. Barnard and P.A. Moss 1982. *Biomass for energy in the developing countries.* Oxford: Pergamon.

Hall, M.M. and F.L. Long 1921. *Rubber content of North American plants.* Washington, DC.: Carnegie Institution of Washington.

Holdgate, M.W., M. Kassas and G.F. White 1981. *The world environment 1972–1982. A report by the United Nations Environmental Programme.* Natural Resources and the Environment Series. Vol. 8. Dublin: Tycooly.

IEA 1979 *et seq. Biomass Abstracts.* IEA Biomass Conversion Technical Information Service. Dublin: Institute of Industrial Research and Standards.

Johnson, J. and H.E. Hinman 1980. Oil and rubber from arid lands. *Science* 208: 460–464.

Karrar, G. 1984. The UN plan of action to combat desertification and the concomitant UNEP campaign. *Environ. Conserv.* 11, 2: 99–102.

Kingsolver, B. 1981. *Euphorbia lathyris* reconsidered; its potential as an energy crop for arid lands. *Biomass* 2: 281–298.

Kohan, S.M. and D.J. Wilhelm 1980. *Recovery of hydrocarbon-like compounds and sugars from Euphorbia lathyris.* US Dept. Energy Contract E4-76-C-03-0115. Stanford, California: SRI Intel.

Lemaine, D. 1982. *Cultivation of hydrocarbon producing plants native to the western U.S. and the whole plant utilization of the oils and by-products.* Washington, DC.: Bio-Energy Council.

Lipinsky, E.S. 1981. Chemicals from biomass – petrochemical substitution options. *Science* 212: 1665–1676.

Malhotra, R.C. 1984. Rural development – national improvement. *Mazingira* 8, 2: 3–8.

Manassah, J.T. and E.J. Briskey (eds) 1981. *Advances in food-producing systems for arid and semi-arid lands.* New York: Academic Press.

Maunder, A. and K. Ohkawa 1983. *Growth and equity in agricultural development.* Aldershot: Gower.

McLaughlin, S.P. and J.J. Hoffmann 1982. Survey of biocrude-producing plants from the south west. *Econ. Bot.* 36: 323–339.

Martin, G. 1944. Competitive rubber plants. *Nature* 153: 212–215.

Neu, H. and D. Bain (eds) 1984. *National energy planning and management in developing countries.* Dordrecht: Reidel.

Odell, P.R. and K.E. Rosing 1983. *The future of oil: world oil resources and use,* 2nd edn. London: Kogan Page.

Pimentel, D. and M. Pimentel 1979. *Food, energy and society.* London: Edward Arnold.

Raisman, J.M. 1984. The future for fossil fuels: a global perspective. In *International energy options conference,* R.G. Taylor (ed.). London: Institute of Electrical Engineers.

Raymond, L.P. 1982. Aquatic biomass as a source of fuels and chemicals. In *Energy, resources and environment,* S.W. Yuan (ed.). Oxford: Pergamon.

Richmond, A. (ed.) 1984. *Handbook of algal mass culture.* Boca Raton, Florida: CRC Press.

Sachs, R.M., C.B. Low, J.D. MacDonald, A.R. Awad and M.J. Sully 1981. *Euphorbia lathyris*: a potential source of petroleum-like products. *California Agric.* July–Aug.: 29–32.

Seegeler, C.J.P. 1983. *Oil plants in Ethiopia: their taxonomy and agricultural significance.* Wageningen: PUDOC.

Simon, J.L. and H. Kahn 1984. *The resourceful earth.* Oxford: Basil Blackwell.

Smil, V. 1983. *Biomass energies.* New York: Plenum Press.

Solo, R.A. 1980. The saga of synthetic rubber. *Bull. At. Sci.* April: 31–36.

Sonalysts 1981. *Assessment of plant derived hydrocarbons.* US Dept. Energy Contract DE-AC01-80ER30006. Waterford, Connecticut: Sonalysts Inc.

Stewart, G.A., J.S. Hawker, H.A. Nix, W.H.M. Rawlins and L.R. Williams 1982. *The potential for production of hydrocarbon fuels from crops in Australia.* Melbourne: CSIRO.

Walsh, J. 1984. The Sahel will suffer even if rains come. *Science* 224: 467–471.

Williams, L.R. and V.N. Home 1984. Extraction of fresh latex producing plants. *Energy Agric.* (in press).

World Bank 1981a. *Agriculture research: sector policy paper.* Washington DC.: World Bank.

World Bank 1981b. *Mobilizing renewable energy technology in developing countries: strengthening local capabilities and research.* Washington, DC.: World Bank.

World Bank 1982. *International development association in retrospect: the first two decades of the IDA.* Oxford: Oxford University Press.

28 Unconventional arid land plants as biomass feedstocks for energy

R. J. Newton[1] and J. R. Goodin[2]

[1]*Department of Forest Science and the Texas Agricultural Experiment Station Texas A & M University, College Station, TX 77843, USA*

[2]*Department of Biological Sciences and the International Center for Arid and Semi-Arid Land Studies, Texas Tech University, Lubbock, TX 79409, USA*

Introduction

As supplies of non-renewable fossil fuels decrease and their costs increase, the need for identification and production of renewable energy resources becomes more urgent. Biomass production from terrestrial plants is renewable, and it could contribute significantly to the world's energy needs. The concept of biomass production involves 'energy plantations' which produce crops that are used: (a) directly in combustion to replace coal in industrial plants or provide fuel wood for heating and cooling, or (b) used indirectly as biochemically-transformed fuels such as alcohol or methane.

A logical question that many researchers have asked is whether native or naturalized plants growing on millions of hectares of rangelands would have any potential use as biomass energy sources. West Texas is of particular interest because it has great potential for biomass production on nearly 30 million hectares of rangeland. Sixteen species indigenous to or naturalized in West Texas have been identified as potential biomass candidates (Newton *et al.* 1980). Four of these were selected for further field and laboratory evaluations (Newton *et al.* 1982). The basis for selection of these four was: (a) reported or predicted yield, (b) energy content, (c) energy cost of production, (d) wide distribution in West Texas, and (e) forage value. *Atriplex canescens, Kochia scoparia, Sorghum halepense* and *Prosopis glandulosa* were selected from the final list of sixteen.

In this communication field data are reported on two of these species: *Sorghum halepense* (johnsongrass) and *Atriplex canescens* (fourwing saltbush). Because these two species showed some promise in the field (Newton *et al.* 1982) preliminary drought/salt tolerance studies were initiated on populations collected throughout the USA utilizing seedlings and tissue culture. Genetic selection and clonal propagation with tissue culture also have been initiated.

The research objectives were to: (a) evaluate the establishment and productivity of *S. halepense* and *A. canescens* in West Texas, and (b) initiate drought and salt tolerance evaluations of these two species.

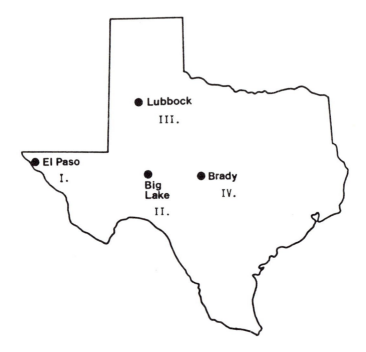

Figure 28.1   Field   research   sites   utilized   for   performance   trials   of   *Atriplex canescens* and *Sorghum halepense.*

Materials and methods

CHARACTERIZATION OF RESEARCH SITES

Four research sites were established in West Texas (Newton *et al.* 1982). These are: I (El Paso in the Trans Pecos), II (Big Lake in the Edwards Plateau), III (Lubbock in the High Plains), and IV (Brady in the Edwards Plateau) (Fig. 28.1). Characteristics of these sites have been described previously (Newton *et al.* 1982). In general, the soils from all four sites are alkaline (pH 〉 8), low in nitrogen and high in calcium, magnesium and potassium. The phosphorous levels are rather high at I and II (Newton *et al.* 1982). The incident solar radiation ranges from 4500 to 5300 kcal/ m²/day providing excellent potential for maximizing biomass production. The annual frost-free period, i.e. the growing season, ranges from 200 to 240 days with the probability of frost occurring first at I and III. Most of the precipitation occurs as rain with snowfall contributing significant amounts at all of the sites except IV. Precipitation in West Texas increases progressively from east to west with a mean of 190 mm at I and nearly 600 mm at IV. The above results are given in Table 28.1.

PROCEDURES

Seedling establishment, plot design, soil characterization and climatological data acquisition have been described previously (Newton *et al.* 1982).

CALLUS INDUCTION AND STRESS TOLERANCE

Atriplex explants were shoot tips and axillary buds approximately 4 mm in length. The tissue surface was sterilized, placed in culture tubes or in petri dishes, and incubated in an environmentally controlled chamber. The culture medium of Murashige and Skoog (MS) (1962) was supplemented with sugars, vitamins, and hormones (Goodin *et al.* 1984). Callus was placed on Heller supports and stress was imposed by adding polyethylene glycol (PEG) and salt (NaC1). There were ten replicates for each treatment.

Sorghum callus was induced from excised mesocotyl tissue derived from young seedlings and placed in a MS medium supplemented with hormones. The tissue was sterilized and cultures were incubated in an environmentally controlled chamber. Salt stress was imposed using procedures described above.

Table 28.1 Climatic description of four test sites in West Texas.

Site	Annual mean solar radiation[a] (kcal/m²/day)	Mean number frost-free days[b]	Mean precipitation[c] (mm)
I. El Paso (Trans-Pecos)	5300	248	190
II. Big Lake (Edwards Plateau)	4600	229	382
III. Lubbock (High Plains)	4750	208	468
IV. Brady (Edwards Plateau)	4500	226	591

[a]Texas Energy and Natural Resources Advisory Council, 1980, Texas Energy Outlook, 1980–2000, Austin, Texas.

[b]Soil Conservation Service, 1981, Temple, Texas, NPMH TX–8.

[c]National Climatic Data Center, Asheville, N. Carolina.

ATRIPLEX GENOTYPE SCREENING

Twenty-eight genotypes were examined for tolerance at the cellular level (Goodin *et al.* 1984). Axillary buds were cultured on Heller supports dipped into the same liquid media described above. Stress was imposed with 8000 MW PEG (-0.9 MPa) and 0.5 M NaCl, ten replicates per treatment. Each replicate was then evaluated based on visual ratings for growth, survival and quality of callus.

SORGHUM WHOLE PLANT SALT STRESS

Plants were initiated from seed in pots containing vermiculite and located in a greenhouse. The limit of salt concentration tolerance was determined by measuring the appearance of leaves, area of the second leaf from the top, and root dry weight from plants growing in an environmentally controlled chamber.

REGENERATION FROM CALLUS

Calli were placed in liquid MS media and rotated under continuous light (Goodin *et al.* 1984). After two weeks, 10 ml aliquots of the suspension were inoculated into 50 ml flasks with varying hormone levels. Cultures were routinely checked for embryonic potential and/or embryos. The embryos were placed on solid media for further development. After further development, embryos were then transferred to 1/2 MS solid media. Plantlets derived from shoot multiplication were placed in 1/2 MS for root formation prior to transfer into vermiculite.

After *Sorghum* callus had been induced on the standard MS medium, the callus was divided and placed on an agar-nutrient medium with reduced levels of hormones.

Results

SEEDLING SURVIVAL

Atriplex and *Sorghum* seedlings were established in a greenhouse in peat pots and transplanted in the field at four test sites in April 1981. All seedlings received one liter of water at time of planting and two additional liters within one month after planting. Three months after field transplanting, *Atriplex* survived only at I and II (Table 28.2).

Table 28.2 Percentage of *Atriplex canescens* and *Sorghum halepense* seedlings surviving after transplanting in spring, 1981[a].

Time	I. El Paso	II. Big Lake	III. Lubbock	IV. Brady
Atriplex canescens				
3 months	79	88	4	0
18 months	83[b]	76	84[b]	0
Sorghum halepense				
3 months	41	95	0	92
18 months	30	2	29[b]	0

Sorghum did not survive at III and survival was only 41 per cent at I. Hot dry winds dehydrated the transplants at III (Newton *et al.* 1982). Nonsurviving transplants were replaced with new transplants in November 1981. Eighteen months after the original transplanting, about 80 per cent of the *Atriplex* survived on I, II, and III, with no survivors at IV (Table 28.2). The complete mortality of *Atriplex* at IV was attributed to flooding and competition from other vegetation (Goodin & Newton 1983).

Similar to *Atriplex*, *Sorghum* did not survive severe dehydration at III, and even after replanting, there were few survivors 18 months after the initial transplanting at II and IV and about 70 per cent mortality at I and III (Table 28.2. Rhizomes allowed these plants to regenerate 24 months after establishment, but their growth was not significant. Further attempts at establishment and monitoring of *Sorghum* growth was abandoned at I, II, and III. However, there was a natural stand of *Sorghum* at IV which provided an excellent study of its biomass potential (See under Biomass yield below).

PRECIPITATION

Precipitation during 1981 (Table 28.3, p. 390), when seedlings were established, was higher than the annual mean (Table 28.1) at all the sites and undoubtedly contributed to the successful establishment of *Atriplex* at three of the four sites (Table 28.2). In 1982, precipitation was similar to mean annual measurement at II and IV, with slightly higher values compared to the mean at I and III (Tables 28.1 & 3). It is speculated that midwinter snows at I and III provided moisture which

[a]192 transplants (48 seedlings/plot X 4 plots) at each site.
[b]Includes replacements during the fall of 1981; percentages based on 192 transplants.

contributed to early spring growth of *Atriplex* (Table 28.4). In August, 1983, summer droughts at I, II, and III presumably inhibited *Atriplex* growth (Table 28.4). Total precipitation at IV was over 500 mm during the first 10 months of 1982 and this was very conducive to biomass production of *Sorghum* (See under, Biomass yield).

Table 28.3 Annual cumulative precipitation (mm), 1981–1983.

Site	1981	1982	1983
I. El Paso	321	270	203
II. Big Lake	515	380	232
III. Lubbock	584	533	514
IV. Brady	670	608	630

Table 28.4 *Atriplex* height (Ht) and basal diameter (BD) at various monthly intervals after transplanting.

Months after transplanting	I. El Paso Ht (cm)	BD (cm)	II. Big Lake Ht (cm)	BD (cm)	III. Lubbock Ht (cm)	BD (cm)
0	8–10		8–10		8–10	
4	17.3±8.6	22.1±10.6	13.8±7.7	12.2±5.4	13.4±3.3	16.4±14.2
18	49.3±19.8	70.4±28.9	25.6±20.1	32.0±26.3	43.7±18.4	36.6±21.5
19	harvest		harvest		harvest	
29	54.4±14.3	70.7±38.5	33.1±21.5	34.7±24.7	63.2±44.8	78.1±41.7

GROWTH

Changes in *Atriplex* height and basal diameter were greater for plants established at I and III compared to II (Table 28.4). The increased precipitation at I and II (Table 28.1 & 3) appeared to have contributed to the increased growth observed. Plants at I and III generally had a higher water content than those at II (Table 28.5), suggesting more turgor maintenance resulting in increased growth. Growth data shown in Table 28.4 reflect the wide variability among the established plants which is also evident from biomass yields (Table 28.5).

BIOMASS YIELD

The naturalized stand of *S. halepense* was utilized to ascertain its biomass production at IV. A one square meter sample of *Sorghum* above-ground biomass shows a maximum dry matter production of 0.64 kg/m² at the end of August (Fig. 28.2, p. 392) which extrapolates to about 6400 kg/ha with precipitation of

Table 28.5 Mean and maximum *Atriplex* biomass yield (individual plant) 19 months after transplanting[a].

Site			No. harvested	Fresh wt (g)	Dry wt (g)	% H_2O
I.	El	Paso				
		Mean	116	342	126	63
		Maximum	1	1782	697	61
II.	Big	Lake				
		Mean	23	353	182	48
		Maximum	1	1700	874	49
III.	Lubbock					
		Mean	62	328	156	52
		Maximum	1	3360	1459	34

600 mm. Although not shown, a maximum fresh weight of 1.4 kg/m² (54% moisture) was obtained in June with a 12 per cent moisture content obtained in September.

To insure survival and subsequent growth of *Atriplex*, all harvested biomass was coppice harvested at 30 cm above the ground (Tables 28.5 & 6). The mean dry matter production for I, II and III for harvests during the autumn of the second growing season was 126, 182, and 156 g/plant, respectively, but the maximum dry wt/plant harvested for each site was 697, 874 and 1459 g/plant (Table 28.5). These data from the largest plants suggest that the potential for biomass production is much greater than the mean values indicate.

Table 28.6 Mean and maximum *Atriplex* biomass yield (individual plant) 31 months after transplanting[a].

Site			No. harvested	Fresh wt (g)	Dry wt (g)	% H_2O
I.	El	Paso				
		Mean	141	319	231	28
		Maximum	1	1400	1008	
II.	Big	Lake				
		Mean	47	327	241	27
		Maximum	1	2450	1778	
III.	Lubbock					
		Mean	141	838	628	25
		Maximum	1	4300	3225	

[a]Yield from plants having a height of 30 cm or more. Some had been harvested previously at 19 months.

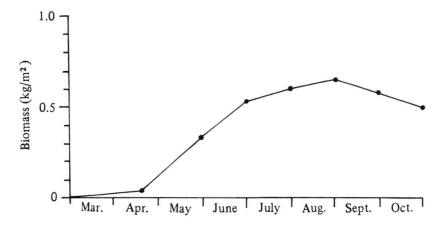

Figure 28.2 Biomass yield of *Sorghum halepense* at Site I (Edwards Plateau, Brady, Texas), 1982.

Similar conclusions can be made from harvests made during the third growing season (31 months after planting) as shown in Table 28.6. The mean dry matter production for I, II and III were 231, 241 and 628 g/plant. The largest plants harvested at each site during the fall of the third growing season were 1008, 1778 and 3225 g/plant, respectively. Again these maximum values reflect the growth potential that could be obtained through careful selection and breeding. Plant water content varied from year to year with 50 to 60 per cent at 19 months (Table 5) and only 30 per cent at 31 months. This is apparently associated with precipitation patterns.

ATRIPLEX GENOTYPE SCREENING

Axillary bud explants from 28 genotypes were observed for explant survival, and growth was observed and ratings were established (Table 28.7). One genotype (#8) was rated as outstanding for drought tolerance and another (#14) was outstanding for salt tolerance (Table 28.7). Three other genotypes (#5, 9, 15) had moderate drought tolerance and two genotypes (#7, 21) were moderately salt-tolerant.

CALLUS STRESS TOLERANCE

Atriplex callus growth is severely inhibited ($\rangle$ 90%) when stressed at -1.2 MPa and its growth is completely inhibited at 0.5 M NaCl (Table 28.8, p. 394). This is in contrast to *Sorghum* callus where mortality is nearly 90 per cent with only 0.2 M

Table 28.7 Drought and NaCl tolerance: *Atriplex canescens* genotype screening using axillary bud explants.

Rating[a]	Genotype	
	Drought (-0.9 MPa)	Salt (0.5 M NaCl)
0 to 0.5	2, 10, 11, 12, 13, 14, 19, 20, 21, 22, 24, 28	3, 5, 6, 9, 11, 12, 15, 16, 17, 20, 22, 23, 24, 25, 26, 27, 28
0.6 to 2.5	1, 3, 4, 6, 7, 16, 17, 18, 23, 25, 26, 27	1, 2, 4, 8, 10, 13, 18, 19
2.0 to 2.5	5, 9, 15	7, 21
4.0	8	14

NaCl in the nutrient medium (Table 28.9, p. 394). It is interesting also to note for *Atriplex* that the water potential of the growth limiting 0.5 molar NaCl medium is -2.2 MPa which is greater than the PEG medium (-1.2 MPa) which limits growth. Thus, it would appear that a matric stress is more detrimental to the tissue than is a solute stress.

SALT TOLERANCE OF SORGHUM SEEDLINGS

Leaf production, leaf growth and root growth of *Sorghum* seedlings as influenced by NaCl were measured (Table 28.10, p. 395). Leaf number/plant was reduced by two with 0.2 M NaCl applied for 47 days and leaf area was reduced by 65 per cent with the same concentration over the same time period. Application of 0.2 M NaCl for six weeks reduced root dry weight over 90 per cent (Table 28.10). These data coupled with callus survival data indicate that *Sorghum halepense* does not tolerate NaCl concentration greater than 0.2 M for prolonged periods of time.

[a] $(1 \rangle 0$ = survives and green, 0 = no growth, 1 = possible growth, 2 = positive growth, $\rangle 2$ = n X original mass.

Table 28.8 *Atriplex canescens* callus growth as influenced by salt (NaCl) and water stress by polyethylene glycol (PEG).

PEG (MPa)	Fresh Weight (g)
0	0.68
-0.3	0.25
-0.6	0.14
-0.9	0.10
-1.2	0.04
NaCl (M)	
0	1.13
0.1	0.42
0.3	0.11
0.5	0.01

Table 28.9 *Sorghum halepense* callus percent[a] survival as influenced by salt stress (NaCl) and time.

NaCl (M)[b]	5 days[c]	10 days[c]	15 days[c]	20 days[c]
0	100	100	100	90
0.1	100	100	78	67
0.15	100	67	56	44
0.20	78	56	49	11

REGENERATION FROM CALLUS

Atriplex cell suspension culture provided many embryos, although many of the embryos had developed within the callus mass as well as on the periphery and were released when shaken in culture. After two weeks cultures were transferred to 1/2

[a]Percentage based on 10 calli/treatment

[b]MS media with 1mg/liter of 2–4–D and 0.5 mg/liter of kinetin

[c]Survival based on visual observation: browning or blackening of tissue, secretion onto filter paper, and growth inhibition

Table 28.10 Influence of salt (NaC1) on growth of callus, leaves and roots of *Sorghum halepense.*

NaCl (M)	Leaf no.[a]	Leaf area[b] (cm^2)	Root dry wt.[c] (g)	Callus survival[d] (%)
0	3.0	56.1	1.2	90
0.05	2.4	39.8		
0.10	1.4	29.0	0.3	67
0.15				44
0.20	0.8	19.4	0.1	11

MS media with various additives. Only one embryo produced a plantlet; this plantlet recallused and produced numerous embryos which subsequently developed into plantlets. The resulting plants were transferred to vermiculite for hardening prior to moving to the greenhouse. Shoots and roots have been formed from *Sorghum* callus two to four weeks after transfer to new media. However, no embryos have been observed.

Discussion

Sites I through IV were selected for these studies because they represented a range of precipitation of 200 to 600 mm annually. The year in which seedlings were established, 1981, precipitation was higher than the mean and this contributed to the successful establishment of *Atriplex* at three of the four sites. The application of small volumes of water (Goodin and Newton 1983) also insured seedling survival. The critical factor in establishment is the soil moisture at the time of transplanting and the first few weeks following. An established transplant has a far greater chance of survival in a relatively dry soil than does a germinating seed (Springfield 1970, Nord *et al.* 1971). *Sorghum* seedlings survived into the second growing season at the two driest sites, but they produced very little biomass. Rhizomes of *Sorghum* are a successful adaptive mechanism for survival in drought-prone environments even though the productivity is low.

[a]Plants were 4 weeks old when treatment began; treated for 47 days

[b]Plants were 4 weeks old when treatment began; area of second leaf from the top after 47 days of treatment

[c]Plants were 4 weeks old when treatment began; treated for 6 weeks

[d]Based on 10 calli/treatment after 20-day treatment period

A significant finding of this study is the demonstration of wide genetic variability exhibited not only by geographically-separated *Atriplex* populations, but even with a collection from a single plant. Van Epps *et al.* (1982) observed a tremendous genetic potential among and within populations of this perennial shrub. Based on maximum plant yield values (g DW/plant) and 5165 plants/ha, biomass yields during the second year after establishment are estimated to be 3500, 4500 and 7500 kg/ha/yr for regions with mean precipitation of 190, 380 and 470 mm, respectively. Estimates of *Sorghum* biomass yields with precipitation of nearly 600 mm were over 6000 kg/ha/yr. It is believed that 6000 kg of dry matter accumulation per year would be required in order to make energy plantations economically feasible (Goodin & Newton 1984). These data suggest that careful selection from individuals within a population could perhaps produce this biomass yield. This could be accomplished by selective inbreeding and cloning by asexual propagation and/or tissue culture.

Our preliminary data suggest that *in vitro* single cell and callus cultures can be used as a selection tool for drought and salt tolerance and maximum productivity on arid lands. Screening of 28 *Atriplex* populations via tissue culture collected from various parts of the western USA have identified several drought and salt tolerant populations. The limitation of rapid clonal propagation of these screened *Atriplex* accessions has been addressed by the demonstration of embryogenesis from callus and single cell suspensions and their development into plantlets. Root and shoot morphogenesis from *Sorghum* callus has also been demonstrated. Once these regeneration procedures have been definitively established, the propagation of salt and drought tolerant plants through tissue culture selection will be realized. Clonal variation of single cells represents another source of variation that can be subjected to salinity and drought screening. The utilization of tissue culture technology combined with conventional plant breeding could result in the production of 6000 kg/ha/yr of biomass in regions with precipitation of 350 to 500 mm. The energy production from this biomass could be very significant.

The energy content of *Atriplex* and *Sorghum* is 4400 kcal/g and 3900 kcal/g, respectively. Using yield values of 6000 kg/ha/yr, the biomass energy is 26 million kcal/ha/yr for *Atriplex* and 23 million kcal/ha/yr for *Sorghum*. This indicates that arid and semi-arid lands producing 25 million kcal/ha would produce 1 Quad of energy on 10 million ha (25.2×10^{10} kcal/Quad $\div 25 \times 10^6$ kcal/ha = 1.0×10^4 ha/Quad). One quad is about one per cent of the total energy requirement of the United States.

The vast land area and intensive solar energy available throughout the arid regions of the world may be attractive for producing fuel and chemical products. It appears that the genus *Atriplex*. with a number of perennial woody shrubs indigenous to many saline regions of the world, provides the best potential for sustainable productivity at a rate of solar conversion sufficient to justify the cost (Branson *et al.* 1976, Goodin & Newton 1983). Careful selection from the perennial species could lead to sustained biomass production for sufficiently harvestable yield to justify costs of land development. As energy costs become greater, renewable resources such as biomass production in arid lands may become more important.

Acknowledgements

The authors are indebted to the Texas Energy and Natural Resources Advisory Council, Exxon Enterprises, Texas Tech University Center for Energy Research, Texas Tech University Water Resources Center, Texas A & M University Center for Energy and Mineral Resources, the University of Texas Lands System, and the U.S. Department of Agriculture for financial support of this work.

References

Branson, F.A., R.R. Miller and I.S. McQueen 1976. Moisture relationships in twelve northern desert shrub communities near Grand Junction, Colorado. *Ecol.* 57(6): 1104–1124.

Goodin, J.R. and R.J. Newton 1983. *Production potential of biomass feedstocks.* Texas Energy Development Fund Tech. Rept. TENRAC/EDF–112, Austin: Texas Energy and Natural Resources Advisory Council.

Goodin, J.R. and R.J. Newton 1984. Establishing halophyte shrubs for fuel production on saline sites. *Reclam. Reveg. Res.* (In press).

Goodin, J.R., N.L. Trolinder and M.H. Kamp 1984. *In vitro techniques of genetic screening for drought and salt tolerance.* Lubbock: Water Resources Center, Texas Tech University.

Murashige, T. and F. Skoog 1962. A revised medium for rapid growth and bioassays with tobacco tissue cultures. *Physiol. Plant.* 15: 473–497.

Newton, R.J., D.R. Shelton and D.J. Reid 1980. *Biomass production in semi-arid regions of West Texas.* TAMURF–3949, Texas A & M Univ., College Station.

Newton, R.J., J.R. Goodin, D.L. Magar and J.D. Puryear 1982. Biomass from unconventional sources in semi-arid West Texas. In *Proceedings of energy from biomass and wastes* VI, D.L. Klass (ed.): 168–219. Buena Vista, Florida: Inst. Gas Tech.

Nord, E.C., P.F. Hartless and W.D. Nettleton 1971. Effects of several factors on saltbush establishment in California. *J. Range Managem.* 24: 216–223.

Springfield, H.W. 1970. *Germination and establishment of fourwing saltbush in the Southwest.* US Dept. Agr. For. Serv. Res. Paper Rm–55.

Van Epps, G.A., J.R. Banker and C.M. McKell 1982. Energy biomass from high producing rangeland shrubs of the intermountain United States. *J. Range Managem.* 15: 22–25.

29 Rubber and phytochemical specialities from desert plants of North America

Eloy Rodriguez

Phytochemical Laboratory, School of Biological Sciences, University of California, Irvine, CA 92717, USA

Introduction

The investigation of under-utilized desert plants, native to the southwestern United States, Baja California and Chihuahua, Mexico, as alternative sources of rubber and phytochemicals, is gaining importance as arid zones increase in area. A significant number of flowering plants, native and common to arid environments, synthesize a variety of organic substances which are suitable substitutes for petroleum-based chemicals (Rodriguez 1980). These natural resources range from rubber polymers and oligomers to biologically active benzofurans, sesquiterpene lactones and phenolics. In North America, about 25 per cent of the United States is semi-arid to arid, while in Mexico, approximately 46 per cent is dryland (Becker *et al.* 1984). Two desert regions which are of great importance to the United States and Mexico, are the Chihuahuan and Sonoran Deserts. Both regions have many unique and endemic plant species that produce significant quantities of organic substances that are economically useful to human communities living in marginal arid lands.

As part of our long-term study to discover and develop new arid land plant species that can be used as sources of agriphytochemicals, we present some of the latest findings on selected desert dominants from North America that are potential crops for organic biochemicals. These species include *Parthenium argentatum* (guayule) and *Flourensia cernua* (tarweed) from the Chihuahuan Desert and *Pedilanthus macrocarpus* (candelilla) and *Larrea tridentata* (creosote bush) from Baja California, Mexico (Sonoran Desert).

Chihuahuan Desert

As previously noted by Campos-Lopez and Roman-Alemany (1980), the Chihuahuan Desert is predominantly in Mexico and the states of Texas and New Mexico (US). The Chihuahuan flora contains over 2000 species included in about 591 genera and approximately 100 families. The most important families are those that produce economically important products such as waxes, steroids, fibers, rubber, gums, resins and medicinal products (Campos-Lopez & Roman-Alemany (1980). Two families that we have studied in detail are the Asteraceae and Zygophyllaceae. In particular, we have investigated *Parthenium argentatum* and hybrids for rubber; *Flourensia cernua* for resin and insecticides and *Larrea tridentata* for antibiotics and photoactive insecticides. Other families from the Chihuahuan Desert that have received considerable attention, but are not discussed in this chapter, include Agavaceae (*Agave* – fiber and food), Cactaceae (*Opuntia* – food, sugars and alkaloids), Euphorbiaceae (*Euphorbia* spp. – wax and rubber), Fabaceae (*Acacia* – tannins, *Prosopis* – proteins), Gramineae (cattle fodder), Liliaceae (*Yucca* – fiber and steroids) and the Solanaceae (*Solanum* – food, medicines and alkaloids).

PARTHENIUM ARGENTATUM

Guayule, *Parthenium argentatum* (Asteraceae), is a shrub endemic to the Chihuahuan Desert (Fig. 29.1). Natural populations of guayule were first exploited for rubber nearly a century ago in Mexico, and more recently, the United States and other countries have initiated research projects to develop guayule as a commercial source of natural rubber. Presently, the natural rubber from *Hevea* costs approximately \$1.30 per kilo, whereas guayule agricultural costs (excluding processing) are estimated at \$2.00 per kilo, assuming a harvest of the entire plant and roots after 36 months and a yield of 2270 kilo/ha. Two ways to make guayule a cost-effective source of rubber are to increase yield (breeding or genetic engineering) and to establish direct seeding techniques. Major objectives of our collaborative research efforts are to increase rubber content, disease resistance and cold tolerance by crossing guayule with other species of *Parthenium*. Presently, the guayule breeding program is being carried out at the University of California at Riverside, with numerous guayule varieties and F_1 hybrids available for detailed chemical analysis. Rubber analysis of the guayule hybrids indicate that all hybrids are capable of producing high molecular weight rubber, with the guayule X *P. rollinsianum* and *P. schotti* hybrids producing up to 2–3 per cent in quality rubber (Fig. 29.2, p. 402). Although guayule produces significant amounts of rubber, two other species, *P. rollinsianum* and *P. alpinum* contain small quantities of high molecular weight rubber (West *et al.* 1984).

Other important products of guayule that we are investigating are the terpenoid constituents present in the stem and leaf resins. We have identified numerous novel sesquiterpene esters and sesquiterpene lactones from guayule, other species of *Parthenium* and F_1 hybrids (Rodriguez 1977). Recently, we established that oxygenated pseudoguaianolides are very effective in deterring feeding by phytophagous insects (Isman & Rodriguez 1983). Also, derivatives of the guayulins (sesquiterpene esters) are effective fungicides and in some cases inhibit larval feeding on guayule treated with the guayulin derivatives (Rodriguez in preparation). Since at the present time the only source of natural rubber is the tropical tree

Figure 29.1 Distribution and potential cultivation sites for Guayule.

Hevea (Euphorbiaceae), the development of guayule and hybrids as desert hydrocarbon crops for the arid zones of the world is necessary and timely.

FLOURENSIA

An important dominant of the Chihuahuan Desert is the genus *Flourensia* (Asteraceae), which is a potential source of resins and insecticides. Thirteen species are distributed throughout the arid zones of Chihuahua, with *F. cernua* (tarweed), a co-dominant with *Larrea* (Zygophyllaceae) and guayule (Dillon 1976). The success

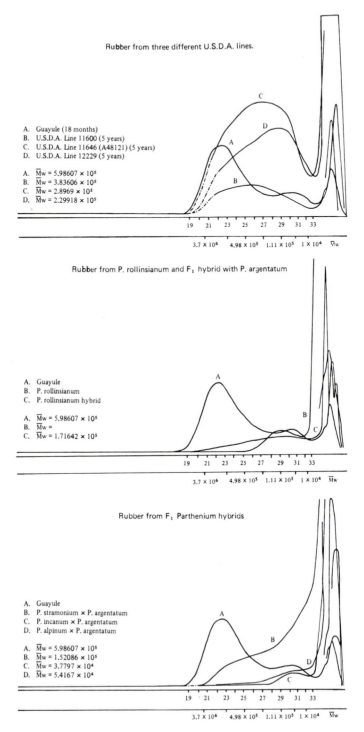

Figure 29.2 Rubber from *Parthenium* hybrids.

of *F. cernua* is in part due to the large quantities of resin present on the leaves. The resin consists of a mixture of secondary metabolites, with the major active constituents identified as benzofurans and benzopyrans (Aregullin & Rodriguez 1984).

An interesting property of the benzofurans is their insecticidal activity. Of these 7-methoxy-2-isopropenyl-5-acetyl-2,3-dihydrobenzofuran-3-ol cinnamate was found to possess antijuvenile hormone activity. Insects treated with the benzofuran underwent precocious moulting, anatomical malformations, retention of juvenile characters and sterility in their second, third and fourth instars (Aregullin & Rodriguez 1984). These observations are analogous to those noted by Bowers (1976) when he treated milkweed bugs with precocene I and II isolated from *Ageratum houstonianum* (Asteraceae).

Further studies of resin producing plants suggest that the benzofurans and benzopyrans are more widely distributed among arid land plants than previously thought. Also, they may be a potential source of insecticides and repellants to be used against desert crop pests.

Sonoran Desert

Baja California is a narrow peninsula stretching southeasterly about 1300 km from the southern boundary of California to its tip at Cabo San Lucas (Wiggins 1980). Baja California is part of the Sonoran Desert, a desert that includes Arizona, New Mexico, California and numerous Western States of Mexico. All in all, it includes a total area of almost 120 000 square miles. The physiography of Baja is characterized by an irregular terrain made up of low desert mountain ranges with extensive bajadas that feed the runoff into washes. The Sonoran Desert has low precipitation and high summer temperatures. In Baja, summer temperatures are highest in the northeastern part of the peninsula where the breezes from the Pacific are cut off by the Sierras Juárez and San Pedro Mártir. Annual rainfall is highest in the mountains, lowest along the desert slopes in their rain shadow (Wiggins 1980). Two different precipitation regimes goven the rainfall of Baja California. The northern half to two-thirds of the peninsula receives almost all of its annual rainfall during the winter months, November to March, with most of it falling between mid-December and the end of February. The major arid subdivisions and dominant plants on Baja California according to Wiggins (1980) are:

(a) Microphyllous desert — consisting of species of *Larrea, Fouquieria, Parkinsonia (= Cercidium), Bursera, Ambrosia* and *Opuntia.*

(b) Sarcocaulescent desert — *Bursera, Jatropha, Ferocactus, Pachycereus* and *Pedilanthus.*

(c) Sarcophyllous desert — *Agave, Ambrosia, Yucca, Fouquieria columnaris (= Idria columnaris)* and *Pachycormus discolor.*

(d) Magdelan region — *Pachycereus pringlei, Lysiloma* and *Machaerocereus gummosus.*

There are approximately 900 genera and 3000 species, subspecies and varieties found in the Sonoran Desert (Wiggins 1980). The most heavily represented family is the Asteraceae which includes 130 genera and 440 species. An important genus of the Sonoran Desert and all low deserts of the Americas is *Larrea* (Zygophyllaceae). In Baja, *Larrea tridentata* is a dominant and probably the most important plant in the arid zones of the United States and Mexico. Other important families include the Euphorbiaceae (*Pedilanthus* and *Euphorbia*), Fabaceae, Anacardiaceae, Fouquieriaceae, Frankeniaceae, Cactaceae and Agavaceae. Many of these families produce organic chemicals ranging from rubber to volatile monoterpenes. The capacity for some plants to produce up to 20 per cent of their biomass in chemical specialities is evident among some of the dominants. Jojoba (*Simmondsia chinensis*) is another important species, native to Baja California, Mexico, that produces an unsaturated oil present in the seeds.

LARREA TRIDENTATA

A common feature of some desert plants of Baja California is the secretion of resinous constituents on the leaf surfaces through specialized glands (Rodriguez *et al.* 1983). *Larrea tridentata* (creosote bush), in the family Zygophyllaceae, represents a dominant component of plant communities in arid regions throughout the southwestern United States and contiguous desert regions of Baja California and Mexico (Fig. 29.3). Copious amounts of a resin coat the leaves and stems of this desert perennial. The chemical composition of this resin has been studied extensively and has been found to contain a complex mixture of phenolics (flavonoids and lignans), saponins and wax esters. Protection of the shrub from pathogens and insect herbivory has been proposed as a function of the resin due to its demonstrated toxicity towards bacteria, fungi, and various insect herbivores (Mabry *et al.* 1977b).

Phenolics
Phenolic-type compounds account for approximately 80 per cent of the resin from the creosote bush, lignans and flavonoids are quantitatively the most abundant phenolics present. Nordihydroguaiaretic acid (NDGA), the major lignan in the resin, can account for up to 5–10 per cent of the dry leaf weight. At least four lignans structurally related to NDGA have also been identified (Fig. 29.4, p. 406). The flavonoid constituents of the resin are characterized by remarkable structural diversity. To date, nineteen flavone, flavonol and dihydroflavonol methyl ethers have been isolated and characterized (Mabry *et al.* 1977b).

Saponins
In terms of quantitative importance, saponins are the second major group of chemicals represented in the resin of the creosote bush. Habermehl and Moller (1974) found these triterpenoid constituents to account for 10–15 per cent of the dry leaf weight.

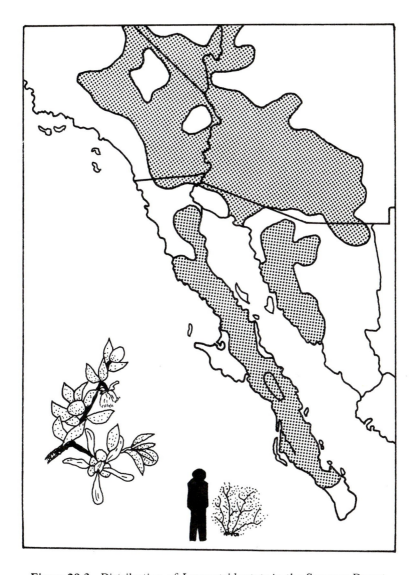

Figure 29.3 Distribution of *Larrea tridentata* in the Sonoran Desert.

Wax Esters
The third and quantitatively least significant (0.1% of dry leaf weight) natural products in *Larrea* resin are the wax esters. These compounds are composed of C-24 to C-30 aliphatic alcohols linked with C-22 to C-32 aliphatic acids in an ester linkage (Mabry *et al.* 1977a and references therein).

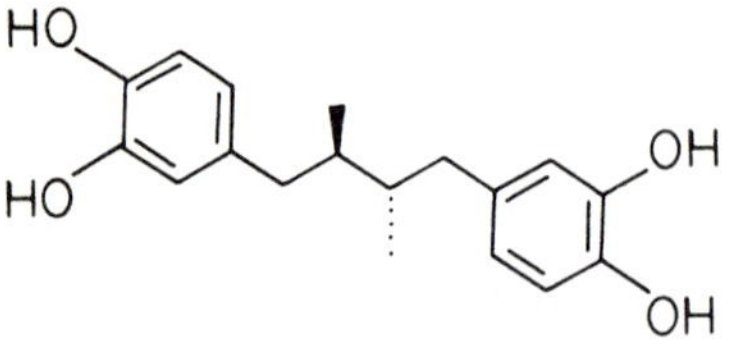

NDGA (Nordihydroguaiaretic Acid) 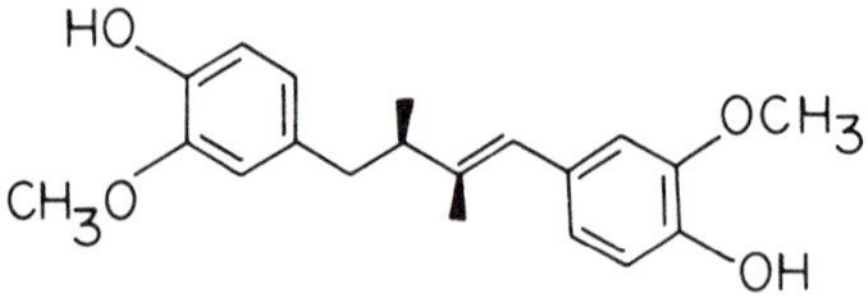   Guaiaretic Acid

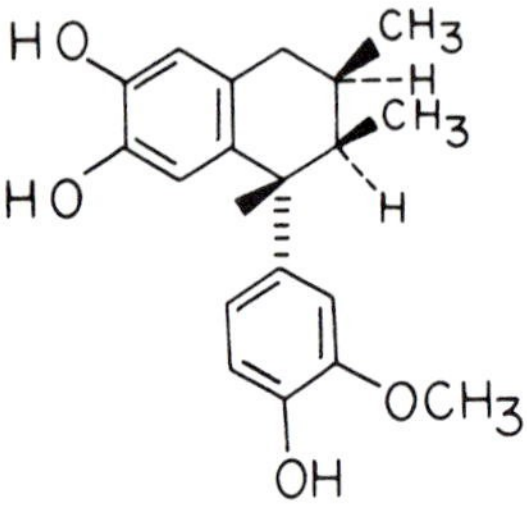

Norisoguaiacin 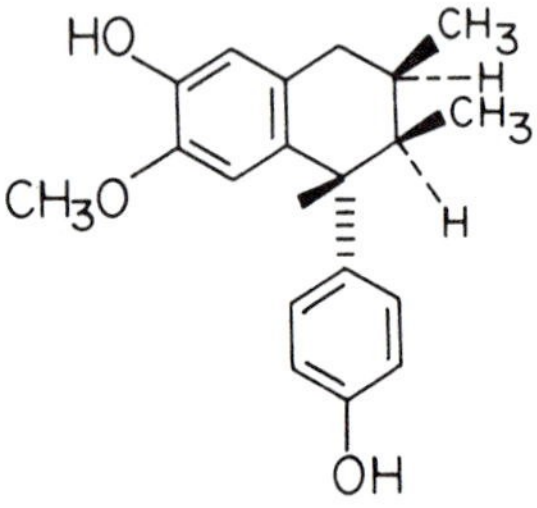   3'- Demethoxyisoguaiacin

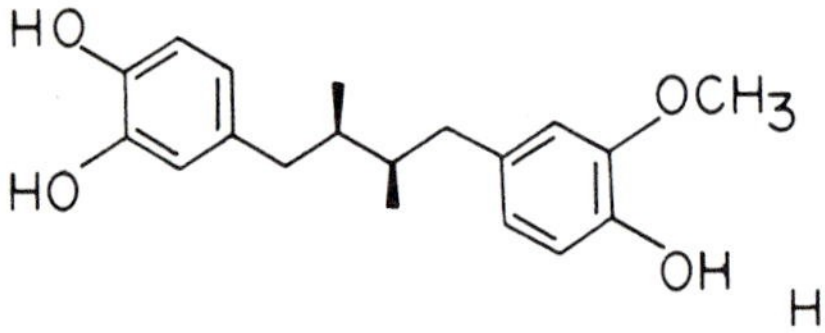

Dihydroguaiaretic Acid

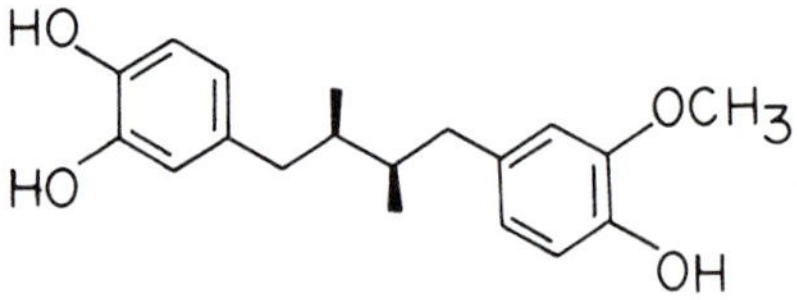

Partially Demethylated
Dihydroguaiaretic Acid

Figure 29.4 Lignans from creosote bush leaf resin.

Toxicology

There is no information available in the literature concerning the biological activity of the saponins or the wax esters of *Larrea.* Considerable information on the biological activity of the phenolic resin constituents, however, is available. NDGA, for example, is reported to be highly toxic to *Salmonella, Penicillium,* and several other pathogens. Lignans and certain flavonoids of the resin have also been shown to be biocidal towards several phytopathogenic fungi including *Rhizoctonia solani, Fusarium oxysporum,* and *Pythium* spp. (Hurtado *et al.* 1979, Florencio *et al.* 1979).

Activity of creosote resin against selected insect herbivores is indicated by the work of Rhoades (1976), who showed that high levels of resin repelled leaf feeding insects. In these same studies, Rhoades found *in vitro* evidence which indicated that elements within the resin complexed with starches and proteins. If this occurred in nature, he suggested, then reduced digestibility of these dietary components might result and thus deter herbivory. Direct toxicity of resin chemicals toward insects was not demonstrated, but the repellant properties suggest possible involvement of resin toxins.

Over the past five years we have screened hundreds of extracts from dominant desert plants of the USA and Mexico in search of new chemicals with interesting and novel biological activities. Recently, we discovered that the antiobiotic activity of a resin extract from *Larrea* was enhanced in longwave ultraviolet light (UV—A; 320—400 nm) (Downum & Rodriguez in preparation). Preliminary research on the chemical(s) responsible for the phototoxic effect has already begun, but the compound(s) have not yet been identified. We have found that the phototoxic fraction is soluble in alcohol (methanol and ethanol), diethyl ether and hexane, suggesting the lipophilic nature of the chemicals. Ultraviolet spectroscopy of an ether extract of the leaf resin revealed a major absorbance at 285 nm with a broad shoulder at 330 nm. Isolation and identification of nordihydroguaiaretic acid (NDGA), the most concentrated resin constituent, showed that this lignan catechol was responsible for the sharp peak at 285 nm, but not the shoulder at 330 nm. Presumably, the absorbance of wavelengths around 330 nm is due to the photo-toxic constituent(s), since many known light-induced toxins absorb in this same region (Smith 1977).

The light enhanced biocidal activity of *Larrea* resin was among the most pro-nounced of all the arid land plants examined over the past five years. Phototoxicity assessment was based on bioassays of gram positive (+) and gram negative (-) bacteria as well as eukaryotic yeast cells (Table 29.1, p. 408). Gram positive (+) bacteria (*Staphylococcus aureus (= S. albus), Bacillus subtilis*) and yeasts (*Candida albicans, Saccharomyces cerevisiae*) were most susceptible to the photoinduced effect. Subsequent studies with purified NDGA revealed that this lignan was responsible for considerable antibiotic activity, and possibly the light induced activity (Downum & Rodriguez in preparation.

It is apparent from the detailed chemical investigations of *Larrea* that if simple and inexpensive extraction procedures are developed for obtaining the bioactive constituents, *Larrea* will be a great resource for the arid zones of North and South America.

Table 29.1 Inhibitory effect of *Larrea tridentata* extract on the growth of micro-organisms exposed and not exposed (dark) to UV—A irradiation (320—400 nm). Zones in mm ± SD. (Downum & Rodriguez, in preparation).

Assay organisms	Inhibitory zones (mm)*	
	UV—A	Dark
Gram (+) bacteria		
Staphylococcus aureus	15 ± 2.5	21 ± 1.5
Bacillus subtilis	11 ± 1.8	20 ± 2.5
Gram (-) bacteria		
Escherichia coli	13 ± 1.3	15 ± 0.5
Pseudomonas aeruginosa	9 ± 1.0	11 ± 1.4
Yeasts		
Candida albicans	8 ± 0.7	12 ± 1.1
Saccharomyces cerevisiae	9 ± 0.9	13 ± 0.9

*Inhibitory zones were measured from edge to edge.

PEDILANTHUS MACROCARPUS

Pedilanthus is a small Mexican and circum-Caribbean genus of the tribe Euphorbieae (Euphorbiaceae). Plants comprising this genus are better known as the slipper spurges (so named for their curious inflorescence – the cyathium), but throughout their Latin American range they are known by such names as gallitos, jamete, pie de nino, zapato del diablo and candelilla (Dressler 1957).

Pedilanthus probably originated on the Pacific slopes of Mexico and the differentiation of the genus commenced in the Tertiary period when hummingbirds entered tropical North America (Dresler 1957, Webster 1967). According to Dressler, it clearly developed from a member of *Euphorbia* subgenus *Agaloma*. This is evident from the petaloid appendages of the spur lobes of *Pedilanthus* which indicate that the ancestor of the genus, if it were still living, would be assigned to the Agaloma. The features found in *Pedilanthus* are not found in any single living *Euphorbia,* but they all may be found in different species of the subgenus *Agaloma.* The closest surviving species to the suspected ancestral type is probably *Euphorbia fulgens,* which has alternate leaves and bright red petaloid appendages (Webster 1967).

All of the 14 species of *Pedilanthus* occur in Mexico today. One species, *P. tithymaloides,* is divided by Dressler into several subspecies and these are distributed throughout southern Mexico, Central America, the northern coast of South America, and through the Caribbean archipelago to the southern tip of Florida (Dressler 1957).

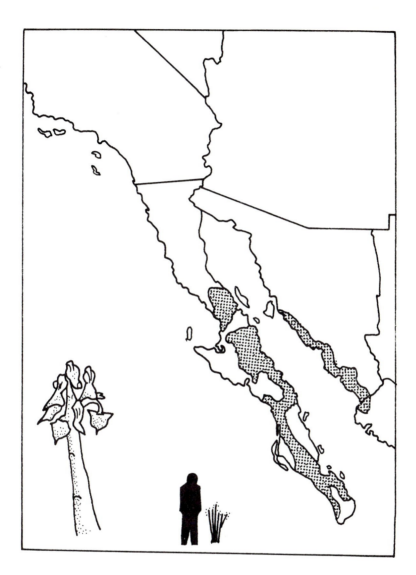

Figure 29.5 Distribution of *Pedilanthus macrocarpus* in the Sonoran Desert.

Species comprising this genus range from tropical deciduous trees to desert succulents. *Pedilanthus macrocarpus* is found in the arid thorn forests of western coastal Mexico and the Sonoran Desert of Baja California (Fig. 29.5; Sternburg & Rodriguez 1982).

Pedilanthus macrocarpus from Baja California is rich in natural rubber and high molecular weight alkanes; both of which are potentially useful compounds (Proksch *et al.* 1981, Sternburg & Rodriguez 1982).

The presence of cis-1, 4-polyisoprene (rubber) was detected and confirmed by H^1-NMR spectroscopy. The spectrum in $CdCl_3$ produced chemical shifts at 1.67 (3H), 2.00 (4H), and 5.12 (1H) ppm. The spectrum is consistent with those obtained from other rubber sources such as *Parthenium argentatum* (guayule) and *Hevea brasiliensis* (Campos-Lopez & Palacios 1976), although the polymer size is somewhat smaller. Purified rubber from *P. macrocarpus* can constitute up to 28 per cent of the dry weight of the latex (Table 29.2).

The normal alkane range in the stem cuticular wax of *P. macrocarpus* was found to be $C_{27}H_{56}$ to $C_{35}H_{72}$. Gas chromatography showed that alkanes with an odd number of carbon atoms predominated; the major components being $C_{31}H_{64}$, $C_{33}H_{68}$ and $C_{35}H_{72}$ with $C_{33}H_{68}$ predominating. Alkanes with even numbers of carbon atoms occurred only in minute amounts. Alkane chemistry varies within *P. macrocarpus* over a wide geographical range. Whether this is due to genetic variability or environmental factor is not known (Sternburg & Rodriguez 1982).

Table 29.2 Total rubber and alkane content in species of *Pedilanthus* (Sternburg, in preparation).

	% rubber dry wt. of latex	% alkanes dry wt. whole plant
P. tithymaloides	0.05	2.4
P. cymbiferus	–	6.0
P. bracteatus	40.0	2.3
P. macrocarpus	28.0	5.0

Although *Pedilanthus macrocarpus* has never been harvested in Baja California for wax or rubber, *Euphorbia antisyphilitica* from the Chihuahuan Desert has long been harvested for its quality wax (Campos-Lopez & Roman-Alemany 1980). The candelilla plant is still collected in parts of Chihuahua, but the rather crude extraction procedures have resulted in low yields and a gradual decline in the production of the candelilla wax. Comparison of the alkanes from *Pedilanthus macrocarpus* and *Euphorbia antisyphilitica* indicate that the waxes are very similar in quantity and quality (Sternburg; in preparation).

Conclusions

The potentials for obtaining a wide variety of phytochemicals from the arid zones of North America and the world have not been realized. Studies of selected desert dominants from the Chihuahuan and Sonoran Desert have established that many species, such as *Parthenium argentatum* and *Pedilanthus macrocarpus* are capable of producing rubber in quantities comparable to that produced by commercially established hydrocarbon crops, e.g. *Hevea*. Also, the potential for obtaining specialized chemicals for agricultural, pharmaceutical and medical use is unlimited. Further research is needed to discover novel and useful pesticides to be used in the development of new crops for the arid lands of the world.

Acknowledgements

I am grateful to the National Science Foundation (PCM–8209100), National Institute of Health (AI 18398 and AI 00472) and the University of California, Irvine, for financial support. I also thank Dr H.T. Huang (NSF), Dr G.H.N. Towers (Canada) and Dr K. Downum (UCI) for their helpful suggestions. I am also very grateful to my research associates, Mr Manuel Aregullin, Mr Charles Sternburg and Mr Jan West who kindly provided unpublished data on the desert plants mentioned in this chapter.

References

Aregullin, M. and E. Rodriguez 1984. Phytochemical studies of the genus *Flourensia* from the Chihuahuan Desert. *Ann. Proc. Chihuahuan Desert.* In press.

Becker, R., R.N. Sayre and R.M. Saunders 1984. Semi-arid legume crops as protein resources. *J. Am. Chem. Soc.* 61: 931–938.

Bowers, W.S. 1976. Discovery of insect anti-allatotropins. In *The juvenile hormones,* L.I. Gilbert (ed.): 394–408. New York: Plenum Press.

Campos-Lopez, E. and A. Roman-Alemany 1980. Organic chemicals from the Chihuahuan Desert. *J. Agric. Food Chem.* 28: 171–183.

Dillon, M.O. 1976. *Systematic study of the genus Flourensia.* Unpubl. Ph.D. dissertation. Austin: University of Texas.

Dressler, R. 1957. The genus *Pedilanthus* (Euphorbiaceae). *Contr. Gray Herb.* 182: 1–188.

Florencio, J.D. and V.G. Pablo 1979. Curative fungicides field evaluation. In *Larrea, Serie el Desierto,* Vol. 2. Mexico: Saltillo (CIQA).

Habermehl, G. and H. Moller 1974. Isolierung und struktur von Larreagenin A. *Justus Liebigs Ann. Chem.* 1974: 169–175.

Hurtado, L., R. Hernandez, F. Hernandez and S. Fernandez 1979. Fungi-toxic compounds in the Larrea resin. In *Larrea, Serie el Desierto,* Vol. 2, E. Campos-Lopez, T.J. Mabry and S. Fernandez (eds). Mexico: Saltillo (CIQA).

Isman, M.B. and E. Rodriguez 1983. Larval growth inhibitors from species of *Parthenium* (Asteraceae). *Phytochem.* 22: 2709–2713.

Mabry, T.J., J.H. Hunziker and D.R. DiFeo (eds) 1977a. *Creosote bush, biology and chemistry of Larrea in New World deserts.* Stroudsburg, Penn.: Dowden, Hutchinson and Ross.

Mabry, T.J., D.R. DiFeo, M. Sakakibara, C.F. Bohnstedt and D. Seigler 1977b. The natural products chemistry of *Larrea.* In *Creosote bush, biology and chemistry of Larrea in New World deserts,* T.J. Mabry, J.H. Hunziker and D.R. DiFeo (eds): 115–134. Stroudsburg, Penn.: Dowden, Hutchinson and Ross.

Proksch, P., C. Sternburg and E. Rodriguez 1981. Epicuticular alkanes from desert plants of Baja California. *Biochem. System. Ecol.* 9: 205–206.

Rhoades, D.F. 1977. Integrated antiherbivore, antidesicant and ultraviolet screening properties of creosote bush resin. *Biochem. System. Ecol.* 5: 281–290.

Rodriguez, E. 1977. The ecogeographical distribution of secondary constituents in *Parthenium. Biochem. System. Ecol.* 5: 207–218.

Rodriguez, E. 1980. Chemical patterns of some desert shrubs of the Chihuahuan and Baja Californian Desert. In *Yucca, Serie el Desierta.* Vol. 3: 243–256. Mexico: Saltillo (CIQA).

Rodriguez, E., P. Healey and I. Mehta 1983. *The biology and chemistry of plant trichomes.* New York: Academic Press.

Sternburg, C. and E. Rodriguez 1982. Hydrocarbons from *Pedilanthus macrocarpus* (Euphorbiaceae) of Baja California and Sonora, Mexico. *J. Am. Bot. Soc.* 69: 214–218.

Smith, K.C. 1977. *The science of photobiology.* New York: Plenum Press.

Webster, G. 1967. The genera of Euphorbiaceae in the southeast United States. *J. Arn. Arb.* 48: 1–23.

West, J., M. Behl and E. Rodriguez 1984. The inheritance of rubber and resin in guayule and hybrids. *Ann. Proc. Chihuahuan Desert.* In press.

Wiggins, I. 1982. *The flora of Baja California.* Stanford: Stanford University Press.

30 Plant information services for economic plants of arid lands

F. A. Bisby

Biology Department, Building 44, University of Southampton, Southampton SO9 5NH, UK

Introduction

Our purpose is to form a picture of the databases and information services available to research and development botanists working on economic plants for arid lands. We need to examine the databases themselves, the means of access to these databases, and the extent to which they act as services giving information beyond that in the database.

There has been a rapid growth in the provision of databases containing information about plants. The well-known databases of BIOSIS, CAB are computerised bibliographic databases containing literature citations and abstracts for retrieval by, for instance, plant names and to a lesser extent by plant uses. A second group of databases which are potentially of much more direct utility, are only now becoming available. These are the various factual databases on plant uses, economic botany and plant breeding: the SEPASAT, IBRA, IBPGR, USDA, and NAPRALERT databases. They provide factual data on plant uses provided by teams that have already gleaned the enormous and dispersed literature and sometimes other sources too. Thirdly there will soon be a number of general purpose plant diversity databases. These will provide, in a few years time, descriptive, chemical, ecological and geographical information of the sort that is essential for biologists considering breeding, introduction or development involving new plants, e.g. the Vicieae/Legume Database, the Seed and Gardens' Records at Kew, or the ROTEM* database in Israel.

Most of these databases provide only indirect access. Enquirers write or telephone their request to the operating team who then put a precise enquiry to the database, and then mail the response back. A few, so far the larger bibliographic databases, provide direct public access by allowing on-line access through the

---

*ROTEM is an acronym for the Hebrew phrase 'Reshet tatzpiot u meida' (network of botanical observations and information); it is also the Hebrew word for *Retama raetam*. ROTEM is equivalent to the Israel Plant Information Centre, which is a joint project of the Society for the Protection of Nature in Israel and the Hebrew University Department of Botany, and is based at the Har Gilo Field Study Center, near Jerusalem.

public telecommunications networks. Alternative methods will appear soon: BIOSIS are providing floppy discs through the mail for use on microcomputers, and other media such as video discs may follow rapidly. We need to think carefully what are the most effective means if dissemination is to be successful in both the developed and the developing nations with arid zones.

General properties of databases

A database is a collection of data stored and organised for multiple uses by many people or in many ways. These days it tends to be stored in a computer and, depending on the organisation, to be stored as several files or data sets within that computer. The multiplicity of uses or users is important as preparing the database will usually require more labour and be more expensive than tabulating the data for the simplest single use. Thus the database itself, rather than any particular analysis, becomes a focus of attention and it is intended to amortise the extra cost of building the database over the economies of multiple uses.

The computer database will often be essentially analogous to a card index or catalogue with various data fields, such as author/title/journal, etc. for a bibliography or name/use/origin, etc. for a plant use. But it differs in important properties: it can be searched on many of the data fields taken individually or in combination, whereas a card index or catalogue is only easily searched on the one field used to order the cards or entries. It can be revised easily very many times. The searches and revisions can extend not only throughout the index or catalogue but also across several indexes or catalogues instantaneously. And because the data are in computer readable form they can be entered to other types of computer activity, such as word-processing or numerical analysis, and transferred to other computers.

Lastly there are two more properties that are consequences of those already listed. We can use the search and extraction properties to extract computer readable subsets of the original database. It may be these, rather than the original full database, that are of great utility to certain classes of user. Secondly, because of the revision facility, we may start to use a database either before it is even filled for the first time, or knowing that it undergoes frequent revision. This in in contrast to most compendia published directly as books where the data is not available until the work is finished and published.

Present and projected databases

Up to the present time, the principal databases to provide assistance in retrieving information about plant uses have been the bibliographic databases. Factual databases containing facts not bibliographic data, with a few exceptions, are only now becoming available or are still in the planning stage.

BIBLIOGRAPHIC DATABASES

There is a well established industry which provides bibliographic guides to the world's literature in the form of searchable computer databases. Typically an organisation screens the literature of a particular discipline and at intervals makes additions to an ever accumulating database. Each entry to the database is a citation of an item in the literature, a book, a pamphlet, or an article in a journal or symposium volume. A selection of the bibliographic databases of greatest interest to workers on plant uses and the plants of arid lands is listed in Table 30.1 together with the producer, the first year's literature to be covered, the approximate number of citations in the database and the approximate annual rate of increase (information from Hall & Brown 1983).

Table 30.1 On-line bibliographic databases relating to the uses of plants of arid lands; 1K = 1000 citations. (Data from Hall & Brown 1983).

Name	Producer	Start	Content	Rate of increase
AGRICOLA	USDA, Beltsville, USA	1970	2 million	144K/yr
AGRIS	FAO, Rome, Italy	1975	850K	180K/yr
CAB ABSTRACTS	CAB, Slough, UK	1972	1.5 million	150K/yr
FOREST PRODUCTS	For. Prod. Res. Soc. Madison, USA	1947	20K	2K/yr
TROPAG	Royal Tropical Inst. Amsterdam, Netherlands	1975	45K	6K/yr
BIOSIS PREVIEWS	BIOSIS, Philadelphia, USA	1969	3.5 million	355K/yr
ZOOLOGICAL RECORD	BIOSIS, York, UK	1978	(300K) (backlog)	60K/yr

AGRICOLA, AGRIS, and CAB ABSTRACTS are the three large databases for the whole of Agriculture. They differ in many details such as the starting year, the coverage and the detail of whether keywords, subject code and abstracts are included. For comparisons see Johnston (1979) and Krabbe (1979). Of these three, CAB ABSTRACTS alone contains full abstracts and additional search terms covering subject area and groups of organisms involved. In coverage CAB ABSTRACTS appears to contain the widest selection of purely biological citations whilst AGRICOLA and AGRIS cover a wider range in economics, management, rural sociology and research methods. FOREST PRODUCTS and TROPAG are just two examples of a number of smaller more specialised databases of less direct interest for uses of plants of arid lands.

Much of the information required by workers on plant uses for arid lands is information on pure biology or biological diversity. These are covered by BIOSIS PREVIEWS and ZOOLOGICAL RECORD. BIOSIS PREVIEWS aims to cover the whole of biology and includes full abstracts whereas ZOOLOGICAL RECORD, apart from being restricted to animals, provides citations indexed by the taxonomic group to which the animals in the publication belong. Similarly indexed botanical citations are available in the *'Kew Record'* a journal not yet available in computerised form.

All of the databases listed in Table 30.1 are available in many countries as on-line services provided by one or more of the database networks such as DIALOG, DATA-STAR, etc. To use such a database one must subscribe for access to the network, and use a computer terminal or microcomputer linked to the telephone system by a modem. Once linked online to the database users can search the database, for example, for citations containing a given plant use, say 'fiber' or 'fibre', or a given plant name, say *Acacia* in the title. A difficulty, of course, is that the target publications will only be retrieved from some of the databases if the plant use or the plant name appear in the title. In CAB ABSTRACTS and BIOSIS PREVIEWS a greater fraction of target publications may be retrieved if the use or plant occurs in the abstract or the additional subject or biosystematic terms. In ZOOLOGICAL RECORD citations are indexed under various taxonomic levels so that, for instance, a paper describing the action of two pest beetles could be retrieved from either of the two specific names, either of the generic names, under Coleoptera, or under Insecta.

FACTUAL DATABASES ON PLANT USES

Factual databases contain factual data carefully assembled possibly by a range of methods and from a range of sources. They have the enormous advantage that for the user the target data is immediately and directly available: the intermediate steps of obtaining the literature and extracting the data have already been performed by the database producers. So our hypothetical enquiry about uses of *Acacia* spp. as fibers can be answered in a single computer enquiry: 'List the scientific and common names of all *Acacia* species used as fibers.'

But factual databases do introduce additional variables many of which may have a significant bearing on the utility of the database. Commonly the data entered is secondary data gleaned by systematically surveying the literature, by recording data from the labels of specimens, or by obtaining opinions or judgements from experts. In all of these the user of the database will want some information about the reliability of the records and possibly the ability to track back to the original sources for further detail. If different data fields are from different sources then this can be achieved by tagging each field with a brief reference to the literature citation, specimen, document, or opinion from which it was obtained. However, doing this may more than double both the size of the database and the labour of producing it. Where whole groups of fields contain data obtained from the same source, e.g. morphological features taken from the same Flora, clearly it is more economical to tag the source for the whole group of fields, and where the same source is used for a particular group of fields throughout the source this can be made clear in a general preamble for users of the database.

The most important need for discussion of reliability and responsibility arises however either when the data assembled is primary data recorded by the database producer's team, say from field trials or living specimens, or where the data to be entered involves a critical opinion or decision, as say in choosing between alternative taxonomic treatments of *Acacia* species. The database user may want to know the principles or method by which the data or opinion was achieved and also be able to check back to the original document or recording sheet used at the time by the worker involved.

A further important variable with factual databases arises from the near impossibility of recording in computer searchable form every single relevant fact known to the database producers. The cost of the very labour intensive data collation and data entry may impose a very restricted limit as to what data fields will genuinely receive sufficient 'multiple use'. A solution frequently adopted is to place in the computer database only the important facts that are predicted to receive heavy use, and to keep the remaining data as documents or photocopies.

There already exist many factual databases with information about plant uses. Because many of these are associated with national or regional development projects and because work may progress for several years before the database is sufficiently filled to be useful, it is difficult to find out about those in other parts of the world. Consequently the list given in Table 30.2 is just a sample of what is available.

Let us take the SEPASAT database at the Royal Botanic Gardens Kew as an example. A team of four have worked for three years building a factual database to cover the uses of non-agricultural plants from arid and semi-arid lands in the tropics. Complete or partial entries have been made for about 5000 species and preparations are in progress to start providing an enquiry service in the autumn of 1984. The data is organised at three levels. 21 descriptors cover names, life form, habitat tolerances, uses and distribution and are fully searchable in the computer database. A further amount of text is held in the computer and can be obtained when requested and manipulated as a word-processor file. Other information, certain original sources and original recording sheets are stored physically in filing cabinets. The principal source of data is the laborious screening of the literature alongside the Index Kewensis and Kew Record teams at Kew. Subsidiary sources are herbarium sheets and the opinions of taxonomists in the herbarium, and collections in the very extensive Kew Museums of Economic Botany.

The enquiry service will allow people to contact the SEPASAT office with requests such as 'What species can be used for dune stabilisation? Do any occur in Somalia? Please list their descriptions and habitat requirements.' A member of the team would put the enquiry to the database using one of the terminals in their office, check on the screen that the response was what was needed, and then have it printed and posted to the enquirer. Duke (1983) illustrates enquiries put to the USDA Minor Economic Plant Species Database covering a range from the nutritional composition of coca leaves (*Erythroxylum coca,* cocaine bush) to the ecological amplitudes of perennial weeds.

Table 30.2 Factual databases on plant uses. (Acronym, name of project, project leader(s), reference, source of funds).

USDA	Minor Economic Plant Species Database J.A. Duke, USDA, Beltsville, USA; Duke 1983; USDA (& formerly Nat. Cancer Inst.) funded
SEPASAT	Survey of Economic Plants for the Arid & Semi-Arid Tropics G.E. Wickens, Royal Botanic Gardens, Kew, UK; Wickens 1984; OXFAM & Kew funded
IBRA	Directory of Important World Honey Sources E. Crane & M.E. Adey, IBRA, Gerrard's Cross, UK; Crane *et al.* 1984; IDRC funded
NAPRALERT	Natural Product Information System N.R. Farnsworth, Chicago, USA; Farnsworth *et al.* 1981; WHO & Nat. Cancer Inst. funded
MUPOTAS	Multipurpose Trees Database P.G. von Carlowitz, ICRAF, Nairobi, Kenya; Carlowitz 1984; ICRAF funded
INIREB	Useful Plants of Mexico S. Avendano, INIREB, Jalapa, Mexico; Gomez-Pompa 1978; INIREB funded

The INIREB Useful Plants of Mexico is the only example from Table 30.2 in which there are plans to make the database available to online enquirers through one of the communications networks, in this case the SECOBI Mexican network.

FACTUAL DATABASES ON PLANT DIVERSITY

Much of the information needed by botanists working on economic plants of arid lands is not of course peculiar to this discipline: the taxonomic, descriptive, chemical, geographical, ecological, and conservation information needed is part of the general 'plant diversity' body of knowledge actively being investigated and disseminated by the rest of the botanical profession. Making this body of information available as databases has lagged behind other disciplines (Bisby 1984a & b) and only now are the first general purpose plant diversity factual databases being developed (Allkin & Bisby 1984).

In Table 30.3 are listed some of the general purpose plant diversity factual databases that are available or in various stages of development. For completeness three

Table 30.3 Factual databases on general plant diversity.

Generic	
GRASSES	L. Watson, CSIRO, Canberra, Australia (Watson & Dallwitz 1981)
CAESALPINIOIDEAE	L. Watson, CSIRO, Canberra, Australia (Watson 1981)
KEW HERBARIUM LIST	R.K. Brummitt, RBG Kew, Richmond, UK (unpublished)
Species	
VICIEAE DATABASE	F.A. Bisby, University of Southampton, UK (Adey *et al.* 1984)
*BIOSIS TRF	M.N. Dadd, BIOSIS UK Ltd, York, UK (Dadd & Kelly 1984)
*ESFEDS	V.H. Heywood, University of Reading, UK (Heywood *et al.* 1984)
**LEGUME DATABASE	F.A. Bisby, University of Southampton, UK (unpublished proposals)
**SOLANACEAE CHECKLIST	R. Lester, University of Birmingham, UK (Lester 1984)
Populations	
PLANT CHARACTER DIVERSITY	S. Weitz-Ingber, Hebrew Univ. of Jerusalem, Israel (Weitz-Ingber 1983)

*These databases are under development and are not yet operational.
**These databases are at planning stage only.

generic databases are included, although most biologists will be more interested in the others that contain information about species, infra-specific taxa or populations. At their simplest, as in the projected Solanaceae Checklist (Lester 1984) or the early stages of the BIOSIS Taxonomic Reference File (TRF) (Dadd & Kelly 1984), they contain carefully collated lists of scientific names cross-indexed to synonyms, and possibly containing further technical data about the publication of the scientific name. Other databases contain or will contain a wide range of data types. The Vicieae Database presently contains morphological descriptions, geographical distributions, and the distributions of secondary substances as well as names and synonyms for most of the 327 species of *Vicia, Lathyrus, Lens* and *Pisum* in the tribe Vicieae (Adey *et al.* 1984). The European Science Foundation's European Taxonomic Documentation System (ESFEDS) project, when fully developed, will contain a mixture of factual data (e.g. on chromosome numbers and geographical distribution) and bibliographical citations to the literature, e.g. on phytochemical and taxonomic issues, for the genera, species and infra-specific taxa of Europe (Heywood *et al.* 1984).

In Table 30.4 some of the more specialised types of databases concerning plant diversity are listed with examples of each. Much the largest class is the germplasm databases run by the germplasm centres. These databases contain up to three main types of data about the individual accessions of seed or other propagules held by the centres: curatorial/management data; passport data about the origin of the propagules; and evaluation data on aspects of the material of interest to breeders (Blixt & Williams 1982). The passport data with details of sites, habitats and conditions and the evaluation data with details of performance, disease resistance and genotype are all of potential interest to a wide range of botanists. And yet the germplasm databases are not easily available to outside users, and do not at present add up to an international system. An enquirer will find that germplasm of any given species is distributed between many international, regional and world centres (Hanson *et al.* 1984), that the centres in question have varied and unconnected database operations, and that even when the database of interest is located, access may not be easy.

Pankhurst (1984) lists only nine major herbaria that have made databases containing data on individual specimens, and these are largely in relatively young and rapidly developing institutions. There are none from the large European herbaria. The databases at the Universidad Nacional Autónoma de Mexico (UNAM) Herbarium in Mexico City and at the Herbarium of the Botanical Research Institute, Pretoria in S. Africa are examples that cover arid lands: they give (or at UNAM will soon give) access to a wealth of distribution, collection and timing data drawn from herbarium collections.

Many nations have set up organisations to collate data on the local distribution and ecology of their flora. Only few of these cover arid zones, but one that does is the ROTEM database covering the distribution, time of flowering and fruiting, environmental and habitat data for wild species in Israel (see footnote p. 413).

Table 30.4 Factual databases on specialised aspects of plant diversity.

GERMPLASM – very many, unconnected	
Izmir Seedbank	INIA, Izmir, Turkey
Kew Seedbank	RBG Kew Wakehurst Place, Ardingly, UK
Pisum Database	DAFS, East Craigs, UK
HERBARIA – rather few	
Pretoria Herbarium (PRE)	National Herbarium, Pretoria, S. Africa
UNAM Herbarium (MEXU)	Herbario Nacional de Mexico, UNAM, Mexico
LOCAL DISTRIBUTION AND ECOLOGY – many, often national	
Biological Records Centre	BRC, Monkswood, UK
ROTEM Project	Dept of Botany, Hebrew Univ., Jerusalem
CONSERVATION	
IUCN database	Conservation Monitoring Centre, RBG Kew

The IUCN Conservation Monitoring Centre runs a database which covers plants and animals of conservation concern, protected areas and wildlife trade (Mackinder, 1984). The CMC database was established in order to provide the international conservation and development agencies with information required to plan their activities. As a consequence the database contains much data of general interest on world distribution, ecological and descriptive data about the taxa covered. At present CMC does not provide an on-line service, access to the database being through CMC staff.

Access and Service

In the context of plant information services for economic plants of arid lands we need to examine carefully the various possibilities for gaining access to databases. Because of the limited funding and technological support for many botanists working in arid zones, we must be prepared to consider only minimal services, and this in turn means being realistic about practicalities. I hope readers from developing countries will forgive me for stressing this point, but it is important if real progress is to be made.

The magnitude of the problems faced by botanists working under difficult conditions in developing countries is illustrated by the figures for subscriptions to long established scientific journals quoted by Woolston (1984). For 10 canadian journals 6.90% of purchases were from developing countries, the remaining 93.10%

being from the OECD, E. European countries and S. Africa. If India, OPEC countries, Brazil, Korea and Israel are also excluded, the developing countries figure falls to 2.70%. The figures for 65 US journals were lower (4.18% and 1.66% respectively) and usage of online databases is thought to be even less. Hence it is no wonder that Woolston claims there is no commercial market for database services in developing countries, and that to reach the appropriate people services must use as low technology as possible, and be funded either externally or internationally.

Just a few of the databases discussed above, the large bibliographic databases, are available online by means of international telecommunications links. This arrangement provides what many consider the ultimate in high technology facilities — both online access to search the database, and this access made to the very latest version of the database. But the facility is achieved at considerable cost, and we should ask in many contexts whether it is essential.

To make this online access the users must be equipped at their own sites with 1) a computer terminal or microcomputer with a terminal emulator, 2) a telephone modem, and 3) reliable telephone services of adequate quality. In addition they must obtain (and pay for) access to one of the international telecommunications networks usually via a node in the capital city of their own country, and obtain (and pay for) access time for the database in question. In practice the instantaneous element is muted in two ways. First, both because costs are high and because searching requires skill, many institutions use professional searchers to do the searching for the real users. Other enquirers without online facilities request national institutions to perform online searches for them, but again with the delays of using intermediaries. Second, a bibliographic database is not like an airline booking system that changes from minute to minute: most are updated at monthly intervals, so that little is gained by catching updates the moment they are made.

A second method of access to a database is to obtain a distributed copy, or more likely a copy of a selected subset. There are few examples of this as yet but in theory it might be as effective as online access, cheaper, and much less demanding of technological support. The one example I know of is the B-I-T-S distribution from the BIOSIS PREVIEWS bibliographic database (Schultz 1983). Users provide search profiles describing the subset of BIOSIS PREVIEWS monthly updates that are of interest to them and details of the disc format for their microcomputer. BIOSIS then supply the selected subset each month as a small database on a floppy disc. On joining the scheme, users are provided with a programme, which enables them to access, search, add to, or delete from the small database using their own microcomputers. If their own microcomputers are of sufficient capacity, say with a hard disc, they can coalesce the monthly files into a larger bibliographic database in their own machines. The B-I-T-S service may well be the forerunner of other services that are distributed physically rather than made available online, particularly if cheap storage media, perhaps video discs, become widely available for use with microcomputers.

An important feature of the B-I-T-S distribution method is that the user needs just a microcomputer (of medium or large size) and funds to purchase the monthly profile on discs. Given the reliability and availability of microcomputers this may, in many places, be relatively easier to organise than telecommunications links for online access. Users of such a system have the advantage of immediate and un-limited access for searching their copies, and with their monthly additions, copies that are updated as frequently as the original.

A third class of access is for a central office to provide an enquiry service as planned for the SEPASAT database. If users send their enquiries by post, absolutely no special equipment nor funds are needed, unless a charge is levied. Alternatively a speedier turn-around time can be achieved simply by using Telex or the new Electronic Mailbox facilities at least for the enquiry if not for the response. If there is no charge, this method effectively transfers the costs back to the organisation funding the database. If the time can be afforded, there are other advantages: the staff may help unskilled enquirers make the best use of the database, and where staff know of further resources not in the database, offer advice on further researches. Despite not being very glamorous this enquiry service method has its advantages and may well be useful for access from developing countries.

Lastly, databases can provide printed catalogues which may be cheaply and rapidly produced directly from the computer, and which may be replaced when the database is revised. Watson's generic databases on grasses and Caesalpinioideae are available in this way on microfiche (Watson 1981, Watson & Dallwitz 1982) as are printed pamphlets generated from the Vicieae Database (Allkin *et al.* 1983). How-ever, while these may be cheap and regularly revised (an advantage over many published compendia), they are no substitute for a fully searchable database as they are fixed in one physical layout and can only be read or searched by eye.

So what are the priorities in access? Clearly these will vary widely with circum-stances, but it is suggested that the following might be borne in mind: (a) access to instantaneously updated versions is usually unnecessary; bibliographic databases are, at best, updated monthly and for many other databases yearly revisions are more likely; (b) to search a database online at will is an important facility of medium priority, but one that must usually be paid for by telecommunications links or by obtaining a copy of the database; (c) away from libraries and great centres of learning, factual databases may prove more valuable than bibliographic databases; (d) given the enormous labour of collating comparative information, even slow access through postal enquiry services should prove valuable, and deserves high priority; (e) printed catalogues revised regularly are useful products from databases, but are no substitute for the ability to search the databases themselves.

Discussion

How might the community of botanists at this conference stimulate improvement of plant information services? Co-ordination, demarcation and funding are obvious candidates.

The wide variety and large number of factual databases described here have mostly arisen when relatively local teams have seen the creation of a database as the solution to relatively restricted, specific information requirements:— the IUCN need for conservation monitoring, the need for Israeli ecologists to understand their flora, the need for Mexican taxonomists to plot distributions from herbarium collection localities. But little effort has so far gone into the larger scale: how can these many parts be brought together into a world plant information system? Clearly the present systems may vary in success, may fill varied niches, and may suffer the full rigour of evolutionary processes:— extinction, radiation, hybridisation, speciation and convergence. So far few have survived beyond the influence of their originators.

There are, however, some positive developments that could lead to further co-ordination. Firstly, the ability to exchange data between databases is probably improving; because of the widespread use of at least some standardised software on medium and large microcomputers, because of the simple rectangular structure of files used in modern relational databases, and because of telecommunication links that can be used to avoid the incompatibilities of tapes and discs. Secondly, (and this is one of the reasons for holding this conference at Kew) taxonomists are at last becoming interested in providing modern general information systems and in standardising descriptions (Hackett 1983) or creating communication formats (Dallwitz 1980). They have always been responsible for a general reference system but there has been a danger that it would be overtaken by modern services for special purposes. Thirdly, international and interdisciplinary communications are so much better than they were that it is increasingly difficult to persist with isolated activities.

Demarcation of expertise and responsibility is another area in which we should look for improvements. Despite the fascinating variety of botanical observation revealed in the papers contributed to this conference, many contain similar classes of information; plant uses, plant diversity (morphology, phytochemistry), and plant ecology (ecology, distribution, environmental limits). In some of these areas, often in plant uses, the author is an expert, but others are peripheral to his expertise. When we turn such data into a database we should provide data in our own areas of expertise, but where appropriate draw on others for their expertise. So by agreeing, for instance, to exchange data between the IBRA, SEPASAT, IUCN and Vicieae databases we can avoid not only duplicated effort but also the dangers of, e.g. the Vicieae Database containing data on plant uses that are less reliable than its taxonomic data. We suppose then that plant information systems of the future will of necessity be co-operative ventures with major elements contributed from different disciplines.

We should discuss funding too. It must be made clear to funding agencies that sensible use of databases does not necessarily mean expenditure on sophisticated computing and telecommunications; indeed, microcomputers are cheap and the principal costs of any database are likely to be in the labour of data gathering. We should check the assumption made all too frequently that, to be justified, databases must be capable of supporting themselves in the commercial market: in the past the plant information services have been largely provided at communal expense and there are good reasons why this should continue. We do, however, need to be sensitive to the difficulty in balancing expenditure on fresh botanical discovery with the collation and dissemination of what is already known. My personal view is that the pendulum may be too far towards discovery, and as a result discoveries go unnoticed or are needlessly repeated. And lastly we need to emphasise the need for services to organise the dissemination of information, as well as the databases themselves.

Acknowledgments

I am grateful to Michael Dadd, Sylvia Fitzgerald, Duncan Mackinder, and Gerald Wickens for very helpful discussions on some of the topics included.

References

Adey, M.E., R. Allkin, F.A. Bisby, T.D. Macfarland and R.J. White 1984. The Vicieae database: an experimental taxonomic monograph. In *Databases in systematics,* R. Allkin and F.A. Bisby (eds): 175–188. London & Orlando: Academic Press.

Allkin, R. and F.A. Bisby (eds) 1984). *Databases in systematics.* London & Orlando: Academic Press.

Allkin, R., T.D. Macfarlane, R.J. White, F.A. Bisby and M.E. Adey 1983. *List of species and subspecies in the Vicieae.* Vicieae Database Project Publication No 1, Southampton: University of Southampton.

Bisby, F.A. 1984a. Automated taxonomic information systems. In *Current concepts on plant taxonomy,* V.H. Heywood and D.M. Moore (eds): 301–322. London & Orlando: Academic Press.

Bisby, F.A. 1984b. Information services in taxonomy. In *Databases in systematics,* R. Allkin and F.A. Bisby (eds): 17–33. London & Orlando: Academic Press.

Blixt, S. and J.T. Williams (eds) 1982. *Documentation of genetic resources: a model.* Rome: IBPGR Secretariat.

Carlowitz, P.G. von 1984. *Multipurpose trees and shrubs, opportunities and limitations – the establishment of a multipurpose tree database.* Nairobi: ICRAF (Working Paper No 17). International Council for Research in Agroforestry.

Crane, E., P. Walker and R. Day 1984. *Directory of important world honey sources.* London: International Bee Research Organization.

Dadd, M.N. and M.C. Kelly 1984. A concept for a machine-readable taxonomic reference file. In *Databases in systematics,* R. Allkin and F.A. Bisby (eds): 69–78. London & Orlando: Academic Press.

Dallwitz, M.J. 1980. A general system for coding taxonomic descriptions. *Taxon* 29: 41–46.

Duke, J.A. 1983. The USDA Economic Botany Laboratory's database on minor economic plant species. In *Plants: the potentials for extracting protein, medicines, and other useful chemicals – workshop proceedings:* 196–214. Washington, D.C.: US Congress, Office of Technology Assessment (OTA–BP–F–23, Sept 1983).

Farnsworth, N.R., W.D. Loub, D.D. Soejarto, G.A. Cordell, M.L. Quinn and K. Mulholland 1981. Computer services for research on plants for fertility regulations. *Kor. J. Pharmacog.* 12: 98–110.

Taxonomic index

Accepted names in roman, synonyms in *italics*.

HIGHER PLANTS

LOWER PLANTS

Mosses

Lichens

General index

Because of limited space subsidiary geographical locations are under the appropriate country; national and international bodies, institutions, etc. are under 'organizations'.

aboriginals or aborigines, Australian 53–68, 200, 298; Pitjantjatjara people 299
Acacia shrublands, Australia 143–158
acetic acid 351
active plants (see also inert plants) 286
adhesive (see also glue and gum) 350, 362
Aegean Islands 132
aeolian sediments 14
afforestation, see reafforestation
Afghanistan 41, 117, 121, 127
Africa 1, 10, 13, 15, 16, 19, 30, 35, 39, 42–3, 45, 90–1, 114, 143, 154, 156, 164, 190, 192, 197, 199, 204, 238, 343, 369, 373, 380
    East 98, 197, 198
    North 36, 117–39 *passim,* 181–2, 193
    South (region) 44, 69–84
    South, Republic of, see South Africa
    West 12, 43, 98, 179, 184, 189
African people, Batswana 235, 237; Bushmen 69, 237; Hottentots 69, 77–8, 81–3; Topnar Hottentots 77; Khoi-Khoin 69; Khoisan 69, 76, 84; Kung San 81; Ndebele 77; Pedi 77, 235; San 69; 75–6, 78, 82–3; Swazi 235; Venda 235; Zulu 235
agaves (*Agave* spp.) 19, 22–4
agricultural tools 88
agriculture (see also crop growing) 93, 370, 372; conventional 96; rainfed 10, 14, 17, 39, 41, 244; runoff 20; silvi-agri-pastoral system 261, 263–4; traditional 95, 177
agriphytochemicals (see also biochemicals) 399, 411
agro-forestry 51, 200
'al'alan (*Juniperus macropoda*) 273, 368
alcohol, potable 23, 24, 193–4, 196, 235, 236
alcohol, fuel 185, 385
Aleppo pine (*Pinus halepensis*) 125, 137
alfalfa (*Medicago sativa*) 107, 109, 165, 182, 205, 210, 224
algae 380–1; blue-green 380
algarrobo (*Prosopis chiliensis*) 106, 108
Algeria 117, 119, 120, 121, 125, 126, 130, 135, 194; Great Eastern Sand Sea 135; Tadmit 126; Tassili N'Ajjers 134
aliphatics 357; aliphatic acids 405; alcohols 405; esters 405
alkali tolerant 183
alkaloids 299, 400
alkanes 409–10
almonds (*Prunus amygdalus*) 16, 73, 78, 81

amaranth (*Amaranthus* spp.) 246; grain 20, 45–6, 246–8; vegetable 247–8
amenity planting (see also ornamentals and landscaping) 166, 183, 290, 293–7, 299–300
America 79, 182, 258
    Central or meso- 26, 29, 98, 164, 246, 408
    Latin 1, 27, 41, 103–4, 106, 164, 408
    North 14, 19, 21, 28, 31, 43, 46, 132, 180, 189, 190, 303, 358, 369, 372, 399, 407
    South 30, 35, 98, 123, 155, 164, 182, 190, 246, 338, 358, 407, 408
American honey locust (*Gleiditsia triacanthos*) 132
American Indians, see Indians
amides 227
amino acids 45, 49, 227, 246, 247, 248; essential 248; cystine 30, 248; isoleucine 248; leucine 248; lycine 247, 248; methionene 248; mimosine 151, 250; phenylalamine 248; proline 49, 50, 225; threonine 248; thryoxin 250; tryptophan 248; tyrosine 248; valine 248
ammonium ions 215, 226, 227
amylase 167; alpha-amylase inhibitors 317
anabolic 257
Andes, mountain range 103–4, 107, 112
Angola 69
animal husbandry, see livestock production
animal products 372; hair 261; hides and skins 107, 261; meat 64, 107, 109, 235; milk and milk products 156, 261; wool 109 156, 261
antimetabolites, see seed antimetabolites
anti-oxidant 24
aphids, see insects
apple sauce substitute 235
Arabia 121, 267
Arabian Sea 267
argan (*Argania spinosa*) 133
Argentina 106–7, 163, 335, 336, 338, 378
aridity coefficient index 9–10, 13
Aristida–Bothriochloa woodlands 144–5
arta (*Calligonum crinatum*) 268
ash 54–63, 71–3, 236, 237, 269
Asia 1, 10, 25, 28, 40, 42, 90–1, 164, 190, 197, 369
    East 189
    Minor 132
    South 192
    South-east 35, 45, 200, 246
    South-west 35
    West 36, 192
asparagus (*Asparagus officinalis*) 77, 79

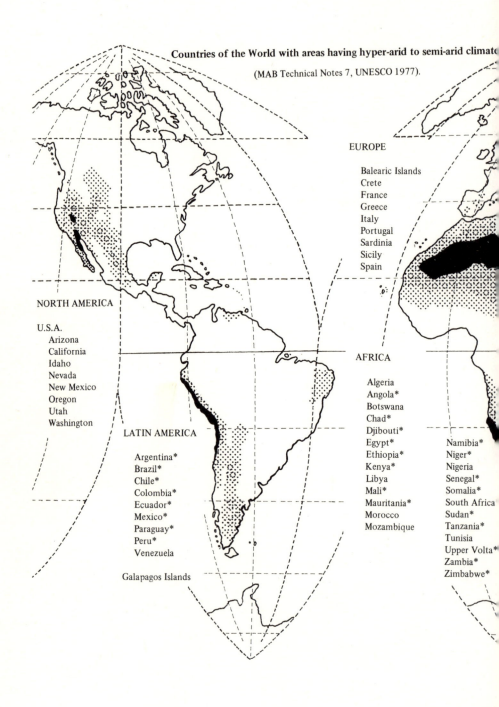

Countries of the World with areas having hyper-arid to semi-arid climate

(MAB Technical Notes 7, UNESCO 1977).

EUROPE

Balearic Islands
Crete
France
Greece
Italy
Portugal
Sardinia
Sicily
Spain

NORTH AMERICA

U.S.A.
  Arizona
  California
  Idaho
  Nevada
  New Mexico
  Oregon
  Utah
  Washington

LATIN AMERICA

Argentina*
Brazil*
Chile*
Colombia*
Ecuador*
Mexico*
Paraguay*
Peru*
Venezuela

Galapagos Islands

AFRICA

Algeria
Angola*
Botswana
Chad*
Djibouti*
Egypt*
Ethiopia*
Kenya*
Libya
Mali*
Mauritania*
Morocco
Mozambique

Namibia*
Niger*
Nigeria
Senegal*
Somalia*
South Africa
Sudan*
Tanzania*
Tunisia
Upper Volta*
Zambia*
Zimbabwe*